ADVANCED CONTROL SYSTEM DESIGN

Bernard Friedland
New Jersey Institute of Technology

Prentice Hall
Englewood Cliffs, New Jersey 07632

Library of Congress Cataloging-in-Publication Data

Friedland, Bernard.
   Advanced control system design / Bernard Friedland.
      p. cm.
   Includes bibliological references and index.
   ISBN: 0-13-0140104 (hard cover)
   1, Automatic control. 2. Control theory. 3. System design.
  I. Title
  TJ213.F74 1996
  629.8--dc20   95-15036
                            CIP

Acquisitions editor: **Tom Robbins**
Production editor: **Bayani Mendoza deLeon**
Copy editor: **Peter Zurita**
Cover designer: **Bruce Kenselaar**
Buyer: **Donna Sullivan**
Editorial assistant: **Phyllis Morgan**

© 1996 by Prentice-Hall, Inc.
A Simon & Schuster Company
Englewood Cliffs, NJ 07632

The author and publisher of this book have used their best efforts in preparing this book. These efforts include the development, research, and testing of the theories and programs to determine their effectiveness. The author and publisher make no warranty of any kind, expressed or implied, with regard to these programs or the documentation contained in this book. The author and publisher shall not be liable in any event for incidental or consequential damages in connection with, or arising out of, the furnishing, performance, or use of these programs.

Printed in the United States of America

10 9 8 7 6 5 4 3 2 1

ISBN 0-13-0140104

Prentice-Hall International (UK) Limited, London
Prentice-Hall of Australia Pty. Limited, Sydney
Prentice-Hall Canada Inc., Toronto
Prentice-Hall Hispanoamericana, S.A., Mexico
Prentice-Hall of India Private Limited, New Delhi
Prentice-Hall of Japan, Inc., Tokyo
Simon & Schuster Asia Pte. Ltd., Singapore
Editora Prentice-Hall do Brasil, Ltda., Rio de Janeiro

*FOR Zita, Barbara,
Irene, and Shelly*

CONTENTS

PREFACE

The world is awash in control theory. There is more of it than a control systems engineer could use in a lifetime of practice. Why still another book on the subject?

There are, I believe, several reasons: Most control engineers are receptive to new theoretical methods and are eager to use them in their work. But the theory is buried in books devoted to exhaustive treatment of individual topics, or in the periodical literature. Most courses, books, and research papers on advanced control theory devote as much space to a topic as it requires for rigorous exposition, rather than to its utility and importance in control system design. Useful but currently unfashionable methods are often passed over in favor of those today in vogue, many of which tomorrow will be consigned to the attic. Graduate courses in control engineering, addressed mainly to students pursuing doctoral studies, also usually cover only a single topic.

To learn about all the methods that could aid your work, you would have to read many books and articles, or take a number of courses. If practical application is your goal, neither option is very appealing. Moreover, some methods that are useful in design of practical control systems do not have much intrinsic theoretical interest; you are unlikely to find them included in advanced books and courses. Hence, even after studying the latest books and completing several courses on advanced methods, you may still be unacquainted with some important concepts and useful design techniques.

As a consequence, many practical control applications are addressed by clumsy and outdated methods. Newly developed methods can be applied to these problems, but often are not, because the engineers in charge of the projects are either unaware of them, or put off by the husk of mathematical theory that must be peeled away to get at the useful kernel within.

This situation motivated the project that started with *Control System Design: An Introduction to State Space Methods* (McGraw-Hill, 1986) and culminates with the present work. My goal has been to describe how the methods of modern control theory can be systematically applied to practical design problems, emphasizing the development of intuition through examples rather than mathematical technicalities. Rather naively, I first thought that the methods developed in *Control System Design* would be sufficient for the design of most practical control systems. But it has become evident that practical control system design by state-space methods requires knowledge of more than the contents of *Control System Design*—not only more theory, but also more acquaintance with the some of the untidy issues that daily confront the practical engineer.

Some readers may take exception to the use of the word "advanced" in the title of this book. I appreciate their misgivings. The book is advanced mainly in the sense that it goes beyond the material covered in *Control System Design*. It has been suggested that a more appropriate title might well be "Practical Control System Design." I resisted taking this suggestion. I don't mean to suggest that other books are "impractical." Also, the word "practical" in the title might give the erroneous impression that the book is concerned with hardware issues rather than design of control algorithms.

Critics may call this a "cookbook," owing to the lack of emphasis on mathematical rigor. It is an appellation at which I take no offense. Just as a skillful chef can prepare a feast without being an expert in the chemistry and physics of cooking, the skillful control engineer should be able to produce an exemplary control system design without having mastered the proofs of the all the theorems used in its development.

I am convinced that the state-space approach is the one most likely to be successful in a wide variety of applications. In dealing with linear systems, this approach is only one of several that could be used. (Other options currently in vogue include the H_∞ and related methods introduced by George Zames and vastly extended by many investigators, and the "quantitative feedback theory" (QFT) method of Isaac Horowitz and his school, both of which rely on frequency domain concepts.) But for nonlinear systems (and what real-world system is not nonlinear), however, there does not appear to be a viable alternative to this approach.

A fundamental concept of the state-space approach is the *separation principle*, by which the design of the control algorithm is divided into the design of an *observer* and the design of a "full-state" feedback algorithm. The observer provides an estimate of the process state, and the full-state feedback is designed on the assumption that all the state variables are directly accessible for measurement. Each design is independent of the other. For linear systems, this "observer-based" design principle has a rigorous theoretical justification and well known optimality properties. In the realm of nonlinear systems, however, this "extended separation principle" is theoretically more tenuous. Nevertheless, it has been my experience that this principle is a valid basis of a systematic and generally reliable method for dealing with a variety of practical design problems.

In selecting topics for inclusion in this book, my objective has been engineering utility, not mathematics. You won't find many theorems, and the proofs of the few that are given probably won't pass muster for rigor. On the other hand, you'll find a discussion of some topics that I think are important, but are not discussed very much, if at all, elsewhere. Examples of such topics are the following:

- How to deal with parasitic phenomena, such as friction and backlash.

- How to combine data from discrete-time and continuous-time sensors.

- How to use zero-crossing and other highly quantized observation data in designing the state observer.

- How the extended separation principle effectively deals with the "integrator-windup."

● How to deal with saturation.

The eclectic selection of topics is bound to raise some eyebrows. Aficionados of the sliding mode will be disappointed with the extent of coverage of that topic; but I have not seen many designs implemented based on this technique and, as explained in Chapter 5, I think it has important practical limitations. Others may be astonished that Pontryagin's Maximum Principle, which occupied so much of control systems research of the past generation, has been compressed to a three-page note at the end of Chapter 5. It's not that I think optimum control theory is not important or useful. (I have even toiled in this field myself.) But, on a utilitarian scale, it does not warrant more space than it has been given. On the other hand, the notion of "the set of recoverable states," which comes from optimum control theory, is accorded more attention than it has received in other recent textbooks. This notion accentuates the hazards of controlling unstable processes with bounded inputs.

Dealing with adaptive control has been perplexing. I have not had much personal experience with the method. To the contrary, I have on occasion voiced the opinion that there is little to distinguish adaptive control designs from those based on other ad hoc methods. On the other hand, I have it on reliable authority that adaptive control methods are used quite extensively in implementing practical control systems, particularly in those applications where it is difficult to develop a good model of the process. Accordingly, I concluded that adaptive control merits serious attention. A long chapter (Chapter 10) is devoted to the subject. Being an outsider in this field, however, my treatment may not emphasize the topics currently viewed by experts as being most important.

Design problems do not have unique solutions. Although some solutions may be more attractive than others, several may be acceptable. It is thus beneficial for you to see and to compare different approaches to the same problem. For this reason, a number "running examples" are presented. These examples have been drawn from various sources. Some (primarily those relating to guidance and flight control) represent applications with which my associates and I have had on-the-job experience. Other examples have been drawn from the published literature (sometimes expanded upon to emphasize practical issues not considered in the original treatment). Still others have been generated to represent systems that could be built, but which to my knowledge have not been.

The importance of simulation as a practical design tool cannot be overemphasized: almost without exception, a simulation study precedes the fabrication of a physical prototype control system. The state-space approach is naturally suited to simulation, since the process model that serves as the basis of the design can also serve as the model for purposes of simulation. On the other hand, the "design model" is sometimes a very much simplified version of the "truth model" that is used in the final stages of an extensive a simulation study. This book addresses the role of simulation and the distinction between different models and their use in control system design.

For whom is this book intended? As I indicated earlier, it is intended for a student or a practicing engineer who is acquainted with space methods for linear, continuous-time systems and who wants to learn how these methods can be applied to other kinds of systems without having to take a number of individual courses. The book is most appropriate for a second MS-level course in a control engineering program. Some of the material could be appropriate for an a senior-level undergraduate course following one that covers the basic state space concepts in linear, continuous-time systems. I would hope that the material is accessible not only to the student, but also to the practicing engineer in search of answers.

This book evolved from the lecture notes for a course on advanced control system design that I gave at the Polytechnic University (Brooklyn, NY) in the late 1980s. Portions of the text were used as supplementary notes for an undergraduate elective, which covered nonlinear and discrete-time systems, at the New Jersey Institute of Technology. A recent version of this text was used in conjunction with a short course that I presented December 1993 in Tel Aviv, under the auspices of the Technion, to a group of engineers employed in Israel's industry. The feedback from participants in these courses was invaluable.

It is a pleasure to acknowledge the contributions of a number of individuals who have had an influence on this work. My former associate and current student, David A. Haessig, Jr. perused the entire manuscript. Not only did he uncover typographical errors, but he made a number of substantive suggestions, many of which were incorporated into the text. Other students who provided useful comments include (now Dr.) Sofia Mentzelopoulou, and Walter Grossman. I am grateful to my NJIT colleagues, Dr. Timothy N. Chang and Dr. Andrew U. Meyer for many stimulating discussions which helped shape the content and style of the book. The text has benefited from critical evaluations of the penultimate draft that were provided by Dr. Farshad Khorrami of the Polytechnic University and Dr. Robert F. Stengel of Princeton University. Many of their suggestions and those of the other reviewers engaged by Prentice-Hall were followed in preparing the final text.

The simulations displayed in the book were done with the aid of a nifty program, *ALSIM*, written by Frank Buran. The results were plotted using the graphics facilities of *Matlab*, Version 4.2, which was also the source of the Nyquist plots, Bode plots, and other curves. David Hur assisted in preparing the graphical output. The block diagrams, visualizations, and other illustrations were executed by Irene Friedland.

I appreciate the indulgence of Prentice-Hall and the editors who patiently waited out the long gestation period of this book. They are probably as surprised as I am that it is finally about to emerge. Thanks also to my wife and daughters who provided much needed moral support.

West Orange, New Jersey
June 1995

Bernard Friedland

Chapter 1

INTRODUCTION

The technology of feedback control systems has come a long way since the field was established as a distinct engineering discipline in the waning days of World War II. (See Note 1.1.) From the crude fire control systems for aiming the guns of warships towards slow-moving targets we have advanced to systems that can drop a missile down the chimney of a building or set a magnetic head on a microscopic spot on a fast-moving disk drive. We can land men on the moon and remotely control the motion of vehicles on Mars. All these modern miracles are witness to the advances of control technology.

These advances, however, are not overly esteemed by the public and often by other engineers. The successes of control technology are all but taken for granted; its failures are disparaged. All but forgotten is the role of control engineering in landing people on the moon and disk heads on tiny spots. But remembered and cursed is the climate control system that freezes one in the summer and roasts one in winter. And the elevator bank in which all the elevators are programmed to go in the wrong direction. That control systems can operate so flawlessly and reliably in countless applications as to go unnoticed and unheralded is a tribute to the technology of automatic control that has been evolving during the past half century.

While there have been remarkable advances in sensor and actuator technology in recent years, most of the strides in the control field are due to the digital computer. Not only is the computer a component in most advanced control systems, but most of the newly-developed techniques for control system design would be inconceivable without digital computers to perform the required calculations and to simulate the performance of designs before prototype systems are built.

Although the attention of researchers is focused on difficult control problems that serve as showcases for the application of advanced theory, most control systems can probably be designed without any theory at all. In many applications, a satisfactory control system can be put together with nothing more than an appreciation of the physical behavior of the process to be controlled, and the right hardware: an actuator by which to make it do what you want it to do and a sensor to measure what it is doing. Applications of this type include thermostats, liquid level controls, simple motion control systems, as well as some of the more sophisticated applications such as pollution control systems. The design challenge in control systems of this type is often in the development of reliable, cost- effective hardware.

At the other extreme are control problems that are too difficult to solve by any theory, in some cases because we don't know how to measure or even what to measure, and in others because there are no feasible actuators. Nuclear fusion and economic processes are examples of this extreme.

The interesting problems of control system design are in applications lying between the two extremes: applications for which control systems are feasible but which need advanced theory for effective solutions. These applications often share some or most of these characteristics:

- The dynamic process to be controlled is complex: nonlinear, high-order, and somewhat uncertain.

- Exogenous disturbances acting on the process are incompletely understood.

- The data is sampled, either because a digital computer to implement the controller, or because only intermittent data is available as output from the process to be controlled.

- The control actuator is limited in "authority": i.e., in the amplitude or the power of the output it can deliver to the process to be controlled.

- Available sensors have limited accuracy, dynamic range, or similar deficiencies.

- It may be necessary to meet more than one requirement on performance quality.

In most applications these are not serious problems. The challenge, however, is not the easy problems but the problems that strain the frontier of possibility. This frontier has several dimensions, only two of which are shown in Figure 1.1. One of the dimensions is dynamic complexity— systems in which high order is combined with large uncertainty. Required performance is another dimension: the necessity of achieving performance (speed, precision) well beyond the ability of existing technology.

The frontier keeps moving outward as technology advances. The difficult problems of the past generation (e.g., flight control of high-performance aircraft, robotic systems, chemical process control) are now within the frontier. Just beyond it are such applications as control of large, flexible space structures and motion control to submicron level for microelectronic manufacturing. And far beyond it are problems in the control of economic, social, and ecological systems; control of nuclear fusion; and antiballistic missile defense.

This book is intended for control system design problems near the frontier. Most problems lie within it. With good fortune you will solve an application outside the frontier and thereby gain the satisfaction and reap the reward of having moved the frontier.

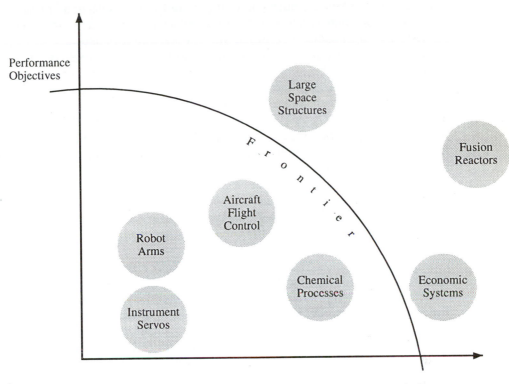

Figure 1.1 The frontier of control systems.

1.1 THE DESIGN PROCESS

Good design eludes definition, but you can usually recognize something that is well designed. One characteristic of good design is that it is exactly right for the specific application. Everything necessary is present, but everything unnecessary is absent.

The skills needed for good design entail experience, intuition, and esthetic sensibility. They are not easily learned from books. The most you ought to expect from any textbook on design (including this one) is to learn about some useful tools.

Like most books on control system design, this book presents tools that can be reduced to mathematical equations: analysis and simulation. Skill in the other aspects of design (such as conceptualization of the overall system, selection of components, dealing with constraints on time and money), which is as important as the mathematical analysis, is acquired and perfected through practical experience.

In practice one learns that most systems develop through evolution: not only biological systems, but also human inventions such as the automobile and the airplane. The sleek, high-performance automobile can be traced to the humble model T; the advanced tactical fighter, still on the contractors' drawing boards, has its roots at Kitty Hawk.

Much of engineering is modifying existing designs. A new model of an existing product is designed to incorporate new technological advances: a new or improved sensor or actuator, a digital processor to replace an analog controller.

Imitation (often called "reverse engineering") is another common design technique. In its most obvious and least creative form you dissect an existing product, and copy the design. The process is legal unless the product copied has patent protection. A more creative use of imitation is to borrow ideas from one application and use them in another. You need to control the level of the liquid in a vessel; consider how it is done in your toilet tank. You must control the temperature of the liquid in the vessel; think of how it is done in your tropical fish tank.

Innovation is often limited by codes, standards, and engineering conservatism. The aircraft industry, for example, took many years to accept the "fly-by-wire" concept, by which the mechanical connections (rods or cables) between the pilot's controls (the stick and pedals) and the moveable aerodynamic control surfaces (the rudder, elevator, and ailerons) are replaced by electric wires carrying signals from the pilot's controls to a flight control computer and from the computer to actuators located at the control surfaces.) While engineering standards may retard progress, they are nevertheless necessary to prevent technological anarchy.

For all these reasons, you rarely get the opportunity to design an entirely new control system. But suppose the opportunity does arise. How would you proceed? The procedure you might follow is depicted in the form of a flow chart in Figure 1.2. As this figure shows, much of the design process falls outside the realm of analysis and simulation.

First you need a physical appreciation of the problem by determining the physical quantity or quantities that must be controlled, and the level of performance required. A consideration that could determine how to proceed might be whether a system similar to the one in question has already been designed. If so, what improvements are you expected to make? If not, is the design perceived as being a difficult problem? Why? What resources (time, money) are at your disposal?

Obtaining a clear statement of the performance requirements is usually more difficult than you might think. In addition to those performance requirements that are

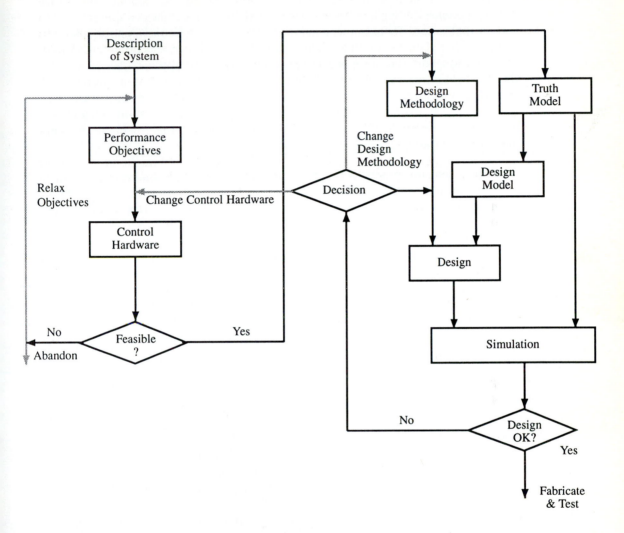

Figure 1.2 How control systems are designed.

explicitly specified (e.g., speed of response, tracking accuracy, disturbance rejection, stability margins), there may also be implicit requirements that are normally met by "conventional" control system designs. Being unaware of implicit performance requirements, you might be tempted to adopt a nonstandard and seemingly innovative design approach leading to a control system that fails to meet these requirements. In certain applications, for example, it may be known that high frequency noise is present from a variety of physical sources. With a conventional design approach it may be appropriate to ignore this noise because experience with this approach has shown that the effect of the noise is negligible. A different (unconventional) design approach that fails to account for the presence of the noise, however, might prove unsatisfactory.

It is often necessary to take nontechnical factors into consideration. Such factors may include economics, esthetics, schedules, or even the internal politics of the manufacturer. Consider, for example, the following scenarios:

- Your company produces a machine that includes a simple analog position servo costing a few dollars. The performance of the servo is adequate for many tasks, but an improved servo would make the machine more versatile and hence potentially worth a higher price. Eager to incorporate "high tech" into the company's product line and knowing that you are knowledgeable in the methods of modern control theory, the director of engineering assigns you the task of designing an improved control system using a microprocessor. You perform a design study (using, for example, the methods of this book) and determine that a microprocessor-based controller can indeed improve performance significantly, but would cost upwards of $100. What would you recommend?

- Suppose you find in your design study that the analog servo was not well designed. In particular, you find that the existing analog servo can be easily modified at no significant cost to yield essentially the same performance as the microprocessor-based system. Now what would you recommend?

- What if the director of engineering, before having been promoted to his present position, was the engineer who designed the original analog servo?

- You are engaged as a control system consultant to a building contractor, who wants to produce a controlled environment. Your fee is a percentage of the cost of the control system. After a detailed design and simulation study you determine that an inexpensive, commercial thermostat will perform nearly as well as a custom computer control costing several thousand dollars. Your recommendation?

As you see, factors other than technical merit can influence engineering design choices.

After satisfying yourself that you understand the technical and nontechnical issues, you make a tentative selection of appropriate hardware. Only after the hardware is selected does the analytic phase of the design process begin.

The first step is to select an appropriate design methodology. (Here a knowledge of the methods discussed in this book would be helpful.) Concurrently you might begin developing a "truth model" of the dynamics of the process—a set of differential and algebraic equations that you are confident adequately represents the behavior of the process. The truth model, which can be used in a simulation for evaluating the performance of the design, is usually too complicated for design purposes. Hence you will probably want to develop a "design model" by simplifying the truth model. The design model should be simple enough to work with but must retain the essential features of the process. Striking the right balance between simplicity and verisimilitude is often a matter of insight and experience.

Having adopted a design methodology and developed a design model, you proceed to perform the design calculations. A generation ago, this step usually entailed laborious hand calculations and curve plotting. Nowadays extensive software is available to assist in performing the calculations, so this step is rarely burdensome.

The final step in the analytical phase of the design is to evaluate the performance of the system by means of a simulation based on the truth model. If the initial design appears to meet the performance and other requirements, fabrication and testing a prototype of the system design would be in order.

Rarely, however, will the starting design satisfy all the objectives, and it will generally be necessary to modify some aspect of the approach. Depending on the perceived deficiencies of the starting design, it might be necessary to change the control hardware. For example, it might be necessary to increase the "authority," i.e., the maximum effort level, of the actuator. Or to add an additional sensor.

On the other hand, it may be found that the design methodology used for the starting design was inappropriate. An example of this would be the decision to base the design on a continuous-time control, with a plan to implement an approximation to the continuous-time control using digital components, and the subsequent discovery that a discrete-time design approach, leading directly to a digital implementation, would have been a better choice of method. After making the appropriate changes, the relevant steps of the design process are repeated as many times as necessary to achieve the desired objectives.

Sometimes you may have to admit defeat: After repeating the design process over and over using different combinations of hardware and design approaches, you may conclude that you don't know how to design a system to meet the requirements. You may have to report that to the best of your knowledge and ability, the required performance requires a breakthrough in hardware or methodology—an invention that does not exist.

As you can see, the selection of the control algorithm is usually only a small part of the overall design process. Experience teaches that the control algorithm is rarely a critical factor in the overall system performance. If you succeed in designing a system (by any method) that works, it is unlikely that a different control algorithm can alone achieve as much as an order of magnitude improvement in performance or reduction in cost. Nevertheless, improvements of less than an order of magnitude may still not be inconsequential and may be worth pursuing.

An unfamiliar process may appear overwhelming unless you are familiar with some of the standard design techniques and conventional approaches. Consider, for example, the problem of controlling the motion of an aircraft so that it proceeds from a given starting point to a specified destination subject to a variety of constraints such as permissible trajectories, available fuel, safety, etc. (Specific instances of this problem include manned aircraft, cruise missiles, and remotely piloted vehicles. Needless to say, each of these applications differs widely in its details from the others.) Problems in aircraft flight control have mostly been solved, but it is instructive to speculate on how you would approach the problem if you didn't know that the solution exists.

In principle, the flight control problem could be formulated as a general optimization problem: to minimize a mathematically specified performance index subject to mathematical constraints. In practice, however, this approach is all but doomed to failure, for many reasons, including the following:

- *Multiple objectives and hidden constraints*: It is rare that everything you want a system to do can be expressed in a single performance criterion to be optimized. Often you only want a system to perform well enough, to be cost effective, and to be reliable. These verbal criteria resist quantitative formulation. Moreover, you may be unaware of all the constraints. The only constraints you may know about are those that have been encountered with similar processes using conventional design approaches. A new design approach, however, may satisfy the known constraints but may create problems not previously encountered.

- *Number of state variables*: The number of state variables in a physical system is often larger than you want or need to deal with.

- *Disparate time scales*: Many processes entail phenomena that occur at widely disparate time scales. In a cruise aircraft application, for example, the position changes on a time scale of minutes or hours; the attitude, on the other hand, changes on a time scale of seconds; and the time scale for the dynamics of the actuators may be in milliseconds.

- *Uncertain dynamics*: The dynamics of the process are generally not as well known as you would like. Although you need some sort of model with which to evaluate performance of the system, your confidence in the model is not high enough to entrust to a brute-force mathematical optimization.

- *Failure to provide insight*: Even with confidence in the dynamics and a mathematically justifiable performance criterion, you may be unwilling to entrust the solution of the problem to a computer-generated optimization, because you may not be able to get an intuitive appreciation of the solution.

For these reasons, among others, a prudent engineer would take a less radical approach. Experienced control system designers make use of such approaches often without conscious knowledge. In some application areas the approaches are virtually "canonical"; no one working in the area would consider any other design approach.

Some of the accepted standard approaches are reviewed and illustrated in the following subsections.

1.1.1   Hierarchical structure

A favorite engineering approach to a complex problem is to break it into several smaller problems. Each of these is further divided into still smaller problems, and the process continued until the smallest problems are easily solved. This hierarchical approach—sometimes called the "top-down" approach—is one of the distinguishing features of systems engineering.

The hierarchical approach is also applicable to the design of feedback controls for complicated systems such as those of the following examples.

Example 1.1   Tactical Missile Guidance and Control

Tactical missile guidance and control illustrates the use of a hierarchical control system architecture. A rigid missile has six mechanical degrees of freedom (three translation variables and three rotation variables). The number of degrees of freedom is even higher when the flexibility of the missile and the dynamics of the actuators and sensors are taken into account. In addition, the motion of the target is a factor to be accounted for. Design of a control system for a tactical missile appears to be a daunting problem. By dividing it into a hierarchy of smaller problems, however, the design process becomes manageable.

A well-known physical characteristic of a tactical missiles is that its speed (i.e., the magnitude of the velocity vector) is governed mainly by the propulsion system and is not significantly changed by the operation of the guidance and control system. Hence it is customary to assume that the speed is a known time function—for design purposes often taken as constant. The guidance and control system changes only the direction of the velocity vector.

Experience further suggests that, for purposes of steering, the missile can be regarded as a point mass. The steering of the velocity vector is accomplished by controlling the lateral acceleration a_L, i.e., the acceleration normal to the velocity vector.

These considerations lead to an overall control system that can be conceived as a hierarchy of two nested subsystems as shown in Figure 1.3, each with distinct functions:

- **The guidance system,** the function of which is to determine a commanded lateral acceleration $(a_L)_C$: the magnitude and direction (in the plane normal to the velocity vector) of the lateral acceleration such that, if achieved, the missile would intercept the target.

- **The autopilot,** the function of which is to make the actual lateral acceleration "track" the desired acceleration with an acceptably small error.

The design of each of the two loops is viewed as a separate task. In the first, only the characteristics of the missile (such as the maximum achievable lateral acceleration) essential for the design of the guidance loop are taken into account; most of details of the missile and its autopilot are crudely approximated. Similarly, detailed knowledge of the guidance system is not required for the design of the autopilot. Needed only is a qualitative description of the acceleration commands and the speed and accuracy that the guidance system expects them to be achieved.

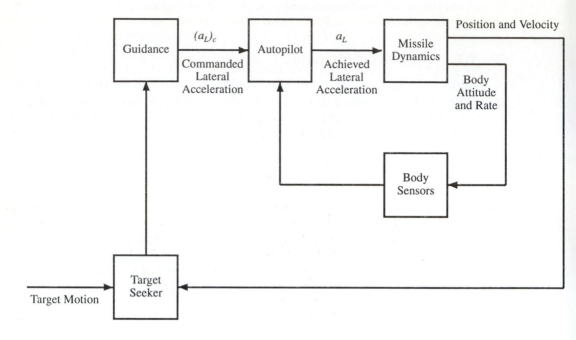

Figure 1.3   Guidance and control of tactical missile uses a hierarchical structure.

The two subsystems, the guidance system and the autopilot, may employ distinct hardware and in many organizations may be designed by different groups of engineers. The guidance group, assuming that the autopilot will perform according to specifications, is concerned with the development of guidance algorithms and with the hardware and software to implement these algorithms. The autopilot group, operating with a detailed knowledge of the missile dynamic behavior, including aerodynamics, is concerned with selection of hardware and design of control algorithms to ensure that the acceleration commands will be achieved at the specified accuracy. (Figure 1.3 shows the use of different sensors in the guidance loop and in the autopilot loop. Many missile systems do employ this architecture. In state-of-the-art systems, however, the inertial sensors (gyros and accelerometers) used in the autopilot can also serve in the guidance loop.)

Separation of the functions of guidance and control has led to many successful designs. Because of the assumptions and approximations that each design group uses, however, it is necessary to verify that the overall system will operate as expected, before it is constructed and tested. The verification is generally accomplished by the use of a detailed simulation using the complete set of dynamics of the missile. Although referred to as a six-degree-of-freedom simulation, the actual number of degrees of freedom is usually much higher, since it takes into account the bending modes of the missile and the dynamics of the actuators and sensors which may have been neglected at the design stage. Development and maintenance of the software for accurate overall system simulation is often the responsibility of still another group in the organization.

Example 1.2 Navigation and Control of an Autonomous Vehicle

Automated factories may use autonomous robotic vehicles to carry such things as tools, materials, work in process, and finished products from one station to another. Usually such vehicles are guided by markings (e.g., tape lines) on the factory floor and are thus confined to follow trajectories defined by the markings. With a more sophisticated approach, however, the vehicles would be able to move freely on any part of the floor.

A hierarchical approach to the design of the control system for the autonomous vehicle would seem appropriate in this application. By such an approach, the overall manufacturing process would be under control of a central supervisory controller, which is cognizant of the objective of the process, that issues high-level commands and monitors the state of the process to determine whether these commands are reasonably executed. The state of the overall process might include more than one robotic vehicle, several stationary robots and other machines. Given all necessary and available information, the controller would prescribe the trajectories to be followed by each of the vehicles. (The trajectories would be subject to change in accordance with possible real-time exigencies, such as a breakdown somewhere in the process.)

Through a suitable communication system, each autonomous vehicle would be sent its own prescribed trajectory, and the control system within the vehicle would endeavor to follow the trajectory. (How the trajectory is to be defined is a question in itself. It could, for example, be specified as a series of points through which the vehicle should pass, or as a series of tangent lines and circular arcs that the vehicle should follow.)

Based on the state (position, velocity, orientation, etc.) of the vehicle and the commanded trajectory, the control system on board the autonomous vehicle guides it along the desired trajectory. One way to accomplish this would be for the guidance system to continuously command the desired instantaneous velocity and vehicle orientation. The errors in these quantities would be zero when the vehicle is on the correct trajectory. But if the vehicle is not on the desired trajectory, the errors will cause the locomotion system to move the vehicle so as to reduce the errors to zero.

The overall system configuration based on this approach is shown in Figure 1.4. Note that it has three loops: the outer supervisory control, the guidance loop, and the "autopilot," which controls the locomotion system.

A type of autonomous vehicle is propelled by two identical independent d-c motors, each attached to its own wheel. The third point of contact with the plane of motion is a "caster" which supports the weight over it but offers no resistance to the motion. A top view of the vehicle is shown in Figure 1.5.

The controls for the vehicle are the voltages u_1, u_2 applied to the motors. These voltages are both limited to U in magnitude.

Based on these controls the vehicle dynamics can be represented by:

$$\dot{x} = v \cos \theta$$
$$\dot{y} = v \sin \theta$$
$$\dot{v} = -\alpha v + \beta(u_1 + u_2)$$

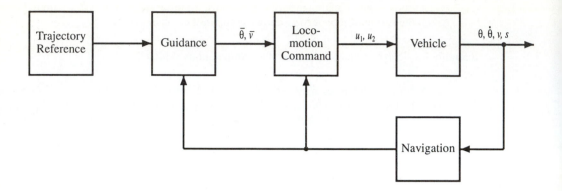

Figure 1.4   Hierarchical structure of control system for autonomous vehicle.

$$\ddot{\theta} = -\gamma\dot{\theta} - \delta(u_1 - u_2)$$

where α, β, γ, and δ depend on the parameters (masses, inertias, torque constants, etc.)
of the vehicle. (See Problem 1.2)

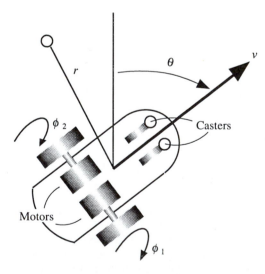

Figure 1.5   Top view of autonomous vehicle.

Possible measurements include the following:

- Intermittent, but very precise, measurements of the coordinates (x, y) of a reference point on the vehicle. (The reference point may be taken to be located at the midpoint of the line between the wheels.) The sampling interval for the intermittent measurements is $T = 0.2$ s.

- Continuous time, but rather noisy, measurements of the total angular rotations ϕ_1, ϕ_2 of the two wheels, each of which can be assumed to roll without slippage on the plane of motion of the vehicle.

1.1.2 Simplification of system dynamics

Example 1.3 Servo with Belt-Driven Load

Motion control systems are frequently designed with a belt drive connecting the servomotor with the load. A typical configuration is illustrated schematically in Figure 1.6. For simplicity, assume that the ratio of pulley diameters is 1. (There is almost no difference in the analysis for other ratios.) If the belt resilience is modeled by an ideal spring, the equations of motion of the system are:

$$J_1\ddot{\theta}_1 = k(\theta_2 - \theta_1)$$
$$J_2\ddot{\theta}_2 = k(\theta_1 - \theta_2) - B\dot{\theta}_2 + Lu$$

where θ_1, θ_2 are the angles of the motor and load shafts, J_1, J_2 are the corresponding inertias, k is the spring rate of the belt, B is the damping torque constant, and L is the control torque coefficient.

These differential equations define a fourth-order system. As the spring constant k representing the belt becomes infinite, the difference in shaft angles $\theta_1 - \theta_2$ vanishes and the model reduces to the second-order system

$$(J_1 + J_2)\ddot{\theta} = -B\dot{\theta} + Lu$$

where $\theta_1 \rightarrow \theta_2 \rightarrow \theta$.

The assumption that the belt is not resilient has reduced the order of the system from 4 to 2. The resonance that would be present for a belt with finite resilience has been eliminated from the model. This is usually permissible, but the resonance could have a parasitic effect as discussed in Chapter 7.

Omitting or simplifying the effects of flexibility is one example that illustrates the reduction of system order in going from the truth model to the design model. Since a reduced-order model has an obvious advantage with respect to design, the subject of reduced-order modeling has received and continues to receive much attention in the literature of control theory. There are two general approaches to order reduction. The first, already illustrated in the previous example, is to use knowledge of the physical characteristics of the process to estimate the magnitude of various effects and omit those that are negligible. The second approach is to use a formal, mathematical technique of order reduction. No single one of the many formal order-reduction methods

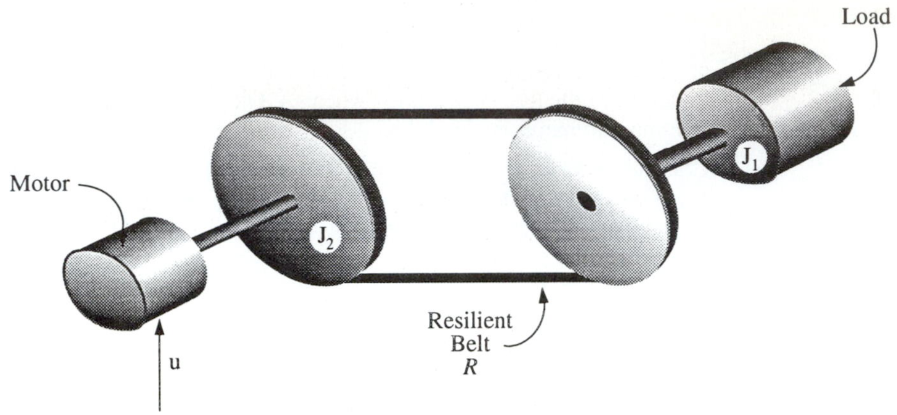

Figure 1.6   Servo with belt-driven load.

is obviously superior to all the others in all applications, so it is probably worth having a repertoire of techniques available for use as needed. A guide to the literature is given in Note 1.2 These techniques are all limited to linear systems.

   Another school of thought on order reduction is to use a high-order model for the control system design, and then to simplify the resulting design by one of the formal techniques of order reduction. Since the latter are limited to linear systems, this design method obviously cannot be used with nonlinear systems. Another disadvantage, even with linear systems, is the loss of physical insight that can be gained by using a reduced-order design model. Nevertheless, since favorable results (see Note 1.3) have been reported in using this approach, it should not be ignored.

1.1.3   Decoupled subsystems

A dynamic system usually can be regarded as comprising a set of coupled subsystems. If the coupling between the subsystems is strong, this way of looking at the system is not helpful. In many applications, however, the coupling between the subsystems is not very strong. In such cases, a feasible approach is to design each subsystem separately, treating as disturbances the coupling between the subsystems.

Example 1.4   Cargo-Moving Gantry

   In this application it is required to control the motion of the payload being moved by a gantry consisting of a truck carrying a motor-driven winch. The truck can move horizontally on a fixed (cantilevered) arm as shown in Figure 1.7.

   The objective of the control system is to move the load from one side of a fixed obstacle to a known destination on the other side, without having the load collide with the obstacle.

   The control variables are the inputs to the independent motors controlling the winch and the truck.

The variables that are readily measured are the cable length l and the position x of the truck. It would also be feasible, if necessary, but less than convenient, to measure the angle ϕ of the cable.

The truth model can be obtained, for example, by use of Lagrange's equations under the following assumptions:

- The cable does not stretch.

- There are no damping effects.

- The motor dynamics are negligible.

The model is given by

$$
\begin{bmatrix} 1 + \frac{m_T}{m_L} & l\cos\phi & \sin\phi \\ l\cos\phi & l^2 & 0 \\ \sin\phi & 0 & 1 + \frac{Jr^2}{m_L} \end{bmatrix} \begin{bmatrix} \ddot{x} \\ \ddot{\phi} \\ \ddot{l} \end{bmatrix}
$$
$$
= \begin{bmatrix} -2\dot{l}\dot{\phi}\cos\phi + l\dot{\phi}^2\sin\phi \\ -gl\sin\phi - 2l\dot{l}\dot{\phi} \\ g\cos\phi + l\dot{\phi}^2 \end{bmatrix} + \begin{bmatrix} \frac{F_T}{m_L} \\ 0 \\ \frac{F_L}{m_L} \end{bmatrix} \tag{1.1}
$$

where

Cable length	:	l
Truck mass	:	m_T
Load mass	:	m_L
Winch inertia	:	J_r
Winch radius	:	r
Truck position	:	x
Cable angle	:	ϕ
Truck control	:	F_T
Winch control	:	F_L

These equations are all coupled and seem to be hopelessly nonlinear. But they can be simplified considerably by making the entirely reasonable assumption that the swing angle ϕ is small enough to permit the approximations $\sin\phi \approx \phi$, $\cos\phi \approx 1$ and that the products of state variables are small.

With these approximations the truth model reduces to

$$
\left(1 + \frac{m_T}{m_L}\right)\ddot{x} + l\ddot{\phi} = \frac{F_T}{m_L} \tag{1.2}
$$
$$
\ddot{x} + l\ddot{\phi} + g\phi = 0 \tag{1.3}
$$

for the truck dynamics and

$$
\left(1 + \frac{Jr^2}{m_L}\right)\ddot{l} = g + \frac{F_L}{m_L} \tag{1.4}
$$

for the length dynamics.

Defining the control variables u_1 and u_2 with F_T/m_L and F_L/m_L, respectively, and using the simplifying approximations has both simplified and decoupled the dynamics.

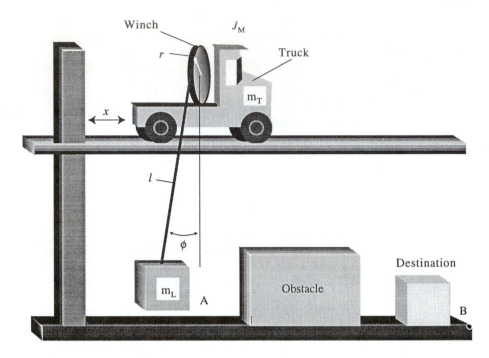

Figure 1.7   Schematic of cargo moving gantry.

Equation (1.4) is nothing more than the equation of vertical motion of a free mass acted upon by a control u_2 and its weight. Equations (1.2) and (1.3) are the equations of a pendulum of length l moving on a car and controlled by u_1.

The control law for u_2 would be the same as that for a free mass acted upon by a constant disturbance.

If the length l were constant, design of the control law for u_1, although more complicated than that for a free mass, would be a fairly straightforward problem. This case is complicated, however, because the length l is not constant over the entire control interval. We can make the control problem easier by deciding to separate the operation into phases: a hoisting and lowering phase, and a translation phase. If we do that, we would be justified in assuming that the length is a known constant during the translation phase.

But it would save time if the two phases of the operation could be done concurrently. Since l is an external variable for the translation problem, (1.2) and (1.3) define a linear but time-varying control problem. If we had an acceptable design approach for this problem, we would have a feasible design for the overall system. One approach to the design of the translation problem is "gain scheduling," described in Chapter 10.

If all the required variables are available for measurement, the control system would have the structure shown in Figure 1.8.

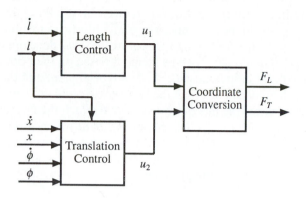

Figure 1.8 System for concurrent control of length and translation.

1.1.4 Use of standard control algorithms

Suggestions of methods for simplifying a problem are presented above. How to solve the simplified problem or problems that result when these methods are used is what this book is all about.

In some applications, however, an acceptable solution can be obtained by simply using one of the standard elementary control techniques appropriate to the situation. Several of these techniques are reviewed here.

Proportional-integral-derivative (PID) control

One of the most common of the standard control algorithms is called "proportional-integral-derivative (PID)" control. It is defined by

$$u(t) = K_P e(t) + K_I \int_{t_0}^t e(\tau)d\tau + K_D \dot{e}(t) \tag{1.5}$$

where e is the system error.

This algorithm is so widespread in some industries (notably process control) that several manufacturers provide ready-made PID controllers. Before the digital computer era, these were analog devices with three knobs, one for each of the scale factors K_P, K_I, and K_D. Nowadays the PID controller is more likely to be implemented digitally, with the knobs replaced by a keyboard—perhaps more accurate, but hardly more sophisticated.

The PID algorithm is particularly useful for processes that are known to be stable and not very oscillatory, but whose parameters are not well known. In such applications it may be difficult to develop a design model, let alone a truth model, and the scale factors are often adjusted with the controller "on-line." A process operator un-schooled in control theory but experienced in observing the behavior of the process can often perform the adjustment by trial-and-error. To minimize the amount of trial-and-error, a recipe for tuning PID controllers was developed by Ziegler and Nichols in 1942 [1]. The Ziegler-Nichols recipe is based on their observation that if a proportional feedback loop is closed around a typical plant and the gain gradually raised, a point will be reached at which the system will oscillate. This is the point at which the root locus crosses the imaginary axis and at which the Nyquist diagram intersects the critical point $s = -1$. Define the loop gain at which this occurs as K_u (termed the "ultimate" gain by Ziegler and Nichols), and let the corresponding period of oscillation be T_u. Then, the proportional (P), integral (I), and derivative (D) gains of the Ziegler-Nichols recipe are given by

$$K_P = \frac{K_u}{2}, \quad K_I = \left(\frac{2}{T_u}\right) K_P, \quad K_D = \left(\frac{T_u}{8}\right) K_P \tag{1.6}$$

From (1.5) the compensator transfer function is given by

$$D(s) = \frac{K_u}{2T_u s}(2 + T_u s + T_u^2 s^2/8) \tag{1.7}$$

which gives a pole at the origin and a double zero at $s = -4/T_u$.

Example 1.5   Temperature Control

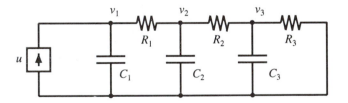

Figure 1.9   Electrical analog of thermal dynamics.

Temperature control systems are representative of many industrial processes for which PID control is often used. A typical temperature control application entails applying the heat at one point in the system and controlling the temperature at a different point. To derive the dynamics of the process, electrical engineers often make use of electrical analogies. With electrical voltage analogous to temperature, current to heat quantity, electrical capacitance to heat capacity, and electrical resistance to thermal resistance, the differential equations governing the system can be written

$$\frac{dv_1}{dt} = -\frac{1}{R_1 C_1} v_1 + \frac{1}{R_1 C_1} v_2 + \frac{1}{C_1} u$$

$$\frac{dv_2}{dt} = \frac{1}{R_1 C_2} v_1 - \frac{1}{R_{12} C_2} v_2 + \frac{1}{R_2 C_2} v_3$$

$$\frac{dv_3}{dt} = \frac{1}{R_2 C_3} v_2 - \frac{1}{R_{23} C_3} v_3$$

where u represents the current (heat) input and

$$R_{12} = R_1 R_2 / (R_1 + R_2) \quad R_{23} = R_2 R_3 / (R_2 + R_3)$$

For the numerical values

$$R_1 = R_2 = R_3 = 1$$
$$C_1 = C_2 = C_3 = 1$$

the process dynamics become

$$\dot{v}_1 = -v_1 + v_2 + u$$

$$\dot{v}_2 = v_1 - 2v_2 + v_3$$

$$\dot{v}_3 = v_2 - 2v_3$$

Suppose that the temperature measured at node 3 is to be controlled, i.e.,

$$y = v_3$$

It is readily established that the transfer function from the heat input u to the output y to be controlled is given by:

$$G(s) = \frac{1}{s^3 + 5s^2 + 6s + 1}$$

which has no zeros and poles on the real axis at $s = $ -0.0198, -1.555, and -3.247. (See Problem 1.4.)

The root locus for the plant, shown in Figure 1.10(a), crosses the imaginary axis at $\omega_u^2 = 6$ at a loop gain of $K_u = 30$. From the frequency of crossing we find $T_u = 2\pi/\omega_u = 2.56$ Using these values in the Ziegler-Nichols recipe (1.6) yields the gains

$$K_P = 15., \quad K_I = 11.72, \quad K_D = 4.80$$

which give a pole at the origin and a double zero at $s = 1.5625$. Note the near coincidence of these zeros with one of the plant poles. The root locus of the compensated system is shown in Figure 1.10(b). Note that it remains in the left half plane for all positive gains. The compensated system thus has an infinite gain margin. The phase margin, however, is only 26 degrees, which is somewhat lower than desirable. The root locus also shows that the dominant poles of the closed-loop system are not very well damped.

The step response shown in Figure 1.11 confirms the prediction of the root locus. The overshoot is more than 50% and the behavior is quite oscillatory.

Depending on the requirements, the design may or may not be satisfactory. But it certainly would be possible to achieve a better design. (See Example 1.7.)

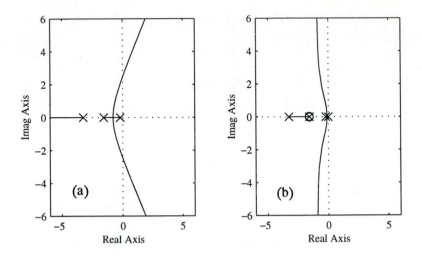

Figure 1.10   Root loci for thermal control. (a) Plant. (b) Plant with PID compensator.

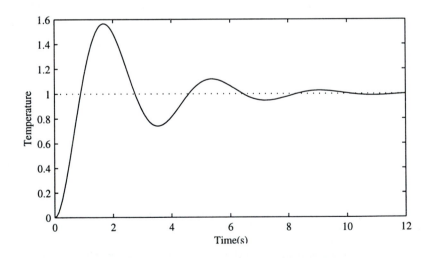

Figure 1.11   Step response of thermal system with PID control.

The foregoing example shows one of the disadvantages of the Ziegler-Nichols recipe: The closed-loop system is often more oscillatory than desirable. Other design methods usually can give better damped response. Another disadvantage is that it is based entirely on the gain and frequency at which the uncompensated system would oscillate in a proportional feedback loop, hence it may not be valid for plants in which the excess of poles over zeros is only two and hence which may have root loci that do not cross the imaginary axis. As a result of parasitic resonances, delays, etc., most processes will oscillate if the gain is extremely high, but the frequency of oscillation will depend on characteristics of parasitic elements, and will not be a suitable quantity on which to base the PID gain settings.

These limitations notwithstanding, however, the Ziegler-Nichols tuning recipe remains popular in the process control industry.

Proportional Navigation

In one of the standard missile-guidance algorithms, called "proportional navigation," the (commanded) lateral acceleration $(a_L)_C$ is made proportional to the velocity of the missile (relative to the target) and the rate of change of the line of sight angle λ from the missile to the target:

$$(a_L)_C = -kv_r\dot{\lambda} \tag{1.8}$$

where $v_r = |V_M - V_T|$. See Figure 1.12.

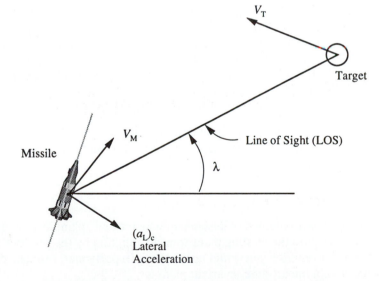

Figure 1.12 Geometry for proportional navigation.

Proportional navigation drives the line-of-sight angle from the vehicle to the target to a constant value. A simple geometric construction shows that if the line of sight angle is constant, the missile and the target are on a "collision course" and the vehicle will intercept the target. (Mariners not aware of this principle will suffer the consequences.)

Since the product of v_r and $\dot{\lambda}$ has the dimensions of linear acceleration, the "navigation constant" k in the proportional navigation law (1.8) is a pure number. It can be shown that any $k \geq 3$ will result in an interception. (See Problem 1.3). Because proportional navigation will operate with a wide range of navigation constants, it is very robust. It will work well with only an estimate of the relative velocity, and is effective even when the target follows a curved trajectory. These are among the reasons that proportional navigation continues to be popular, even though other algorithms may theoretically be able to give superior performance under special circumstances.

Ease of implementation is another advantage of proportional navigation. Many seekers operate at their greatest accuracy when the target is centered in their relatively narrow field of view. Since the missile body axis may not always point to the target, the seeker is mounted in a set of gimbals that permit the sensor axis to point to the target regardless of the missile orientation. If the motion of the seeker is perfectly isolated from the motion of the missile, so that it always points directly at the target, it will be rotating at precisely the line-of-sight rate. A set of gyros attached to the seeker will thus measure the projections of the (vector) line-of-sight rate onto the directions of their "input" axes, fixed relative to the seeker. Hence these gyros directly measure the line-of-sight components of line-of-sight rate required by the proportional navigation algorithm. When their outputs are zero, the missile and the target are on a collision course. This is an important feature, particularly in systems that lack sophisticated computer facilities that may be needed to implement other guidance laws.

Using proportional navigation, it is not possible to control the angle at which the trajectory of the missile crosses that of the target. Sometimes this angle is important. If so, it is possible to include a term in (1.8) proportional to the line-of-sight angle:

$$(a_L)_C = -k_1 v_r \dot{\lambda} - k_2 g \lambda \qquad (1.9)$$

where g is the acceleration of gravity and is included in (1.9) to make k_2 dimensionless.

If (1.8) is called proportional navigation, (1.9) might deserve to be called "proportional plus integral" navigation.

Plant Inversion

Another obvious method of design might be termed "plant inversion." In this method you compensate the existing plant transfer function by the product of its inverse and the transfer function you would rather have. In particular, suppose the actual transfer function of a linear, time-invariant plant is

$$G(s) = \frac{N(s)}{\Delta(s)}$$

and you would prefer that it were

$$\hat{G}(s) = \frac{\hat{N}(s)}{\hat{\Delta}(s)}$$

It would seem that all you need to do would be to make the compensator transfer function

$$D(s) = \hat{G}(s)[G(s)]^{-1} = \frac{\hat{N}(s)}{\hat{\Delta}(s)} \frac{\Delta(s)}{N(s)}$$

If this were a generally applicable method, control system design would be trivial (and books like this would be superfluous). Unfortunately (or fortunately), however, the method as stated almost never is applicable. There are a number of reasons:

- The transfer function of the compensator may be more complicated than necessary.

- The degree of the numerator of $D(s)$ is usually higher than its denominator, which necessitates differentiation of the plant output, an undesirable operation in the presence of noise.

- If the plant has zeros or poles in the right half plane, the compensator will have corresponding poles or zeros and this will result in an unstable closed loop system.

While the first two of these limitations may be tolerable, the last is fatal. In order for a plant pole or zero in the right half plane not to appear in the forward loop transfer function it must be cancelled by a corresponding zero or pole. Although the pole-zero cancellation appears to take place in the frequency domain, all that happens is that it is made uncontrollable or unobservable. This can be interpreted as the presence of a branch of the root locus, of zero length, between the cancelled pole and zero. Since this branch is in the right half plane, the closed loop system must be unstable. An exogenous disturbance will cause an unbounded response.

Although the creation of unobservable or uncontrollable pole-zero pairs occurs for pole-zero cancellations occurring in the left-half plane, the result is benign; the response due to an exogenous disturbance will decay harmlessly.

Poles or zeros in the left half plane near the imaginary axis can create practical difficulties. If the pole or zero of the plant that is to be cancelled by the zero or pole of the compensator were to drift, owing to environmental factors for example, a root-locus branch of nonzero length would result. If this branch were to swing into the right half plane, as shown in Figure 1.13, the system could become unstable.

In most applications the drastic step of inverting the entire plant is not necessary. Often the desired effect may be achieved by inverting only one factor of the plant transfer function, leaving in place the part of the transfer function that is acceptable. This technique is illustrated by the following example.

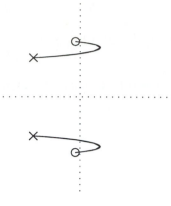

Figure 1.13 Danger of pole-zero cancellation near imaginary axis.

Example 1.6 Temperature Control (Continued)

The transfer function for the temperature control problem of the previous example
can be written with its denominator in factored form:

$$G(s) = \frac{1}{(s + 0.0198)(s + 1.555)(s + 3.247)}$$

In order for the closed-loop system to follow a step with zero steady-state error, the loop
transmission must have a pole at the origin. Suppose, also, that we want a rapid response
such as would be characteristic of an open-loop system with a pole at $s = -3.247$. This
would suggest making the loop transmission

$$F(s) = D(s)G(s) = K\frac{1}{s(s + 3.247)}$$

which is achieved with a compensator having the transfer function

$$D(s) = K\frac{(s + 0.0198)(s + 1.555)}{s} = K\frac{s^2 + 1.5748s + 0.0308}{s}$$

This transfer function is that of a PID control, but with gains chosen to place zeros over
the undesirable poles and hence very different from those given by the Ziegler-Nichols
recipe. The step response for $K = 2$ is shown in Figure 1.14. The response is both faster
than that of the Ziegler-Nichols response and free of overshoot. Clearly this design is
superior.

1.2   THE "DESIGN MODEL" AND THE "TRUTH MODEL"

The ultimate verification of a control system design is its performance in the real
world. Nothing is better than building and extensively testing a prototype of the proposed system.

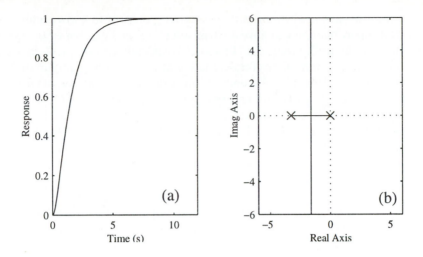

Figure 1.14   Temperature control response with compensator designed by partial inversion of the plant. (a) Step response. (b) Root locus.

But there are many reasons why this is not always desirable or even possible. First there is the matter of safety. The risk of catastrophic consequences of failure in an aircraft flight control system or a chemical process control system prevent construction and testing of the system unless it is all but certain that the control system will not fail even under conditions of extreme and extraordinary stress. Even when safety is not a concern, construction of a prototype may not be an efficient use of engineering time and effort, particularly when the design is to be selected from among several alternatives.

An alternative, or a preliminary step, to construction and testing a prototype is *simulation*. Organizations (such as aircraft manufacturers) that are concerned with the development of control systems for complex processes often have entire departments dedicated to the development of simulations of control systems and supporting technology. There is an advantage to separating the operation of performance simulation from the operation of control system design: there is much less likelihood of the same error going undetected in two independent operations.

When using a simulation as an alternative to a physical prototype, it is important to include models of all the phenomena that may conceivably influence the behavior of the process; the more accurately that the overall model of the system captures the characteristics of the actual system, the better will be the performance predictions based on the simulation. The simulation model that includes all the known relevant characteristics of the real system is often called the *truth model* of the system.

Notwithstanding every effort to incorporate all that is known about the system into the truth model, danger lurks in identifying the truth model with the real system. You almost never can know or model everything there is to know about a system. More-over, many phenomena have a trivial effect on the behavior of the system. Including them in the truth model could make it intractably complicated. In a mechanical system of finite extent, for example, there is always a certain amount of flexure, even if it is built of nominally rigid members. Accurate representation of such a system would entail the use of partial differential equations and that would lead to a truth model of infinite order. If it were necessary to represent every mechanical system by an infinite-order model, it would be all but impossible to design a control system. Nevertheless, short of building a prototype, there is no avoiding the need for a simulation based upon a thoughfully developed truth model

The truth model, however, is often too complicated for use in control system de-sign. For the latter purpose, a simpler model, often known as the *design model*, is fre-quently used. Such a model captures the important features of the process for which a control system is to be designed but omits the details which you believe are not sig-nificant. Knowing which details must be included and which can safely be omitted is a skill grounded in physical understanding and honed by experience.

The cargo-moving gantry considered in Example 1.4 illustrates the distinction be-tween a truth model and a design model. The nonlinear, coupled equations of motion are the beginnings of a truth model. Greater verisimilitude would result from includ-ing the stretching of the cable, the dynamics of the drive motors, and friction effects. Including out-of-plane motion and aerodynamic effects would yield an even more ac-curate truth model. A design model, on the other hand, developed by eliminating all unpleasant nonlinear effects while capturing all the essential features of the process, is much more amenable to use for the design of a control system. How do we know that the omitted nonlinear effects are in fact negligible? By testing the control system design using the truth model that incorporates all the effects worth considering.

Sometimes the complexity of the model of a process can be reduced by simply omitting small terms from the model. Candidates for omission in a design model are terms that are not entirely negligible—else they should not have been included in the truth model—but that experience or physical insight suggests would be small enough to omit in the design model. There is no firm rule on how small a term must be in order to justify omitting it. (The customary rule of thumb is 10 percent.) Sometimes a much smaller term is retained because it is no trouble to do so, and sometimes a much larger term is omitted because its presence is inconvenient. If you elect to drop a term that you suspect should not be dropped, it might be prudent to "cover" this term in the design procedure by assuming the presence of an appropriate amount of noise. (Chapter 7, which deals with parasitic effects, has a more extensive discussion of this point.)

A complex system often consists of a number of relatively simple subsystems that are only weakly coupled. By assuming for purposes of design that the subsystems are entirely uncoupled, a design model breaks down into a number of separate pieces, each of which can be individually controlled. This will result in what is termed a "decentralized" control configuration. This configuration may aid implementation: a number of relatively simple, localized control processors might be preferred over one large central processor. But even when the control is implemented by one central processor, it might be preferable to design the control algorithm as if the subsystems were uncoupled.

A real-world application will usually entail several types of complexity; thus it will require you to use a combination of several of the above techniques to obtain a suitable design model. In practice, you may need a hierarchy of models of varying complexity: a very detailed truth model for final performance evaluation, several less complex truth models for use in evaluating particular effects, and one or more design models.

1.3 SIMULATION TECHNIQUES

Analog computers were used for simulation of dynamic systems before the advent of modern digital computers. But virtually all dynamic systems are now simulated by means of digital computers, which provide greater flexibility, ease of maintainability, and cost effectiveness. All that now remains as the heritage of analog computers is the block-diagram graphical language used for representing the behavior of dynamic processes.

Two basic types of simulations are used in control engineering: mathematical and physical. In a mathematical simulation the characteristics of all the components of the system are represented mathematically by logical operations, algebraic equations, differential equations, and difference equations, which are solved concurrently within the digital computer. The veracity of the simulation depends on how well the equations embody the true behavior of the system and how carefully they are programmed. The purpose of the simulation is primarily to evaluate the performance of the entire system under the assumption that the characteristics of all its components are known.

Physical simulation, on the other hand, is intended to create a environment in which the performance of a real component is to be tested. To evaluate the behavior of a component in the presence of acceleration, a centrifuge may be used to create a centrifugal acceleration to simulate an acceleration that would occur by some other physical mechanism.

Mathematical Simulation

Mathematical simulation of a dynamic process is relatively simple: all you need is a reliable numerical integration package and the ability to express the equations of the process in an appropriate computer language. Writing a simulation program for a simple process is scarcely more than an exercise in computer programming.

Instead of developing a separate computer program for each process that you may want to simulate, it is more efficient to use a commercial simulation program, which is essentially a "shell" that facilitates inputting the description of the process and does the tasks common to all simulations.

A number of popular simulation shells are commercially available for mainframes and personal computers; more appear each year. The description of the process dynamics in these programs are inputted to these programs in two formats: algebraic or graphical. In the algebraic-style programs, you input the equations representing process dynamics in the form of computer program statements, either in a standard computer language or in a language invented by the program developer for this purpose.

Example 1.7   Truth Models for Cargo-Moving Gantry

For a simulation program in which the process dynamics are coded in the C language, the process could be represented by:

```
/*
**  Truth Model for Gantry Crane
*/
#define MLOMT    fpar[1]
#define JR2OML   fpar[2]
#define MT       fpar[3]
#define ML       fpar[4]
#define G        fpar[5]

derv(double t, double *x, double *dxdt)
{
double mass[9]={0., 0., 0., 0., 0., 0., 0., 0., 0.};
double rhs[3]={0., 0., 0.};

/********* Plant Dynamics **********/
dxdt[1] = x[4];
dxdt[2] = x[5];
dxdt[3] = x[6];

/*** Compute mass matrix ***/
mass[0] = 1 + MLOMT;
mass[1] = x[3]*cos(x[2]);
mass[2] = sin(x[2]);
mass[3] = mass[1];
mass[4] = x[3]*x[3];
mass[5] = 0.;
mass[6] = mass[2];
mass[7] = 0.;
mass[8] = 1 + JR2OML;

/*** Compute right-hand sides ***/
rhs[0] = -2.*x[6]*x[5]*cos(x[2]) + x[3]*x[5]*x[5]*sin(x[2]) + u[1];
rhs[1] = -G*x[3]*sin(x[2]) - 2.*x[3]*x[6]*x[5];
rhs[2] = G*cos(x[2]) + x[6]*x[5]*x[5] + u[2];

/*** Invert mass matrix ***/
```

```
matinv(mass,3);

/*** Calculate accelerations ***/
dxdt[4] = mass[0]*rhs[0]+mass[1]*rhs[1]+mass[2]*rhs[2];
dxdt[5] = mass[3]*rhs[0]+mass[4]*rhs[1]+mass[5]*rhs[2];
dxdt[6] = mass[6]*rhs[0]+mass[7]*rhs[1]+mass[8]*rhs[2];
}
```

The previous fragment is the actual code used in the simulation of this process in one simulation shell. Symbols in capital letters are constant data elements read into the program as the initial step in the simulation. Data for the simulation are supplied to the global array `fpar[ ]` from a data file supplied by the user. Elements of the global arrays `x[ ]`, `dxdt[ ]`, `u[ ]` and others are computed during the simulation and read into a file for plotting and other post-simulation analysis.

Note the use of the computer function `matinv` to numerically invert the 3×3 mass matrix after its elements are computed.

In graphic-style programs, you input the process dynamics by drawing the corresponding block diagram, selecting elements (e.g., integrators, summers, gain elements) from a library provided by the program author. The software converts the interconnected blocks to the proper algebraic, differential, and difference equations. These simulation shells, *tours de force* of software engineering, are fun to use and do not require a knowledge of computer programing. They may be less convenient to use than an algebraic-style program, however, because the block representation of a system of moderate complexity can easily get out of hand. (Try drawing a block diagram representation of the dynamics represented by the foregoing example.)

Some simulation shells give you the advantage of a graphic interface without sacrificing the ability to express complex dynamics in computer code. Such shells permit you to define a two-input, three-output block with inputs F_L and F_T and outputs l, x_T, and ϕ to which you can connect the control system you would want to simulate.

Building your own simulation shell is a feasible project and may be cost effective as well as educational. Among the factors to be considered either in developing your own software or acquiring commercial software are the following:

- Numerical integration algorithm(s)

- Capability for mixing various process modeling techniques.

- Convenience of varying initial condition and process parameters.

- Tabular and graphical output
    - Produced during running of program
    - Files for post-processing

- Ability to interface with other software (e.g., control system design software).

- Ability to interface with hardware.

Physical Simulation

The most spectacular physical simulations are used to evaluate one particular system component, a *human being*. A single aircraft or spacecraft flight simulator can cost tens of millions of dollars and can make use of dozens of computers running simultaneously to simulate the visual and motion inputs that a pilot would experience in performing a typical task. Development and manufacture of such simulators is a substantial industry.

Similar "person-in-the-loop" simulators are used in the process control and nuclear power industries. The purpose of these simulators is generally training the operator, not verifying the operation of a control system, although such simulators are sometimes used for the latter purpose.

Person-in-the-loop simulations are examples of applications in which the goal is to interface real hardware components with simulations of other components of the system. One such application would be to test the operation of a dedicated ("stand-alone") computer that would be used to implement a control system. The dedicated computer would be interfaced to a general-purpose computer through the input and output ports of both computers. If the dedicated computer is designed to handle analog data, the general-purpose computer would have to generate this data by means of a suitable digital-to-analog converter. Similarly, if the dedicated computer generates an analog signal to drive an actuator, for example, the general-purpose computer would need an analog-to-digital interface.

Real-time operation is a key requirement of a hardware-in-the-loop simulation, in contrast to mathematical simulation in which this requirement is absent. Not only must the dynamic variables have the correct numerical values, but they must also occur at the same instants as they would in the physical process. This imposes a severe speed requirement on the computer used to perform the simulation and is one application that may demand a supercomputer or one with a massively parallel architecture. Speed, however, is not a basic requirement of a mathematical simulation, as long as you are willing to wait long enough to obtain the desired results.

1.4   NOTES

Note 1.1   History

Although the recognition of control technology as a distinct discipline is relatively recent, feedback control systems date back to ancient times. The industrial revolution in eighteenth and nineteenth century England stimulated the invention of many control systems, Watt's ball governor being perhaps the best known. The behavior of these systems, notably stability, engaged the attention of some of the most famous mathematical physicists of the era: Hook, Maxwell, Airy, and others. The early twentieth century saw the development of more ambitious control systems such as aircraft autopilots and ship stabilizing systems.

Until World War II, however, there was no systematic approach to control system analysis and design; each problem in stability or performance necessitated the development of new analytical techniques to elucidate the problem and the invention of hardware to correct it.

Assembled during the 1940s to aid in the war effort, a group of engineers, scientists, and mathematicians at the MIT Radiation Laboratory worked on control problems related to pointing guns and radar antennas, and developed systematic techniques for the analysis and design of feedback control systems, or "servomechanisms" as they were then called. A highlight of this activity was the use of frequency domain methods (including the Laplace transform) which had theretofore been used only for the analysis and design of electrical networks for use in communications. The results of the wartime effort at MIT were summarized in the famous monograph of James, Nichols, and Phillips, [2] which was the "bible" for many control engineers of the early postwar era.

The successes of frequency-domain methods for the analysis and design of control systems for linear, continuous-time control systems motivated the development of similar transform methods (i.e., the $\mathcal{Z}$-transform) for discrete-time systems. The development started during World War II and reached its climax during the mid-1950s.

Preoccupation with transform methods and hence with linear, time-invariant systems, however, created the erroneous impression that only such systems deserved serious consideration, and all but precluded (at least in the United States) systematic consideration of other methods, particularly state-space methods that had been developing in the USSR. A few voices, such as those of R.E. Kalman and J.E. Bertram (who brought Liapunov's Second Method [3] to the attention of the US control community), were heard without very much interest.

A shift in attention to nonlinear systems, state-space methods, optimum control theory, and other Soviet mathematical developments in dynamics systems was probably a consequence of the Soviet launch of *Sputnik* in 1957. This event turned all eyes eastward. The work of Soviet scientists and mathematicians was translated into English, dissected, and devoured. Names that had never before been heard (Letov, Tsypkin, Pontryagin, Aizerman, Feldbaum, and many others) became almost as familiar as those of Western control scientists and engineers. By consensus, Moscow was the natural venue for the first Congress of the International Federation of Automatic Control in the summer of 1960.

In the US, under the aegis of the Research Institute of Advanced Studies (RIAS), Solomon Lefchetz organized a group of scientists and engineers to conduct research in dynamic systems and control. This group, which included such notable figures as J. LaSalle, R.E. Kalman, R. Bucy, R. Bass, and H. Kushner, produced a number of important theoretical contributions of practical value and established the mathematical style that is characteristic of contemporary control theory. (At the demise of RIAS, the group moved to Brown University.)

The center of gravity of control system research moved westward during the 1970s and 80s and is again firmly rooted in the United States. The American Control Conference (held annually in early summer under the sponsorship of the American Automatic Control Council) and the Conference on Decision and Control (held annually in early winter under sponsorship of the IEEE Control System Society) are widely regarded as the world's most important.

Note 1.2   Order reduction methods

Use of your knowledge of the physical principles of operation of a system to eliminate the unimportant phenomena is an effective way of reducing the order of a dynamic system. There may be instances, however, when physical knowledge and insight is inadequate: the model may, for example, be specified in the form of a high-order transfer function with no accompanying explanation of the significance of the state variables. Or the model may be the result of a finite-element analysis, in which the state variables have physical significance, but there are just too many of them.

The earliest (and still popular) method of reducing the order of a transfer function is to expand it in partial fractions and simply omit the contributions from the poles farthest from the origin, adjusting the overall gain, if necessary to maintain the correct d-c response. One problem with this method is that it may give a reasonable approximation for the open-loop behavior of the plant but be inadequate for predicting the closed-loop behavior.

In an effort to avoid this problem, Hutton and Friedland [4] developed an approximation method in which the transfer function is expanded in continued fractions with the expansion being terminated at the proper stage to give the desired order approximant. This method is similar to a Padé approximation, which has been studied by a number of investigators.

These methods have state-space counterparts. The state-space equivalent of partial fraction expansion is transformation to normal coordinates (e.g., the Jordan normal form), which results in a number of first- or second-order systems in parallel. Omitting the subsystems having the eigenvalues at the highest frequencies is equivalent to omitting the contributions from the poles farthest from the origin. The Routh approximation method also has a state-space counterpart.

Another state-space order-reduction method was developed by B. Moore [5] based on singular value analysis of the controllability and observability matrices.

Note 1.3   Controller simplification

Instead of simplifying the process model for purposes of design, B.D.O. Anderson and his colleagues [6] suggested designing the controller for the actual model and then simplifying the resulting controller. Since the computational complexity of designing the controller increases as some power of the order of the process, the practicality of this approach depends upon the availability of computers and software tools with which to perform the needed computations. Rapid strides in computer hardware and software make this approach increasingly attractive. One advantage that this approach has over simplifying the process model for design is that the design for the more complete model is available, so you can see how much performance is lost by using a simpler control law.

1.5   PROBLEMS

Problem 1.1   Dynamics of motor-driven robot arm

A robot arm is driven by a d-c servomotor as shown in Figure 1.15. Show that the process is governed by the following differential equations:

$$\dot{\theta} = \omega \tag{1.10}$$

$$\dot{\omega} = -\alpha\omega - \Omega^2 \sin(\theta) + \beta u \qquad (1.11)$$

Find α, β, and Ω^2 in terms of the physical parameters of the system, namely

K : torque-to-current ratio
J : motor moment-of-inertia
R : armature resistance
m : mass of moment arm (assumed concentrated at the end)
l : length of moment arm

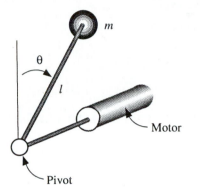

Pivot

Figure 1.15 Servomotor-driven robot arm.

Problem 1.2 Parameters of autonomous vehicle

Consider the autonomous vehicle described in Example 1.2. Each of the wheels of the vehicle is directly driven by its own d-c motor.

(a) Determine the parameters α, β, γ, and δ in terms of the physical parameters of the vehicle:

J_m : Motor moment of inertia
K : Motor torque-to-current ratio
R : Motor armature resistance
r : Radius of wheel
J_w : Moment of inertia of wheel
m : Total mass of platform, including motors, wheels, etc.
J_p : Moment inertia of platform about vertical axis
d : Distance between wheels

(b) Determine the numerical values of the parameters for the following physical data:

J_m : 2×10^{-5} Kg-m^2
K : 0.015 N-m/amp
R : 2 Ω
r : 0.06 m
J_w : 10^{-4} Kg-m^2
m : 10. Kg
J_p : 0.1 Kg-m^2
d : 0.3 m

Problem 1.3   Proportional navigation

Consider the missile navigation problem described in Section 1.1.4. Show that the control law (1.8) with $k > 3$ will result in interception of a target moving at constant velocity. *Hint:* Express the equations of motion of the missile in a coordinate system attached to the target and then resolve the acceleration vector into components along and normal to the velocity vector.

Problem 1.4   Transfer function for temperature control

Starting with the differential equations in Example 1.5 , derive the transfer function from the input to the designated output.

1.6   REFERENCES

[1] J.G. Ziegler and N.B. Nichols, "Optimum Settings for Automatic Controllers," Trans. ASME, Vol. 64, No. 8, pp. 759–768, 1942.

[2] H.M. James, N.B. Nichols, and R.S. Phillips, *Theory of Servomechnaisms* (MIT Radiation Laboratory Series, Volume 25), McGraw-Hill, New York, 1947.

[3] R.E. Kalman and J.E. Bertram, "Control System Analysis and Design via the Second Method of Lyapunov," Trans. ASME, J. Basic Engineering, Vol. 82, pp. 371–399, 1960.

[4] M. Hutton and B. Friedland, "Routh Approximations for Reducing Order of Linear, Time Invariant Systems," IEEE Trans. on Automatic Control, Vol. AC-20, No. 3., pp. 329–337, June 1975.

[5] B.C. Moore, "Principal Component Analysis in Linear Systems: Controllability, Observability, and Model Reduction," IEEE Trans. on Automatic Control, Vol. AC-26, No. 1., pp. 17–27, January 1981.

[6] Y. Liu, B.D.O. Anderson, and U.-Y. Ly, "Coprime Factorization Controller Reduction with Bezout Identity Induced Frequency Weighting," Automatica, Vol. 26, No. 2, March 1990.

Chapter 2

QUALITATIVE BEHAVIOR AND STABILITY THEORY

Everyone has an intuitive grasp of the concept of stability: the ability of a system to return to equilibrium after a small disturbance or perturbation. Is a knowledge of the mathematical theory of stability indispensable in order to design a satisfactory control system? Probably not. If it were, there would not be many working control systems.

You can empirically determine whether a prospective control system is stable by simulating its behavior after a disturbance. Running a simulation is prudent and should be done if possible, but it may not answer all questions. Only a finite number of cases can be investigated by simulation; it is possible (by Murphy's Law even likely) that one of the cases not investigated will be the one that results in disaster. Moreover, you are generally concerned not only with the stability of a nominal design, but with its robustness, i.e., the ability of the design to weather the vicissitudes of the environment, aging, sloppy fabrication, etc. In other words, you want to know how much punishment the design can accept without losing stability.

Simulating all possible departures from the nominal design, however, is not realistic. It is desirable to predict the qualitative behavior of a system by some means other than simulation. At the minimum, you want to know whether the system will be stable. It would be helpful to have some theory that could be used for this purpose.

One of the goals of this chapter is to make the intuitive concepts of stability somewhat more precise (rigorous). The only way of reaching this goal is via a series of definitions, theorems, and proofs. Although this approach runs counter to the spirit that informs this book, it appears to be unavoidable. If they turn you off, feel free to just skim this chapter and accept it on faith or authority.

The objective of the mathematical theory of stability is to determine whether the process is stable *without* needing to solve the differential equations that describe the process, either analytically (which may be impossible) or by numerical integration, i.e., simulation. This is the goal that was addressed by the Russian mathematician, A.M. Liapunov, [1] nearly a century ago. For about fifty years Liapunov's work was scarcely noticed in the technical community of the West. His work was brought to the attention of control engineers in the United States and Western Europe only in the 1950s, largely through the efforts of S.Lefschetz and his colleagues. (See note [1]).

As long as one is dealing with linear, time-invariant systems, there is little need for a precise definition of system stability, since all definitions lead to the same criterion for stability, namely the nature of the eigenvalues, or poles of the system. If all the system poles are in the left half plane, the system is (asymptotically) stable; if one or more poles are in the right half plane, the system is unstable. Poles on the imaginary axis, which is the line dividing the region of asymptotic stability from the region of instability, are a complication of no great consequence. Stability theory for linear, time-invariant systems thus reduces to the algebraic theory of the location of the roots of polynomials.

You can readily appreciate that the situation with regard to nonlinear systems is not nearly as simple. It is not only that the nonlinear systems do not have eigenvalues to characterize their behavior, but even more fundamentally, that various definitions of stability may not lead to the same stability criteria: a system may turn out to be stable under one definition but not under another. (As a practical matter, a system that is classified as unstable by *any* criterion is likely to be unsatisfactory.)

The two basic classes of stability are:

1. Stability of equilibrium state(s)

2. Input-Output stability

Stability of equilibrium states is concerned with the question of whether a system will return to equilibrium after a disturbance. In input-output stability the question is whether a particular class of inputs (usually those that are bounded in magnitude) will produce a bounded output. This is usually called *bounded-input, bounded-output* (*"BIBO"*) stability.

2.1   EQUILIBRIUM

Physically, a system is in equilibrium when it remains in one state in the absence of any external (exogenous) input. This means that the time derivative $\dot{x}$ is zero.

---

[1] Also spelled "Lyapunov."

For an unforced (time-invariant) dynamic system governed by

$$\dot{x} = f(x) \tag{2.1}$$

the equilibrium state (or states) is the solution (or solutions) to the algebraic equation

$$f(x) = 0 \tag{2.2}$$

In a linear system $f(x) = Ax$ so the origin, $x = 0$, is an equilibrium state of every linear system. Moreover, if A is nonsingular, the origin is the *only* equilibrium state. If A is singular, the linear subspace that is defined by $Ax = 0$ constitutes an equilibrium region.

As you might imagine, matters are not so simple in a nonlinear system. Since $f(x)$ may not have even one *real* solution, the system may have no real equilibrium states. (The significance of *complex* equilibrium states in the dynamic model for a physical system is problematical.) Moreover, the origin may or may not be one of the equilibrium states. Still further: unlike the situation in a linear system in which multiple equilibrium states lie in one continuum, in a nonlinear system the equilibrium states, if any, can be isolated points, continuums (i.e., curves, surfaces, etc.) or combinations thereof. Finally, actually finding the equilibrium states, i.e., solving $f(x) = 0$ may be an awesome task.

Example 2.1   Pendulum

The classic nonlinear system is the pendulum, the motion of which, in the absence of damping, is given by

$$m\ddot{\theta} + g\sin\theta = 0$$

which, in state-space form, are

$$\frac{d\theta}{dt} = \omega$$

$$\frac{d\omega}{dt} = -(g/m)\sin\theta$$

As is well known, there are two equilibrium states

$$[\omega = 0, \quad \theta = 0] \quad \text{and} \quad [\omega = 0, \theta = \pi]$$

The first is stable, the second is unstable.

Example 2.2   Aronowitz Equation for Ring-Laser Gyro

Nonlinear systems without an equilibrium state are not mathematical curiosities, but may represent important physical systems. As an example consider the system governed by

$$\dot{\phi} = \omega_i - \omega_L \sin \phi$$

This equation, known in the field of ring-laser gyros as the "Aronowitz equation" [1] governs the behavior of a ring-laser gyro that exhibits the "lock-in" phenomenon. The output of the gyro is proportional to ϕ. In an ideal gyro, for which ω_L would be zero, the output rate $\dot{\phi}$ would be proportional to the input rate ω_i. In the presence of lock-in ($\omega_L \neq 0$) the output rate would not be equal to the input rate. In fact, if

$$|\omega_i| < \omega_L$$

the Aronowitz equation would *have* an undesired equilibrium solution: The gyro would be locked-in. The desired condition is

$$|\omega_i| > \omega_L$$

in which case the gyro would not have an equilibrium state.

Time-varying systems cause other problems. In particular, for a time-varying system

$$\dot{x} = f(x, t)$$

to have an equilibrium state, the equation $f(x,t) = 0$ must have a solution

$$x = \text{const.} \quad \text{for all } t, \tag{2.3}$$

since, if the solution to (2.3) were to depend on t, its time derivative would not be zero.

2.2 TRAJECTORIES IN STATE SPACE

For understanding the behavior of a nonlinear system it is often convenient to consider the "trajectories" of the system in state space. These trajectories are curves that show the evolution of the state of the system in time. An arrow on the trajectory points in the direction taken by the trajectory as time increases. Pictorial representations are most useful for second-order systems, in which the state space is a plane. With the aid of graphical techniques, trajectories for third-order systems can be visualized on a two-dimensional surface. Trajectories for higher order systems can be imagined but not readily depicted. (Since the state space of a first-order system is a line, its trajectories do not make very interesting pictures.)

Typical trajectories of second-order time-invariant linear systems are shown in Figure 2.1. There are six generic trajectory types, depending on the nature of the eigenvalues of the system, as listed in Table 2.1

Table 2.1   Trajectories of Second-Order Linear Systems

Definition of Equilibrium State	Nature of Eigenvalues
Stable Node	Both real and negative
Unstable Node	Both real and positive
Stable Focus	Complex, with negative real parts
Unstable Focus	Complex, with positive real parts
Center	Imaginary
Saddle Point	Real and of opposite sign

The terminology used in Table 2.1 was originated by the French mathematician H. Poincaré at around the beginning of the twentieth century. (See Note 2.1).

Since the origin is not the only equilibrium state if one of the eigenvalues of the system is zero, it is not as meaningful to represent typical trajectories.

The behavior of second-order nonlinear systems can be substantially more complicated than the behavior of linear systems. First of all, there may be more than one isolated equilibrium point. The trajectories in the vicinity of each equilibrium point might resemble those shown in Figure 2.1. But at distances remote from any of the equilibrium points, the trajectories may have no resemblance to those of linear systems. One special type of trajectory that is not found in linear systems is the "limit cycle," a closed, periodic trajectory of fixed shape, as shown in Figure 2.2. Although a limit cycle and the trajectory around a center are both closed and periodic, there is an important distinction between the two. If the initial state is perturbed from the trajectory around a center it will move to a new closed trajectory. On the other hand, if the initial state is perturbed away from a limit cycle, the resulting trajectory will either return to the limit cycle (if the limit cycle is stable), or diverge entirely from the limit cycle. A limit cycle typifies the behavior of an oscillator with a fixed amplitude. Such an oscillator is often desirable in technical applications and represents one situation in which a nonlinear system is required.

For most of the century, it was thought that limit cycles represented the most complex behavior likely to occur in nonlinear systems. Although the possibility of more complex, chaotic behavior of nonlinear systems was suggested by the work of Poincaré, hardly anyone gave much heed to the possibility. The situation changed very rapidly with the personal computer revolution of the 1970s.

2.3 CHAOS

Nothing in the theory of dynamic systems has aroused more excitement in the scientific community than the topic of *chaos*. Not only has it engendered a large number of papers in the mathematics and systems theory literature, but it has also spawned a number of books written for the general educated reader. One of these books [2] has even made it to *The New York Times* best-seller list.

Chaos is said to be implicated in such diverse phenomena as turbulence, erratic behavior of the stock market, biological population growth and decay, and the diversity

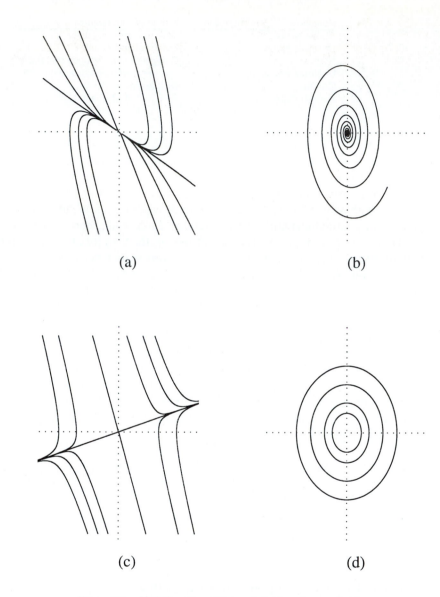

(a)   (b)

(c)   (d)

Figure 2.1   Varieties of equilibrium behavior of second-order linear systems. (a) Stable node. (b) Stable focus. (c) Saddle point. (d) Center.

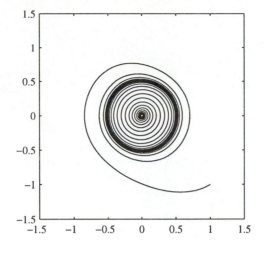

Figure 2.2 Limit cycle.

of geological formations.

One would expect that a system governed by "orderly" differential equations, i.e., differential equations with right hand sides that are smooth functions, would exhibit orderly behavior. Trajectories would be expected to converge to equilibrium states, diverge to infinity in a predictable manner, or to well-defined limit cycles. That such systems, even of relatively low order (2 or 3), can exhibit chaotic behavior came as a surprise.

The earliest investigation of chaotic behavior was reported in a now famous paper [3] by Lorenz published in a relatively obscure journal on meteorology. Lorenz hypothesized a very simple model for studying an atmospheric phenomenon. This model is recognized to be a gross simplification of the true phenomenon, but when the work was done in the 1960s computer resources were scarce. The simple model proved to be a blessing in disguise, because it shattered the notion that simple systems are incapable of complex behavior. Lorenz's model consists of the third-order system:

$$\dot{x} = 10(-x + y)$$
$$\dot{y} = 28x - y - xz \tag{2.4}$$
$$\dot{z} = -8z/3 + xy$$

It is hard to imagine that a dynamic system as innocuous as this could be responsible for a whole new branch of systems science, but it was.

The striking result that Lorenz exhibited in 1962 was obtained by a computer simulation of (2.4)—a simulation that by now has probably been repeated thousands of times. A typical trajectory (see Figure 2.3) has the appearance of a butterfly: it seems to wander around one wing for a substantial time and then, without warning, take off and go to the other wing. This process recurs from time to time, but it is not periodic.

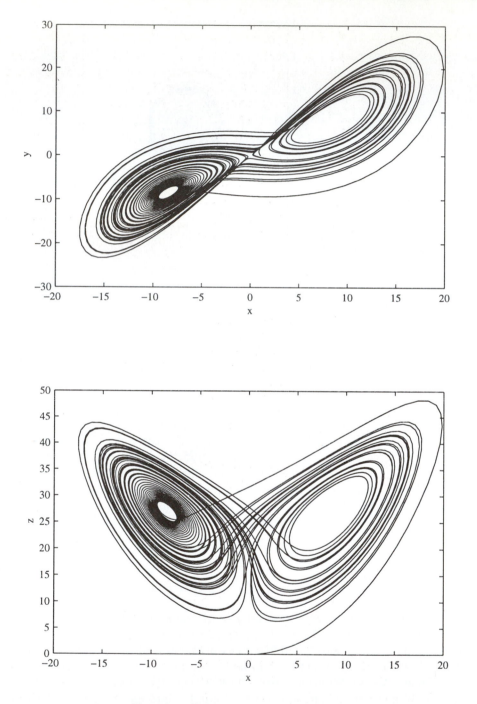

Figure 2.3   Trajectories of Lorenz attractor.

Lorenz's butterfly is an example of a "strange attractor," a term devised by Rouelle and Takens [4] to characterize this type of bounded, aperiodic motion characteristic of chaotic systems. Proof that a given system has a strange attractor or exhibits other chaotic behavior is devilishly difficult, entailing a combination of tedious geometric constructions and use of some of the most sophisticated ideas of differential geometry.

Insights into chaotic phenomena are frequently the results of computer simulations of dynamic systems, in which such objects as "strange attractors" are visualized on the screen of the computer terminal. You might think that the observed chaotic behavior is an artifact of the computer simulation—numerical integration and finite precision arithmetic. But the same behavior can be produced with physical analog components so the chaotic behavior must be accepted as a physical reality.

What interest does chaos hold for the control engineer? First, by recognizing the possibility of chaotic behavior, you are warned to be on guard against it. Second, you may have a better appreciation of the challenge you are up against when called upon to control a chaotic system.

2.4   STABILITY IN THE SENSE OF LIAPUNOV

Stability in the sense of Liapunov is concerned with the behavior of a system in the vicinity of an equilibrium state. Qualitatively one would say that an equilibrium state is stable if the system returns to the equilibrium state from an initial state near the equilibrium state. Since a nonlinear system may have a number of equilibrium states, some stable and others not stable, it would not make sense to define this type of stability as a property of the system rather than of its equilibrium states.

As noted earlier, the origin $x = 0$ may or may not be an equilibrium state of a nonlinear system. To simplify definitions and results, however, it is convenient to consider systems in which the origin is the equilibrium state of interest. If the origin is not the equilibrium state, it is always possible to translate the origin of the coordinate system to that state, so no generality is lost in assuming that the origin is the equilibrium state of interest.

The concept of stability in the sense of Liapunov is contained in the following:

Definition 2.1 (Liapunov Stability) *The origin of the system $\dot{x} = f(x, t)$ is stable if and only if for any $\epsilon > 0$ there exists a $\delta(\epsilon)$ such that $\|x(t)\| < \epsilon$ for all $t > \tau$ for any initial state $x(\tau)$ that satisfies $\|x(\tau)\| \leq \delta(\epsilon)$.*

If this "epsilonics" definition leaves you cold, think of a contest between you, the system designer, and an adversary (nature?). The adversary picks a region in state space of radius ϵ and challenges you to find another region, of radius δ, such that if the initial state starts out inside your region, it remains inside his region. If you can do this, your system design is stable. The situation is depicted in Figure 2.3.

It is important to note that the regions in the definition must be of the same dimension as the state space. In a third-order system, for example, the regions must be

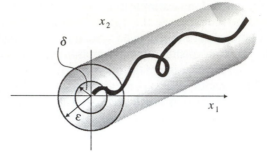

Figure 2.4   Stability definitions.

three-dimensional regions. The regions are usually k-dimensional hyperspheres:

$$\|x\| = (\sum_{i=1}^{k} x_i^2)^{1/2}$$

where k is the dimension of the state space. But it is sometimes desirable to use regions defined by more general norms, such as

$$\|x\| = \max_i |x_i|$$

or,

$$\|x\| = \sum_{i=1}^{k} |x_i|$$

The difference between a limit cycle and trajectories around a center illustrates this definition. A center is a stable equilibrium point, because no matter how small a region your adversary picks, you can always pick an initial condition to keep the trajectory of your system inside his region. If your system has an equilibrium state surrounded by a stable limit cycle such as shown in Figure 2.5, then states inside the limit cycle will lie on trajectories that converge to the limit cycle. Your adversary has only to pick his region anywhere inside the limit cycle to prevent you from finding an initial state that will result in trajectories that remain inside his region.

The possibility that trajectories can remain arbitrarily close to an equilibrium state without actually getting to that state (as in the case of a center) is precluded when the equilibrium state is asymptotically stable:

Definition 2.2 (Asymptotic Stability) *The origin is asymptotically stable if it is stable and if $x(t) \to 0$ as $t \to \infty$.*

Isn't it sufficient, you might ask, that $x(t) \to 0$? Why does the definition of asymptotic stability require stability? In time-invariant linear systems, of course, a system

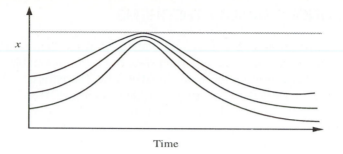

Figure 2.5   Unstable equilibrium state.

for which $x(t) \to 0$ has eigenvalues with negative real parts, so it must be stable. But this is not necessarily the case in a nonlinear system. Figure 2.5 shows a situation in which the state tends initially away from the origin but subsequently returns. In this case it will not be possible to guarantee staying within your adversary's region, if he selects that region to be smaller than the limit L.

The above definitions of stability are "local": you are free to select your region of initial states as small as you need to in order to stay inside your adversary's region. The largest region of initial states that you can choose around the equilibrium point, is called the "region of attraction" or, more simply, the stability region. If the stability region is the entire state space, the equilibrium state is said to be "globally stable." The origin of a stable linear system is globally stable. But in nonlinear systems global stability is a rarity. In particular, a nonlinear system having more than one equilibrium state cannot have any states that are globally asymptotically stable, since if the region of initial states around one equilibrium state includes states in the region of attraction of another, the trajectory of the system will converge to the latter and hence cannot remain arbitrarily close to the former.

The above definitions are probably adequate for most practical stability analysis. There are more definitions, however, which can be used to draw finer distinctions between types of stability in time-varying systems. Such systems may be thought of as having dynamic characteristics that change with time. The eigenvalues of $A(t)$ in the linear system $\dot{x} = A(t)x$, for example, may move with time. If they start in the left half plane but move to the right half plane and stay there, the system is doubtless unstable. If they start in the right half plane and move to fixed locations in the left half plane, the system is asymptotically stable, but the transient behavior starts out looking like that of an unstable system. To further classify the type of behavior characteristic of time-varying systems, the above definitions may be qualified by the term "uniform." If the stability is uniform, the region inside which the initial state must start can be selected independent of the starting time.

2.5   LIAPUNOV STABILITY THEOREMS

The basic problem that Liapunov addressed in the theory that bears his name is to determine whether an equilibrium state of a system is stable *without* solving the differential equations of the system analytically. The nature of the eigenvalues of a linear, time-invariant system completely determines whether it is stable. Hence it is not necessary to solve the differential equations of the system to determine whether it is stable.

The absence of an analog for the eigenvalues complicates the problem immensely in a nonlinear system. Nevertheless, Liapunov succeeded in developing a remarkably simple and elegant criterion for stability. (See Note 2.1 [9].) This criterion is embodied in the theorem which in its basic form can be stated as follows:

If there exists a positive-definite function $V(x)$ that is never increasing, then the origin is stable.

The theorem is suggested by the physical concept of the energy in a system. Energy in a physical system is a positive definite function of the state; in a conservative (stable) system it remains constant; in a system with dissipation the energy always decreases. The energy can increase in a physical system only if it is unstable.

Reasoning physically on the basis of the energy in a system is not satisfactory because in a complex system with components of many different types (mechanical, electrical, chemical, etc.) it may not be possible to physically identify the energy in terms of the state variables. Hence it is necessary to use a function that has the mathematical properties of the energy in a system, even if it is not the actual energy in the system. Remarkably enough, the only property of energy that is needed to establish stability is the fact that it is positive-definite. To emphasize: If any positive definite function of the state with a non-positive derivative can be found, the origin is stable.

Stated formally, Liapunov's stability theorem is:

Theorem 2.1 (Liapunov Stability Theorem) *If there exists a continuously differentiable function $V(x)$ such that*

1. $V(0) = 0$
2. $V(x) > 0$ for all $x \neq 0$
3. $\dfrac{dV}{dt} = \left(\dfrac{\partial V}{\partial x}\right)' f(x) = \displaystyle\sum_{i=1}^{n} \dfrac{\partial V}{\partial x_i} f_i(x) \leq 0$ for all x

then the origin of the time-invariant system

$$\dot{x} = f(x)$$

is stable.

Before we proceed with the proof of this theorem, a few preliminary remarks are in order.

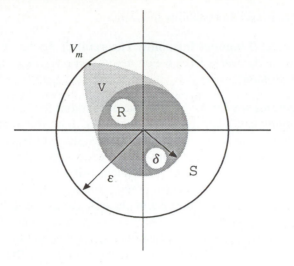

Figure 2.6 Proof of Liapunov's stability theorem.

- A function $V(x)$ satisfying conditions 1 and 2 above is called positive definite.

- Condition 3 is the mathematical statement of the requirement that the function $V(x)$ is non-increasing. The vector $\partial V/\partial x = [\partial V/\partial x_1, \ldots \partial V/\partial x_n]'$ is called the gradient of $V(x)$.

- A function $V(x)$ that satisfies all three conditions of the theorem is called a *Liapunov function*. Hence the stability of the origin of a dynamic system is assured if a Liapunov function for the system can be found.

To prove the theorem we must show that the state $x(t)$ can be kept inside a region $\mathcal{S} := \{x : \|x\| < \epsilon\}$ by starting with the initial state inside a region $\mathcal{R} := \{x : \|x\| < \delta\}$. These two regions are shown in Figure 2.6. Somewhere in $\mathcal{R}$ the function $V(x)$ has a least upper bound (supremum), say V_m. Consider the region $\mathcal{V} := \{x : V(x) \leq V_m\}$. Inside this region lies another region $\mathcal{S} := \{x : \|x\| \leq \delta\}$. Since $V(x)$ is a non-increasing function, then for any x starting in $\mathcal{S}$, it cannot become larger than V_m. Thus the state is confined to $\mathcal{V}$, which is a subset of $\mathcal{R}$. Hence *a fortiori* x is confined to $\mathcal{R}$.

If $\dot{V}(x) = dV(x)/dt$ is not merely non-increasing, but is actually negative definite (i.e., $-\dot{V}(x)$ is positive-definite) then $V(x)$, which is the integral of $\dot{V}(x)$, will ultimately become negative unless $x(t) \to 0$. Since $V(x)$ is postulated to be positive-definite, it is not allowed to become negative. Thus we conclude that $x(t)$ must necessarily approach zero with increasing time. This argument is summarized by the following theorem.

Theorem 2.2 (Asymptotic Stability Theorem) *If there exists a positive-definite function $V(x)$ having a negative-definite derivative $\dot{V}(x)$, the origin is asymptotically stable.*

There is also an instability theorem:

Theorem 2.3 (Liapunov Instability Theorem) *If there exists a positive-definite function* $V(x)$ *whose derivative* $\dot{V}(x)$ *is non-negative in a region containing the origin, then the origin is unstable.*

Theorems similar to the above can be stated for time-varying systems and using time-varying candidate Liapunov functions. The theorems are quite similar to the above, but the time-varying nature of the system introduces technical problems that deflect us too far from our main course. (See Note 2.2.)

Analyzing the stability of a nonlinear using the above theorems has come to be known as Liapunov's Second (or Direct) Method. (His first, or indirect method entails studying the behavior of the system linearized at the equilibrium state, to be discussed below.) The essence of this method is *finding* a Liapunov function. It is easy enough to specify a positive-definite function; any positive-definite quadratic form $x'Qx$ with a positive-definite matrix Q, for example, will do. If the derivative of this function is non-positive (negative-definite) in an entire region containing the origin, then the origin is stable (asymptotically stable). On the other hand, if the derivative is non-negative in a region containing the origin, then the origin is unstable. In either case the analysis is conclusive. More than likely, however, the derivative of an arbitrary positive-definite function will be sign *indefinite* in the vicinity of the origin, i.e., it will take on both positive and negative values. In this case the analysis is inconclusive; it provides no information about the stability of the origin.

For a period of time in the late 1950s and early 1960s a considerable amount of research went into trying to develop control system design techniques based on Liapunov's Second Method. This entailed finding both a control law *and* a Liapunov function. Although these efforts met with limited success, they were soon superseded by research into optimum control, which really was an outgrowth of the work on Liapunov's Second Method. After another hiatus of a decade or so, Liapunov's Second Method reappeared as an analytical tool for proving the stability of certain adaptive control algorithms and for studying robustness. These applications are discussed in later chapters. Note 2.1 provides more of the historical background.

One application of Liapunov's Second Method is to derive a stability criterion for linear systems of the form

$$\dot{x} = Ax \tag{2.5}$$

Consider the following integral as a candidate for a Liapunov function:

$$V := \int_t^\infty x'(\tau)Qx(\tau)d\tau \tag{2.6}$$

where Q is a positive-semidefinite matrix. On differentiating V as defined by (2.6) with respect to the lower limit t we have

$$\dot{V} = -x'Qx \tag{2.7}$$

which is obviously negative semidefinite.

Since V is the integral of a non-negative quantity, it cannot be negative. But it may be zero, depending on the nature of $x(\tau)$. Moreover, if the system that generates $x(\tau)$ is unstable, the integral will not even exist, i.e., it will become infinite as its upper limit becomes infinite. But we can show that the linear system is stable if and only if V is a Liapunov function. In the process, we obtain an algebraic stability criterion.

The solution to (2.6) is

$$x(\tau) = e^{A(\tau-t)}x(t) \tag{2.8}$$

Thus, if V exists it is given by

$$
\begin{aligned}
V &= x'(t)\left(\int_t^\infty e^{A'(\tau-t)}Qe^{A(\tau-t)}d\tau\right)x(t) \\
&= x'(t)Mx(t)
\end{aligned}
\tag{2.9}
$$

where

$$
\begin{aligned}
M &= \int_t^\infty e^{A'(\tau-t)}Qe^{A(\tau-t)}d\tau \\
&= \int_0^\infty e^{A't}Qe^{At}d\tau
\end{aligned}
\tag{2.10}
$$

The time derivative of V can be calculated from (2.10) with the aid of the chain rule:

$$
\begin{aligned}
\dot{V} &= \dot{x}'Mx + x'M\dot{x} \\
&= x'A'Mx + x'MAx
\end{aligned}
\tag{2.11}
$$

On equating this expression for $\dot{V}$ with 2.7, and noting that these expressions are valid for any $x = x(t)$, we obtain

$$A'M + MA = -Q \tag{2.12}$$

Equation (2.12) is called a *Liapunov equation* for obvious reasons. It is known to possess a unique solution if and only if A has no eigenvalues that are the negatives of other eigenvalues [5]. If A corresponds to an asymptotically stable dynamic system, its eigenvalues all lie in the left half plane and thus none can be negatives of others. Thus, we may conclude that if the linear system $\dot{x} = Ax$ is asymptotically stable, a unique solution M to (2.12) exists. Moreover, V is a Liapunov function and hence M is a positive-definite matrix. Conversely, if the solution M is positive definite, the corresponding dynamic system is stable.

Note that this result holds for any positive-definite matrix Q. In fact, by an extension to the basic theorem, Theorem 2.1, Q can even be positive, semidefinite, provided that $\dot{V}$ does not become identically zero for any initial state (as is usually the case). Thus one way of determining whether a linear system is stable is to solve the Liapunov equation (2.12) for any convenient positive-definite matrix Q (for example, $Q = I$) and test the resulting solution for positive-definiteness. In principle, solving the Liapunov equation and testing the solution for positive-definiteness require a finite number of steps, while finding the exact eigenvalues of a matrix requires an infinite number of steps. With up-to-date numerical methods, however, the computation of the eigenvalues of a matrix is not usually a very difficult problem, and the Liapunov equation has no special advantage. (It should be noted that if the characteristic equation of A, i.e., $|sI - A| = 0$ is known, the stability can be determined using the Routh-Hurwitz algorithm. See Note 2.3.)

2.6   STABILITY OF LINEARIZED SYSTEMS; LIAPUNOV'S FIRST METHOD

The most frequently asked question is the behavior of a nonlinear system in the immediate vicinity of an equilibrium state: If the dynamics are linearized about the equilibrium state, and perturbations away from equilibrium are small, can we predict whether the behavior will be stable? Is it sufficient to examine the dynamics of the linearized system? These questions can be answered affirmatively by using Liapunov's *First* Method to be developed in this section.

For simplicity, assume that the equilibrium point to be tested for stability is the origin. Thus, in the nonlinear system

$$\dot{x} = f(x) \tag{2.13}$$

we have

$$f(x) = 0 \tag{2.14}$$

Suppose that the elements of the Jacobian matrix

$$A = \begin{bmatrix} \frac{\partial f_1}{\partial x_1} & \cdots & \frac{\partial f_1}{\partial x_n} \\ \cdots & \cdots & \cdots \\ \frac{\partial f_k}{\partial x_1} & \cdots & \frac{\partial f_k}{\partial x_k} \end{bmatrix} \tag{2.15}$$

exist and are continuous at the origin. Then, by consequence of the multidimensional Taylor theorem [6] we can write

$$f(x) = Ax + g(x) \tag{2.16}$$

where

$$\lim_{x \to 0} \frac{\|g(x)\|}{\|x\|} = 0 \tag{2.17}$$

As a candidate Liapunov function we select $V(x) = x'Mx$ where M is the solution of a Liapunov equation (2.12) with

$$Q = I$$

Then

$$\dot{V} = (x'A' + g'(x))Mx + x'(Ax + g(x))$$
$$= -x'Qx + g'(x)Mx + x'Mg(x) \qquad (2.18)$$

Now, by virtue of (2.17), $g(x)$ approaches zero faster than x. Thus, by keeping x sufficiently small, we can keep $\|g'(x)Mx + x'Mg(x)\|$ smaller than $x'Qx = x'x$. Hence, we can conclude:

Theorem 2.4 (Stability of Linearized System) *The origin of the nonlinear system $\dot{x} = f(x)$ is asymptotically stable if the Jacobian matrix (2.15) has all of its eigenvalues in the left half plane (excluding the imaginary axis).*

The Liapunov instability theorem can be invoked to show that the origin is unstable if the Jacobian matrix A has eigenvalues in the right half plane. But if the linearized system has eigenvalues on the imaginary axis, the stability in the vicinity of the origin depends on the nature of the higher-order terms, i.e., $g(x)$ in (2.16).

2.7 REGION OF ATTRACTION

Asymptotic stability is a local property. It holds in the neighborhood of the equilibrium state of which we are investigating the stability. In a linear system, local asymptotic stability implies global asymptotic stability: If the trajectories starting at states near the origin all approach the origin, then trajectories starting from states arbitrarily far from the origin also approach the origin. But this is often not the case in nonlinear systems. Suppose, for example, that the nonlinear system has two asymptotically stable equilibrium states. This means that trajectories starting in neighborhoods of the equilibrium states will approach these equilibrium states asymptotically. These neighborhoods, however, cannot encompass the entire state space; as both neighborhoods grow from the equilibrium states at their midst, they must ultimately intersect. Trajectories starting at points in the intersection will have to go to both equilibrium states, which would contradict the assumption of unique solutions of the differential equations of the system.

In a typical nonlinear system the region around an equilibrium state within which all trajectories approach the equilibrium state is called its *region of attraction*. The size of the region of attraction is often one of the important factors in the performance of a control system. You want to be sure that noise or temporary faults cannot bring the state of a system outside the region of attraction of the desired equilibrium state.

In second-order systems, the region of attraction is usually bounded by an unstable limit cycle. (Why unstable?) This limit cycle can be determined by simulation. Since the limit cycle is unstable, the best way to find the limit cycle is to reverse the direction of time in the numerical integration; this will turn the unstable limit cycle into a stable one.

Since a limit cycle is a one-dimensional curve, it cannot serve as the boundary of the region of attraction in a system of third or higher order, and hence simulation is not so useful a tool for estimating the size of the latter. For an alternative method we can turn to Liapunov theory.

Suppose the Liapunov function $V(x)$ is globally positive-definite. Its derivative $\dot{V}(x)$ is negative-semidefinite in the vicinity of the origin. (Otherwise V would not be a Liapunov function.) If $\dot{V}(x)$ is globally negative-semidefinite, the origin is globally asymptotically stable. If not, however, there will be a region $\mathcal{U}$ for which $\dot{V}(x) \leq 0$. You might be tempted to regard $\mathcal{U}$ as an estimate of the region of attraction. But it would not be a very good estimate, because there is no guarantee that trajectories starting in this region will remain in this region. It is possible that a trajectory that starts in this region will leave the region of attraction, even though $V(x)$ starts out decreasing. A valid estimate of the region of attraction would be a set of points $\mathcal{V} \subset \mathcal{U}$ for which $V(x) \leq C$. Starting in $\mathcal{V}$, every trajectory will go to states of smaller values of V and cannot escape from $\mathcal{V}$. Any $\mathcal{V}$ is a conservative estimate of the region of attraction; the actual region of attraction can of course be larger. To get the least conservative estimate of the region of attraction, you would try to maximize the size of $\mathcal{V}$ by making its boundary tangent to the boundary of $\mathcal{U}$. The constant C that achieves the largest value is given by

$$\bar{C} = \min_{\dot{V}(x)=0} V(x)$$

Having found $\bar{C}$, the estimate of the region of attraction is

$$\bar{\mathcal{V}} = \left\{ x : V(x) \leq \bar{C} \right\}$$

as shown in Figure 2.7

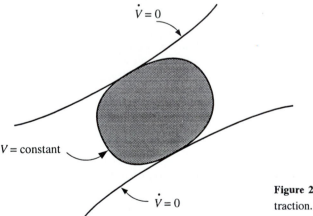

$\dot{V} = 0$

$V = \text{constant}$

$\dot{V} = 0$

Figure 2.7   Estimating the region of attraction.

2.8   NOTES

Note 2.1   Background

The subject of differential equations has occupied applied mathematicians for two centuries. One endeavor has been the development of analytical methods for obtaining exact or approximate solutions to specific differential equations. Interest in this aspect of differential equations has waned with the increasing feasibility of obtaining accurate numerical solutions by use of a digital computer.

Interest continues, however, in the qualitative behavior of differential equations, especially in view of the recent discovery of chaotic phenomena that defy numerical integration methods. Investigation of this topic started in the nineteenth century with the work of H. Poincaré [7] and A.M. Liapunov [9], among others. Fundamental research into the behavior of second-order systems was carried out by Poincaré and Bendixson [8], who introduced the terminology and classification of the different types of trajectories in the phase plane. The foundations of stability theory was laid by A.M. Liapunov [9] at the turn of the century.

The theory of differential equations was studied by engineers in the field of applied mechanics, but was paradoxically of little interest to control engineers in the United States until the mid 1950s. The field of automatic control was dominated by electrical engineers, who successfully adapted frequency-domain methods, developed originally for communications systems, to the solution of a variety of control problems. Frequency domain methods, however, were not entirely successful in dealing with nonlinear systems. One of the few works that addressed some of the engineering issues of nonlinear control was the wartime monograph of N. Minorsky [10]

Interest of control engineers in the theory of differential equations, linear and nonlinear, kindled at least in part by the Soviet achievements in aerospace, began around the time of the launching of the Sputnik in 1957. Mathematicians who had been engaged in the study of differential equations (notably S. Lefschetz and J. LaSalle) joined in a partnership with the control engineering community to initiate the development of modern control theory. A very influential tutorial paper of the period was the Kalman and Bertram survey of Liapunov theory [11] which opened the eyes of many engineers to the burgeoning state-space approach.

The importance of the theory of differential equations, especially qualitative behavior, to the field of automatic control is now universally recognized. Several monographs on nonlinear dynamic systems with an orientation towards problems of interest to control theory have recently appeared [12], [13], [14].

Note 2.2   Time-varying Liapunov functions

For a time-varying function $V(x, t)$ to be a Liapunov function, it must satisfy

$$\alpha_1(||x||) \leq V(x, t) \leq \alpha_2(||x||) \tag{2.19}$$

where $\alpha_1(r)$ and $\alpha_2(r)$ are strictly increasing functions of the scalar argument r. The left hand side of this inequality comports with our customary understanding a positive-definite function. The right-hand side is a requirement that $V(x, t)$ not become large too rapidly in time. Further details can be found, for example in [13], [14].

For a quadratic candidate Liapunov function

$$V(x, t) = x'M(t)x$$

(2.19) requires that $M(t)$ must never become singular nor infinite.

Note 2.3   Routh-Hurwitz criteria

In a linear, time-invariant system the locations of the roots of the characteristic polynomial (eigenvalues) determine whether the system is stable. The Routh-Hurwitz tests are algebraic algorithms for determining whether a polynomial with real coefficients is a *Hurwitz polynomial*, i.e., whether its roots lie in the left half plane (excluding the imaginary axis), without actually finding the roots, and thereby determining whether the underlying dynamic system is stable. Pencil-and-paper calculation of the roots of a polynomial of even moderate degree is a daunting task; by contrast, the calculations entailed in the Routh-Hurwitz algorithms are relatively modest. Discovery of the algorithms in the late nineteenth century were important contributions to the theory of dynamic systems.

The Routh algorithm [15] is presented, usually without proof, in most elementary textbooks on control theory. The Hurwitz algorithm [16] which has some advantages over the Routh algorithm, is given less frequently. The theoretical foundation of these algorithms is devoid of any relationship with Liapunov stability theory. Fairly recently, however, Parks [17] succeeded in developing a state variable transformation which established the connection between the Routh-Hurwitz algorithms and Liapunov stability theory. An account of the algorithms and their relationships is given in [18].

2.9   PROBLEMS

Problem 2.1   Stability of motor-driven robot arm

From physical considerations it is clear that the equilibrium state $[\theta, \omega] = [0, 0]$ of the motor-driven robot arm of Problem 2.1 is *almost* globally asymptotically stable, i.e., it is asymptotically stable in the entire state space except for the equilibrium state $[0, \pi]$.

Demonstrate this by use of Liapunov's second method.

2.10   REFERENCES

[1] F. Aronowitz, "The Laser Gyro," in M. Ross (Ed.) *Laser Applications* Vol. 1, pp. 133-200, Academic Press, New York, 1971.

[2] J. Gleick, *Chaos: Making a New Science*, Viking Press, New York, 1987.

[3] E.N. Lorenz, "Deterministic Nonperiodic Flow," J. Atmospheric Science, Vol. 20, pp. 130–141, 1963.

[4] D. Rouelle and F. Takens, "On the Nature of Turbulence," Communications on Mathematical Physics, Vol. 20, pp.167-192, 1971.

[5] F.R. Gantmacher, *Matrix Theory*, Chelsea Publishing Co., New York, 1959.

[6] T.M. Apostol, *Mathematical Analysis*, Addison-Wesley Publishing Co., Reading, MA, 1957.

[7] H. Poncaré, *Oeuvres*, Gauthier-Villars, Paris, 1928.

[8] I. Bendixson, "Sur les courbes définies par des équations différentielles," Acta Mathematica, Vol. 24, pp. 1–88, 1901.

[9] A. M. Liapunov, *Problème Général de la Stabilité du Mouvement*, Princeton University Press, Princeton, NJ, 1947. (Reprint of Russian monograph of 1892.)

[10] N. Minorsky, *Introduction to Nonlinear Mechanics*, J.W. Edwards, Ann Arbor, MI, 1947.

[11] R.E. Kalman and J.E. Bertram, "Control System Analysis and Design via the Second Method of Lyapunov," J. Basic Engineering, Vol. 82, pp. 371-399, 1960.

[12] R. Mohler, *Nonlinear Systems* 2 Volumes, Prentice-Hall, Englewood Cliffs, NJ, 1991.

[13] H.K. Khalil, *Nonlinear Systems*, Macmillan Publishing Co., New York, 1992.

[14] M. Vidyasagar, *Nonlinear Systems Analysis*, 2nd Ed., Prentice-Hall, Englewood Cliffs, NJ, 1993.

[15] E.J. Routh, *A Treatise on the Stability of a Given State of Motion*, MacMillan, London, 1877.

[16] A. Hurwitz, "Über die Bedingungen, unter welchen eine Gleichung nur Wurzeln mit negativen reelen Teilen besitzt," Math. Ann. Vol. 146, pp. 273–284, 1895.

[17] P.C. Parks, "A New Proof of the Routh-Hurwitz Stability Criterion Using the 'Second Method' of Lyapunov," Proc. Cambridge Philosophical Society, Vol. 58, Pt 4., pp. 694-720, 1962.

[18] R.J. Schwarz and B. Friedland, *Linear Systems*, McGraw-Hill, New York, 1965.

Chapter 3

GRAPHICAL METHODS FOR STABILITY ANALYSIS

Before desktop computers made simulation a practical engineering tool, it was not easy to predict the behavior of a nonlinear control system without building it. A few graphical methods, developed by extension of methods developed for linear systems, were among the few available tools. Nowadays it is easy to predict how a system will behave by simulation, and the graphical methods discussed in this chapter are no longer essential. Nevertheless, they continue to be useful for several reasons:

- Graphical methods can sometimes provide insights into the behavior of a system that may be difficult to obtain from a simulation alone. For example they can predict the existence of a limit cycle, and provide an estimate of the amplitude of the limit cycle.

- The insights obtained from an indirect, graphical analysis can assist in the definition of the parameters to be used in a simulation study. For example, if a limit cycle is predicted by a graphical analysis, choice of the initial conditions for the simulation would be determined by the predicted amplitude of the limit cycle.

- Simulations are susceptible to blunders—misplaced decimal points, sign errors, etc. By making a qualitative comparison of the results of a simulation with the prediction of a graphical analysis, you can often discover a blunder that would be difficult to detect by any other method.

The earliest graphical method is the Describing Function Method. It represents an attempt to predict limit cycles on the basis of the assumption that if such limit cycles exist the plant output is likely to be nearly sinusoidal. Under this assumption, the nonlinear function is represented by a linear element the gain of which adjusts itself to sustain the oscillation.

While the Describing Function Method is usually reliable, its rigorous justification is weak. Examples have been found in which the method gives the wrong answer: It can fail to predict limit cycles that actually exist in a system, and it can predict limit cycles that do not exist.

The "circle criterion" and its more general form— Popov's criterion—were developed in response to the need for graphical methods that are rigorously justified. These methods are based on Liapunov's Second Method, which was presented in Chapter 2. Because Liapunov's Method tends to be conservative, the graphical methods based upon it are also conservative: A system that passes a Liapunov test is guaranteed to be stable, but one that fails is not necessarily unstable. Hence the Liapunov- based graphical tests are not definitive.

Since no single method, graphical or simulation, answers all needs, a you are well advised to study the behavior of the system by more than one method.

The "standard" nonlinear control system configuration consists of a linear dynamic plant having the transfer function $G(s)$ and preceded by a single, static (i.e., "zero-memory") nonlinear element, as shown in Figure 3.1. The nonlinear characteristic is designated by

$$u = \phi(e) \tag{3.1}$$

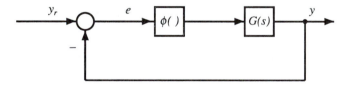

Figure 3.1   Standard nonlinear control system configuration.

Of course not every nonlinear control system fits the configuration of Figure 3.1. But in most cases you are interested in asymptotic stability of the unforced system: $y_r = 0$. In this case, the representation reduces to a single loop containing one non-linear element and one linear dynamic plant. Any system having only one nonlinearity can be represented in this form; all you need to do is to calculate the transfer function from the output of the nonlinearity to its input. You can use any convenient method, such as block diagram manipulation, and the resulting transfer function is the negative of $G(s)$ in Figure 3.1.

Example 3.1   Appraising the Effect of Friction on Motor Control System

Friction is a nonlinear force that depends upon velocity. Its action in a control system is represented by the nonlinearity $f(\ )$ in Figure 3.2. The force due to friction can be expressed as

$$u_f = -f(v) \qquad (3.2)$$

(The minus sign is used in (3.2) because friction acts to oppose increase of velocity.) To use the graphical methods to be developed it is necessary to obtain the transfer function from u_f to v. From Figure 3.2 it is observed that

$$v = \frac{1}{s + \alpha}\left[\frac{-D(s)}{s}v - u_f\right] \qquad (3.3)$$

Solving (3.3) for v in terms of u gives

$$v = -\frac{s}{s(s + \alpha) + D(s)}u_f$$

Hence, for the purpose of appraising the effect of friction on the stability of the control system, you would use the transfer function

$$G(s) = \frac{s}{s(s + \alpha) + D(s)} \qquad (3.4)$$

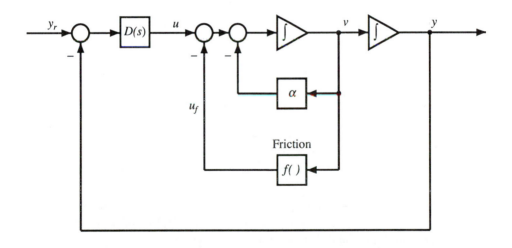

Figure 3.2   Control system for motor with friction.

3.1   DESCRIBING FUNCTION METHOD

Consider the standard nonlinear control system of Figure 3.1 except with the nonlinearity replaced by a variable gain K. Imagine the root locus for the system as K is increased. If the system can become unstable, the root locus will cross the imaginary axis. Except in very unusual circumstances when the root locus passes through the origin, it will cross the imaginary axis at some frequency ω. If the gain is fixed at the value at which the root locus crosses the imaginary axis, the system will oscillate at this frequency, which is also the frequency at which the Nyquist diagram passes the critical point $s = -1$.

Suppose now that the nonlinear element can be considered as a gain element with a variable gain. If one of the values that this gain can assume is the gain at which the root locus crosses the imaginary axis, it is reasonable to suppose that the system will exhibit a tendency to oscillate at the frequency ω. So our task is to define the gain of a nonlinear element in such a way that it can help predict whether the system will oscillate.

One of the characteristics of an oscillating system is the presence of a periodic signal at the input and the output of every component of the system. In particular, the output $u(t)$ of the nonlinear element is a periodic signal, and the output $y(t)$ of the plant is also periodic at the same frequency. A typical plant, however, has a "low-pass" frequency characteristic, meaning that its attenuation increases with increasing frequency. The consequence of this property is that although the input to the plant (i.e., the output of the nonlinear element) may be highly non-sinusoidal—rich in harmonics—the output of the plant is likely to be nearly sinusoidal, the harmonics present in the input having been attenuated ("filtered out") by the plant. Thus the input $e(t)$ to the nonlinear element, which as Figure 3.1 shows is the negative of the plant output, is also likely to be nearly sinusoidal. It is not unreasonable to assume that $e(t)$ can be approximated by a pure sinusoid:

$$e(t) \approx A \sin \omega t \tag{3.5}$$

The output of the nonlinearity will also be periodic, but probably not very sinusoidal. Nevertheless, it can be represented by a Fourier series:

$$u(t) = U_0 + \sum_{n=1}^{\infty} B_n \sin(n\omega t) + C_n \cos(n\omega t) \tag{3.6}$$

But the assumed low-pass characteristic of the plant implies that the harmonics of the input beyond the first (i.e., the fundamental) are filtered out by the plant and do not appear in its output. Hence, for the purpose of assessing stability, we can approximate the input to the plant by

$$u(t) \approx U_0 + B_1 \sin \omega t + B_2 \cos \omega t$$

Moreover, the assumption of (3.5) that the plant output has no constant (*d-c*) term implies that

$$U_0 = 0$$

Thus, finally, we have the approximation

$$u(t) \approx B_1 \sin \omega t + B_2 \cos \omega t \tag{3.7}$$

Subject to the assumptions stated above, we can assert that there is evidence that the oscillation will occur at the frequency ω if the plant output

$$y(t) = -e(t) = -A\sin\omega t \tag{3.8}$$

when the input to the plant is given by (3.7) where B_1 and B_2 are the amplitudes of the in-phase and quadrature components of the output of the nonlinearity whose input is $A\sin\omega t$. Thus an oscillation is predicted if a set of real numbers A, B_1, and B_2 and a frequency ω can be found consistent with (3.7) and (3.8). If these numbers can be found, the predicted oscillation will occur at the frequency ω and will have an amplitude of A at the output of the plant.

If you reflect on the reasoning used above, you will see that it is based on tracing the path of a sinusoid around the loop, disregarding any higher harmonics. Another viewpoint is that of "balancing" the first harmonic of the output of the nonlinearity with the output of the plant. This viewpoint justifies calling the method of analysis the "harmonic balance" method, a term especially popular in Europe.

3.1.1 Definition and Calculation of Describing Functions

The artifice of ignoring the harmonics of the output of the nonlinearity on the assumption that the plant will remove them anyway, permits us to use a frequency-domain representation of the nonlinearity. Since the input to the nonlinearity is

$$e(t) = A\sin\omega t = \Im[Ae^{j\omega t}]$$

(where $\Im[\ ]$ denotes the *imaginary part* of the bracketed expression), its output is approximated by

$$u(t) = B_1\sin\omega t + B_2\cos\omega t$$
$$= \Im[\mathcal{N}(A)Ae^{j\omega t}]$$

where

$$\mathcal{N}(A) = N(A) + j\hat{N}(A) = \frac{B_1 + jB_2}{A} \tag{3.9}$$

is defined as the *describing function* of the nonlinearity $\phi(\ )$.

The describing function has the same role at the single frequency as the transfer function of a linear system at a single frequency; it is a complex number, which is multiplied by the phasor representing the input to get the phasor representing the output.

In principle, the describing function has a real part $N(A)$ and an imaginary part $\hat{N}(A)$. As we shall see, however, the imaginary part of the describing function, which represents phase shift through the nonlinearity, is absent for a single-valued nonlinearity. Also in principle, the describing function can be a function of frequency; hence a more descriptive notation would display the frequency dependence by using the symbol $N(A,\omega)$ instead of $N(A)$. The practical instances in which the describing function depends on frequency, however, are rare, and the unencumbered notation seems more appealing.

(The notation for the describing function is, regrettably, not standardized. The symbol N seems to be the most common, but some authors use other symbols. The symbols used to denote the real and imaginary parts of complex describing functions are even less standarized. Hence, in reading a book or technical paper that uses the Describing Function Method, you have to be sure of the symbols the author is using.)

The real and the imaginary parts of the describing function are simply the coefficients of the fundamental (i.e., first harmonic) terms in the Fourier series expansion of the output of the nonlinearity, divided by the amplitude A of the input. Thus, in accordance with the theory of Fourier series, the real and imaginary parts of the describing function of the nonlinear function $\phi(\ )$ are given by

$$N(A) = \frac{1}{\pi A} \int_{T}^{2\pi+T} \phi(A \sin \omega t) \sin \omega t \, d(\omega t) \tag{3.10}$$

$$\hat{N}(A) = \frac{1}{\pi A} \int_{T}^{2\pi+T} \phi(A \sin \omega t) \cos \omega t \, d(\omega t) \tag{3.11}$$

The starting time T in the integrals can be chosen arbitrarily.

To evaluate these integrals, the integration interval $[-\pi/2, 3\pi/2]$ is especially convenient. Also break each integral into two parts. The reasons will soon be evident.

$$N(A) = \frac{1}{\pi A} \left[\int_{-\pi/2}^{\pi/2} \phi(A \sin \theta) \sin \theta \, d\theta + \int_{\pi/2}^{3\pi/2} \phi(A \sin \theta) \sin \theta \, d\theta \right] \tag{3.12}$$

$$\hat{N}(A) = \frac{1}{\pi A} \left[\int_{-\pi/2}^{\pi/2} \phi(A \sin \theta) \cos \theta \, d\theta + \int_{\pi/2}^{3\pi/2} \phi(A \sin \theta) \cos \theta \, d\theta \right] \tag{3.13}$$

where

$$\theta = \omega t$$

Introduce the change of variable:

$$x = \sin \theta$$
$$dx = \cos \theta \, d\theta = \pm\sqrt{1 - x^2} \, d\theta$$

Note that $\cos \theta \geq 0$ for $\theta \in [-\pi/2, \pi/2]$ and $\cos \theta \leq 0$ for $\theta \in [\pi/2, 3\pi/2]$. This is the reason why the defining intergals were broken into two parts.

With this change of variable

$$N(A) = \frac{1}{\pi A} \left[\int_{-1}^{1} \phi(Ax) \frac{x \, dx}{\sqrt{1 - x^2}} + \int_{1}^{-1} \phi(Ax) \frac{-x \, dx}{\sqrt{1 - x^2}} \right]$$

$$= \frac{2}{\pi A} \left[\int_{-1}^{1} \phi(Ax) \frac{x \, dx}{\sqrt{1 - x^2}} \right] \tag{3.14}$$

while

$$\hat{N}(A) = \frac{1}{\pi A} \left[\int_{-1}^{1} \phi(Ax)dx + \int_{1}^{-1} \phi(Ax)dx \right] = 0 \qquad (3.15)$$

Thus it is seen that the imaginary part $\hat{N}$ of the describing function vanishes.

The above formulas for the describing function were derived on the basis of the assumption that $\phi(x)$ is a single-valued nonlinearity. When hysteresis is present, however, $\phi(x)$ is *not* a single-valued nonlinearity. The typical representation of a hysteresis nonlinearity is as shown in Figure 3.3. The nonlinearity has one value for x increasing ($\dot{x} > 0$) and another value for x decreasing ($\dot{x} < 0$). Thus we should express a hysteresis nonlinearity as a function of both its input and the time derivative of the input:

$$\phi(x, \dot{x}) = \begin{cases} \phi_1(x), & \dot{x} > 0 \\ \phi_2(x), & \dot{x} < 0 \end{cases} \qquad (3.16)$$

Using this representation of hysteresis, the analysis used to obtain (3.14) and (3.14) yields:

$$N(A) = \frac{1}{\pi A} \left[\int_{-1}^{1} [\phi_1(Ax) + \phi_2(Ax)] \frac{x\,dx}{\sqrt{1-x^2}} \right] \qquad (3.17)$$

$$\hat{N}(A) = \frac{1}{\pi A} \left[\int_{-1}^{1} [\phi_1(Ax) - \phi_2(Ax)]dx \right] \qquad (3.18)$$

Note that the imaginary part of the describing function is proportional to the *area* under the hysteresis curve. (This fact was apparently first observed in Hsu and Meyer [1].)

Example 3.2 Describing Function for Deadzone

Use of the integral formulas (3.17) and (3.18) is illustrated by calculating the describing function for a "deadzone" nonlinearity, defined by

$$\phi(x) = \text{dez}(x, \delta) = \begin{cases} 1, & x > \delta \\ 0, & |x| \le \delta \\ -1, & x < -\delta \end{cases}$$

Since this is a single-valued nonlinearity, we know that the imaginary part of the describing function is zero. The real part is calculated using (3.17). By symmetry, we can calculate the describing function using only the positive part of $\phi()$ and double the result. Thus:

$$N(A) = \begin{cases} 0, & A < \delta \\ \dfrac{4}{\pi A} \displaystyle\int_{\delta/A}^{1} \frac{x\,dx}{\sqrt{1-x^2}} = \frac{4}{\pi A}\left[-\sqrt{1-x^2}\right]_{\delta/A}^{1} = \frac{4}{\pi A}\sqrt{1-\left(\frac{\delta}{A}\right)^2}, & A > \delta \end{cases}$$

$$(3.19)$$

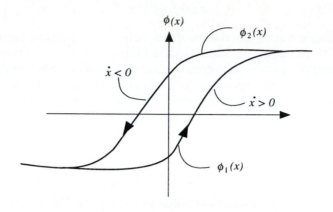

Figure 3.3   Hysteresis nonlinearity.

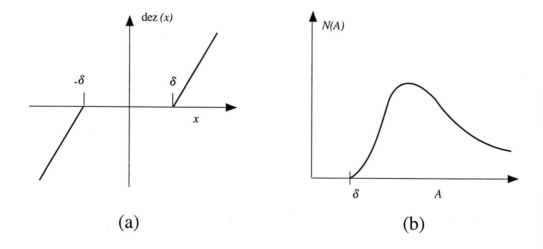

Figure 3.4   Deadzone nonlinearity and its describing function.
(a) Nonlinearity. (b) Describing function.

Note that the describing function is zero for an input amplitude A smaller in magnitude than the width δ of the deadzone. As the input increases beyond the width of the deadzone, the describing function increases to a maximum value and then decreases. It approaches zero asymptotically, because the output amplitude remains bounded as the input amplitude increases without bound.

It is readily established that the maximum value is $2/\pi\delta$ which is attained when $A = \sqrt{2}\delta$.

When $\delta = 0$, the deadzone nonlinearity becomes the "sign" function:

$$\phi(x) = \text{sgn}(x) = \begin{cases} 1, & x > 0 \\ -1, & x < 0 \end{cases}$$

the describing function of which is

$$N(A) = \frac{4}{\pi A}$$

Example 3.3   Simple Hysteresis

As a second example we consider the calculation of the describing function of the hysteresis function shown in Figure 3.5(a). The two functions ϕ_1 and ϕ_2 are shown individually in (b) and (c). Their sum and difference are shown in (d) and (e). With the aid of these figures we see that

$$N(A) = \begin{cases} 0, & A < \delta \\ \frac{4}{\pi A}\sqrt{1 - \left(\frac{\delta}{A}\right)^2}, & A > \delta \end{cases} \tag{3.20}$$

$$\hat{N}(A) = \begin{cases} 0, & A < \delta \\ -\frac{4}{\pi A^2}\delta, & A > \delta \end{cases} \tag{3.21}$$

Note that the real part of the describing function is the same as that of a deadzone of the same width; the imaginary part of the describing function is proportional to the width of the deadzone.

Extensive tables of describing functions have been calculated and tabulated in several textbooks, e.g., [1], [2] so in most cases it is not necessary for you to perform this messy calculation.

3.1.2   Use of the Describing Function

In using the describing function for stability analysis, the goal is to determine whether there is an amplitude A and a frequency ω for which the characteristic equation:

$$1 + \mathcal{N}(A)G(j\omega) = 0 \tag{3.22}$$

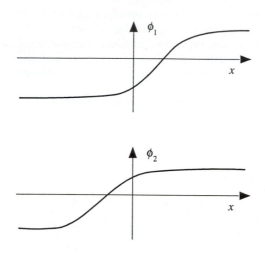

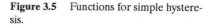

Figure 3.5 Functions for simple hysteresis.

(This is the same equation that would be used to test for the stability of a linear system with $\phi(\ )$ having the transfer function N.)

If (3.22) has a solution (A, ω), then a limit cycle may exist with (approximately) this amplitude and frequency. It may happen that more than one solution to (3.22) can be found. If so, each solution represents a possible limit cycle, and has to be tested separately as explained below.

The characteristic equation (3.22) is equivalent to

$$-\frac{1}{\mathcal{N}(A)} = G(j\omega) \tag{3.23}$$

which is often solved graphically. The imaginary part of the right-hand side plotted vs. the real part is simply the Nyquist diagram of $G(j\omega)$. The left-hand side is a plot of $-1/\mathcal{N}(A)$ as a function of A. If the two curves intersect, the values of A and ω at which the intersection occurs determine a possible limit cycle of corresponding amplitude and frequency.

A possible plot of $G(j\omega)$ superimposed upon a plot of $-1/\mathcal{N}(A)$ is shown in Figure 3.6. (For illustrative purposes, a hypothetical describing function was selected that has an imaginary part and intersects the Nyquist diagram at more than one point; in many applications the describing function is not as complicated.) There are two intersections (A_1, ω_1) and (A_2, ω_2); hence two limit cycles are possible. The nature of these limit cycles can be assessed by the describing function method by reasoning as follows:

The map of the right hand of the s-plane lies to the right of the Nyquist diagram. If the system were to operate at the intersection (A_1, ω_1), a slight increase in gain would drive the system into the unstable region, causing a still larger oscillation. Hence intuition would suggest that the limit cycle predicted at (A_1, ω_1) is unstable. On the

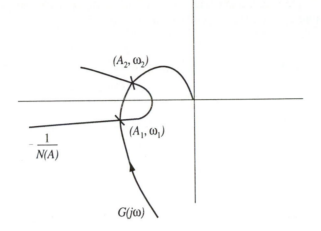

Figure 3.6   Stability analysis by describing function method.

other hand, if the system were to operate at (A_2, ω_2), a slight increase in gain would drive the system into the stable region. To sustain the limit cycle, the system adjusts itself to recover this operating point. This suggests that the limit cycle (A_2, ω_2) is stable. This intuitive reasoning is usually reliable.

Instead of finding the intersection of the inverse of the describing function with the Nyquist plot, you may find it more convenient to find the intersection of the describing function itself with the "inverse Nyquist plot":

$$- \mathcal{N}(A) = G^{-1}(j\omega) := \frac{1}{G(j\omega)} \tag{3.24}$$

This form is especially convenient with complex describing functions for which the expression for $1/\mathcal{N}(A)$ may be messy.

Example 3.4   Temperature Control

Consider the customary control law for a heating system. When it is too cold, the heat is turned on; when it is too hot, the heat is turned off. Can you imagine a system that is simpler? Nevertheless, from personal experience you know that this type of control usually works quite well. But sometimes it doesn't. If the plant is third-order or higher and the temperature sensor is physically separated from the source of heat, for example, this control law can result in an undesirable limit cycle. The situation is illustrated by the third-order temperature control system considered in Chapter 1, which has the transfer function

$$G(s) = \frac{1}{s^3 + 5s^2 + 6s + 1}$$

Under the (somewhat unrealistic) assumption that heat can be removed or added, the simplest control law is based simply on the difference e between the reference temperature v_R and the measured temperature y

$$u = K \operatorname{sgn}(e) \qquad (3.25)$$

where

$$e = y_R - y \qquad (3.26)$$

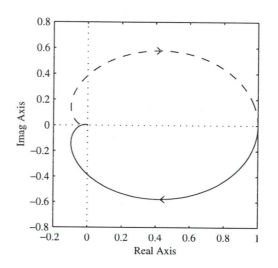

Figure 3.7   Nyquist diagram for temperature control.

It is easiest to analyze the case in which the reference temperature is zero. The Nyquist diagram for the transfer function $G(s)$ is shown in Figure 3.7, which exhibits a phase shift of 180 degrees at $\omega = \sqrt{6} = 2.44$ rad/sec. At this frequency, with $K = 1$, the transfer function is

$$G(j\sqrt{6}) = -\frac{1}{29}$$

As determined earlier, the sgn nonlinearity has the describing function $N(A) = 4/\pi A$. Solving,

$$G(j\sqrt{6}) = -\frac{1}{29} = -\frac{1}{N(A)} = -\frac{\pi A}{4}$$

or $A = 4/29\pi = 0.0420$.

The simulated transient response of the system with a small, but nonzero, initial condition, is shown in Figure 3.8. A limit cycle is indeed present. The amplitude and frequency are found, respectively, to be $A = 0.0448$ and $\omega = 2.39$ rad/sec. These simulated values agree quite well with the corresponding values predicted by the describing function analysis.

The presence of a reference input has an effect upon the amplitude and frequency of the limit cycle. Since the largest control amplitude that can be generated by the control is 1, and the d-c gain of the plant is also 1, the output cannot exceed 1 in amplitude, no

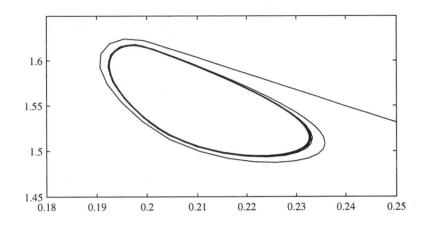

Figure 3.8 Limit cycle in response of temperature control system with output feedback.

matter how large the reference input. Hence it is only meaningful to study the behavior of the system for reference between 0 and 1 in magnitude. If the reference input is exactly 1, the control input will be exactly 1 and there will be no limit cycle. As the reference input is decreased, limit cycles of increasing amplitude and frequency will occur, as the simulation results of Figure 3.9 show. These results are summarized in Figure 3.10. The frequency and amplitude of the limit cycle is fairly close to the value predicted by the describing function analysis for reference inputs less than about 0.5 in magnitude. The ability of the describing function method to predict the amplitude and frequency of the limit cycle diminishes rapidly as the reference input increases above 0.5 in magnitude. Among the reasons for this is that the approximation of the output of the system by a sinusoid degrades seriously as the reference input approaches 1 in magnitude. In addition, the presence of the d-c signal in the loop implies that the describing function used in the analysis should be calculated for an asymmetrical element with a nonzero d-c value, rather for the sgn function. But the d-c value that you need to use in calculating the describing function cannot be determined until you do the describing function analysis. Hence the describing function and the d-c value must be determined concurrently. The theory for performing the analysis is elucidated in texts, such as Hsu and Meyer [1], that deal with the *dual-input describing function*

In some noncritical applications a limit cycle may be tolerable. If not, however, it can easily be eliminated. One way of eliminating the limit cycle would be to use a linear control region, as shown in Figure 3.11(a). To get a linear region, however, you would need a heater with an adjustable power output, which could be expensive. A more economical option would be to use a deadzone with a width δ large enough to ensure that its describing function does not intersect the Nyquist diagram. From the analysis in Example 3.2 the maximum value of the describing function for a deadzone is $2/\pi\delta$. The Describing Function Method predicts that there will be no limit cycle if $\delta > 2/29\pi = .021$, since the describing function will not intersect with the Nyquist plot of the plant. Simulations of

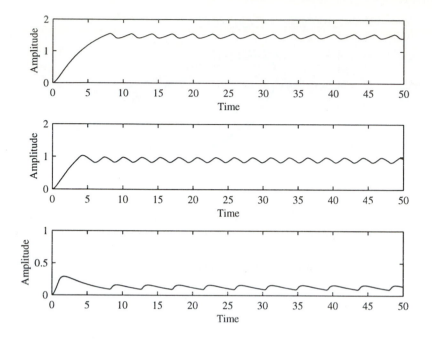

Figure 3.9   Response of temperature control system with sgn
nonlinearity to several levels of reference inputs.

the transient response for two deadzone widths, one narrower than 0.021, and one wider, are shown in Figure 3.12. Observe: when deadzone is too narrow, the limit cycle is sustained; but when the deadzone is sufficiently large, the oscillation decays to zero. The amplitude and frequency of the limit cycle ($A = 0.0375, \omega = 0.390$) are consistent with the values predicted by the describing function analysis.

Alas, the deadzone does not eliminate the limit cycle when the reference input is nonzero. Simulations indicate that a limit cycle is present even when the deadzone is as large as 0.25 (ten times the value needed to eliminate the limit cycle when the reference input is zero). The major effect of increasing the size of the deadzone is to increase the steady state error.

The effect of using a saturating linear control instead of a deadzone is the subject of Problem 3.3.

Example 3.5   Temperature Control with Hysteresis

The analysis of the effect of hysteresis on a closed-loop system is illustrated replacing the deadzone nonlinearity of the previous example by the simple hysteresis function considered in Example 3.3 .

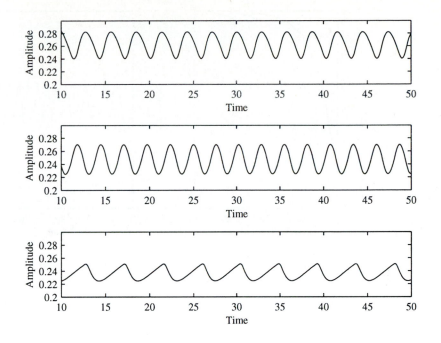

Figure 3.10 Variation of limit cycle amplitude and frequency with reference input.

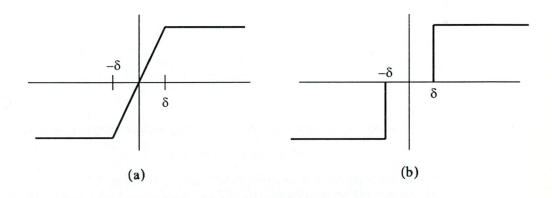

Figure 3.11 Possible modifications of nonlinearity to avoid limit cycles. (a) Linear region (b) Deadzone.

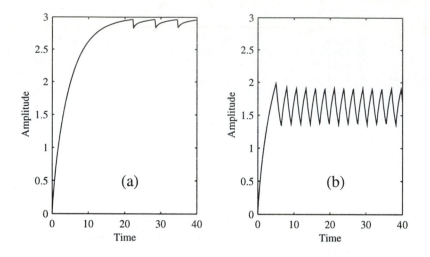

Figure 3.12   Transient response of temperature control system with deadzone nonlinearity. (a) $\delta = 0.25$. (b) $\delta = 0.20$.

Table 3.1   Comparison of Predicted and Simulated Limit Cycles with Hysteresis.

δ	Predicted		Simulated	
	Amplitude A	Frequency ω	Amplitude A	Frequency ω
0.00	0.045	2.40	0.048	2.39
0.05	0.086	1.52	0.110	1.48
0.10	0.089	1.21	0.161	1.18

In this example it is somewhat easier to work with the inverse Nyquist plot, i.e.,

$$G^{-1}(s) = s^3 + 5s^2 + 6s + 1$$

and the describing function, as given by (3.20) and (3.21). Figure 3.13 shows the inverse Nyquist plot and the describing functions for (i) $\delta = 0.05$ and (ii) $\delta = 0.10$. For these cases describing function analysis predicts limit cycles of amplitudes and frequencies as shown in Table 3.1. The amplitudes and frequencies of the limit cycles determined by simulations are shown for comparison. The agreement between the limit cycle frequency predicted by the describing function analysis and the simulation is quite good. The agreement of the amplitudes is only fair.

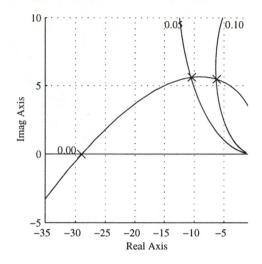

Figure 3.13 Inverse Nyquist plot and describing functions for hysteresis.

3.1.3 Limitations of the Describing Function Method; Absolute Stability

On the basis of the describing function analysis you might be led to conclude that there will be no limit cycle if the Nyquist diagram and the negative of the inverse describing function do not intersect. Unfortunately, *this conclusion is wrong.* Fitts's counterexample, given in Example 3.6 below, demonstrates this.

If the Describing Function Method fails to predict a limit cycle, perhaps there is a more reliable method of analysis. Indirectly, this is one of the issues that concerned a number of investigators during the 1950s and 1960s in the US and the USSR. They were interested in "absolute stability." In particular they wanted to know the conditions under which a system with an imperfectly known nonlinearity would be globally asymptotically stable. The only thing known about the nonlinearity is that it lies in a sector of the first and third quadrant, as shown in Figure 3.14.

Mathematically, a sector-bounded nonlinearity is defined by:

$$a \leq \frac{\phi(x)}{x} \leq b \qquad (3.27)$$

A famous conjecture of the 1950s, known as *Aizerman's conjecture* [3], was that if the closed-loop system with a sector-bounded nonlinearity were stable for all gains in the sector, i.e., if $\phi(e)$ in Figure 3.1 were replaced by Kx, and the resulting linear feedback system were stable for $a \leq K \leq b$, then the nonlinear system would be asymptotically stable for any nonlinearity in the sector. If this conjecture were true, it would be very useful, since many nonlinearities in practice are sector bounded. Moreover, it would serve as vindication of the describing function method, because it is relatively

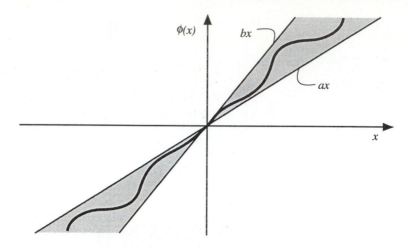

Figure 3.14　　Sector-bounded nonlinearity.

easy to show that the describing function for a sector bounded nonlinearity lies in the sector (see Problem 3.4):

$$a \leq N(A) \leq b$$

Hence, if the describing function did not predict a limit cycle, and Aizerman's conjecture were true, the feedback system would be asymptotically stable.

But *Aizerman's conjecture is false.* Several counterexamples to conjecture have been discovered. One of the earliest is the instructive example by Fitts [4]:

Example 3.6　Fitts's Counterexample

The linear plant in Fitts's counterexample is

$$G(s) = \frac{s^2}{[(s + 0.5)^2 + (0.9)^2][(s + 0.5)^2 + (1.1)^2]}$$

and the nonlinearity is

$$\phi(e) = e^3$$

Clearly the nonlinearity lies in the first and third quadrant. It is readily established that the linear system having the return difference

$$1 + KG(s)$$

is asymptotically stable for all $K \in [0, \infty)$. Hence, by Aizerman's conjecture, the closed loop system ought to be stable for $\phi(e) = e^3$.

The describing function for $\phi(e)$ is readily determined to be

$$N(A) = \frac{3}{8}A^2$$

Hence

$$-\frac{1}{N(A)} = -\frac{8}{3A^2}$$

which lies entirely on the negative real axis. The Nyquist plot for $G(j\omega)$ is shown superimposed upon the describing function in Figure 3.15. It is seen that the Nyquist plot and the describing function intersect only at $\omega = 0, \infty$. (By moving the zeros of $G(s)$ slightly to the right of the origin, the Nyquist plot would lie to the right of the imaginary axis at $\omega = \infty$ and there would be no intersection of the Nyquist plot with the describing function.) The describing function method thus also does not predict a limit cycle.

Nevertheless a limit cycle is present. The limit cycle happens to be unstable, so it is difficult to find by simulation. By carefully selecting the initial conditions, however, it is possible to exhibit a number of oscillatory cycles before they decay to zero or depart to infinity. The control input and the plant output are shown in Figure 3.16. Note that the frequency of oscillation is approximately 1.7 rad/sec. Nothing as yet discussed in this chapter would lead you to suspect a limit cycle at this frequency.

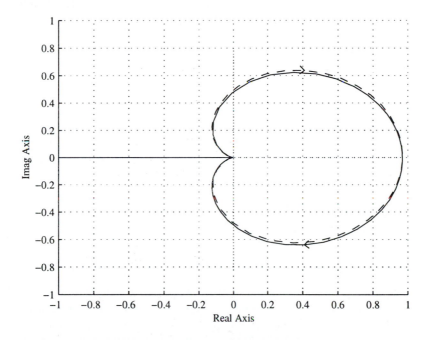

Figure 3.15 Nyquist plot and describing function for Fitts's counterexample to Aizerman's conjecture.

Another famous conjecture was that proposed by Kalman [5]: if, for the nonlinear function $\phi(x)$,

$$a \le \frac{d\phi(x)}{dx} \le b \tag{3.28}$$

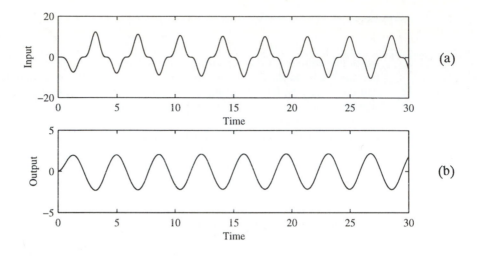

Figure 3.16   Simulated input and output for Fitts's counterexample to Aizerman's conjecture. (a) Input to plant. (b) Output
of plant.

for all x, the origin is asymptotically stable. Although the conditions (3.28) of Kalman's
conjecture are stronger than those of Aizerman's, Kalman's conjecture is also false,
as Fitts's counterexample shows. The failure of the Describing Function Method and
the conjectures of Aizerman and Kalman motivated a search for reliable conditions
for asymptotic stability and nonexistence of limit cycles. Liapunov methods provide
these conditions.

3.2   LIAPUNOV METHODS

3.2.1   Passive Electrical Networks and Positive-Real Functions

The title of this section suggests that there is some connection between electrical networks and the stability of nonlinear systems. What is the connection?

One property of a linear, passive electrical network (i.e., a network comprising only
resistors, capacitors, and inductors) is that it dissipates the energy that is delivered to
its terminals. As electrical engineers know, this means that the input resistance—the
real part of the input impedance— of a passive network is always *positive*. Experience
and intuition suggest that if a nonlinear "resistor" is connected in series with such a
passive network, as shown in Figure 3.17, the resulting combination will remain passive: in the absence of an input excitation (i.e., a short-circuit in place of the input
voltage v_i in Figure 3.17), all the energy stored in the capacitors of the linear network
will ultimately dissipate in its internal resistors and in the nonlinear external resistor.

Surprisingly, this commonly known fact about electrical networks, suitably generalized, can become the basis of several important and useful graphical methods applicable to all nonlinear systems, not only electrical networks. To see how this principle may be generalized note that the equations defining the operation of the network of Figure 3.17 are

$$v_i(t) - v(t) = e(t) \tag{3.29}$$
$$e(t) = \phi(i(t)) \tag{3.30}$$
$$V(s) = Z(s)I(s) \tag{3.31}$$

where $V(s)$ and $I(s)$ are the Laplace transforms of $v(t)$ and $i(t)$ respectively. A block diagram representation of these relations is given in Figure 3.18. It is seen that the block diagram has the general form of a single-loop feedback system with a nonlinearity in the forward path.

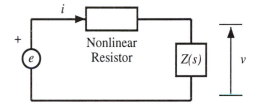

Figure 3.17   Series combination of passive, linear network and a nonlinear resistor.

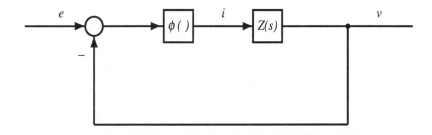

Figure 3.18   Block diagram representation of series combination of passive, linear network and nonlinear resistor.

To exploit this intuitive idea we must generalize the concepts of

- Nonlinear resistor

- Passive network

The concept of a nonlinear resistor is intuitive: it is a nonlinear zero-memory element whose input-output characteristic lies in the first and third quadrant. This is a statement of the fact that the power going into the resistor is non-negative, i.e., that it *absorbs* power:

$$ei = i\phi(i) \geq 0$$

The property of a linear network that makes it passive is that its impedance $Z(s)$ is a *positive-real (PR) function*, one for which

$$\Re[Z(j\omega)] \geq 0 \quad \text{for all } \omega$$

where $\Re[\ ]$ denotes the real part of the bracketed expression. This simple property implies a number of other properties, some of which are not immediately evident:

1. $Z(s)$ has no poles or zeros in the right half of the s plane; it is a stable, minimum phase transfer function.

2. $Y(s) = 1/Z(s)$ is a positive-real function.

3. The degree of the numerator of $Z(s)$ cannot differ from the degree of its denominator by more than 1.

A consequence of the property that the real part of $Z(j\omega)$ is positive is that its phase shift must lie between ± 90 degrees. Its Nyquist diagram must consequently lie to the right of the imaginary axis of the s plane, as shown in Figure 3.19. As you know from experience with various systems, the transfer functions of very few linear systems possess this property. (Even in *electrical networks* the positive-real property applies only to ("driving point") impedances and admittances, and not, in general, to other network functions such as transfer impedances or voltage ratios.) By manipulation of the block diagrams of systems with a single nonlinearity, however, it is possible to obtain useful results for such systems, even when the linear part is not a positive-real function. The circle criterion, to be developed in Section 3.2.3, is an example of this sort of manipulation.

3.2.2   Kalman-Yacubovich Lemma and Absolute Stability

The failure of the Describing Function Method to predict limit cycles in all cases and the counterexamples to Aizerman's and Kalman's conjectures attest to the danger of relying on intuition alone. A more rigorous foundation is needed. In this instance, the needed rigor is provided by a famous result known as the "Kalman-Yacubovich Lemma."

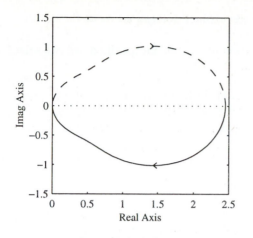

Figure 3.19 Nyquist diagram of positive real function must remain to the right of the imaginary axis.

Although the Kalman-Yacubovich lemma is usually stated in a slightly more general form to account for poles and/or zeros on the imaginary axis (corresponding to ideal resonances or anti-resonances), if we exclude the possibility of such poles for simplicity we can state the lemma as follows:

Kalman-Yacobovich Lemma. Let

$$Z(s) = c'(sI - A)^{-1}b$$

be the transfer function of a linear, time-invariant system

$$\dot{x} = Ax + bu \qquad\qquad\qquad (3.32)$$

$$y = c'x \qquad\qquad\qquad (3.33)$$

The transfer function is positive-real if and only if there exist positive definite matrices P and Q that satisfy

$$PA + A'P = -qq' - \epsilon Q \qquad\qquad (3.34)$$

$$c = Pb \qquad\qquad\qquad (3.35)$$

with ϵ sufficiently small.

Oddly, the proof of the Kalman-Yacubovich lemma is omitted in several recent books on nonlinear control [6][7][8]. A slightly more general form of the lemma is proved, using the method presented by Kalman [2], in the venerable monograph of Lefschetz [10].

Owing to the importance of the lemma, an indication of the proof is presented here. In the interest of saving space, only the more difficult part of the lemma is outlined: the fact that if $Z(s)$ is a positive real function, then (3.34) and (3.35) are satisfied. To save a few steps, we take $\epsilon = 0$. (Proof of the more general form of the Kalman-Yacubovich lemma follows the path of the development that follows and is left as an exercise. See Problem 3.5.)

The development will result in the explicit determination of the vector q and the matrix P.

Since the state-space representation of $Z(s)$ is not specified, we can select it to be in the controllable companion form:

$$A = \begin{bmatrix} 0 & 1 & 0 & \cdots & 0 \\ 0 & 0 & 1 & \cdots & 0 \\ & & \cdots & & \\ & & \cdots & & \\ -a_n & -a_{n-1} & -a_{n-2} & \cdots & -a_1 \end{bmatrix} \qquad b = \begin{bmatrix} 0 \\ 0 \\ \vdots \\ 1 \end{bmatrix} \qquad (3.36)$$

$$c' = \begin{bmatrix} c_1 & c_2 & \cdots & c_n \end{bmatrix} \qquad (3.37)$$

With this representation

$$Z(s) = \frac{N(s)}{D(s)} = \frac{c_1 s^{n-1} + c_2 s^{n-2} + \cdots + c_n}{s^n + a_1 s^{n-1} + \cdots + a_n} \qquad (3.38)$$

Note that because $Z(s)$ is positive real, it is necessary that the coefficients

$$c_i > 0$$

The inequality is strict; none of the coefficients may be zero.

Now consider

$$\begin{aligned} Z(j\omega) + Z(-j\omega) &= 2\Re[Z(j\omega)] \\ &= \frac{N(j\omega)}{D(j\omega)} + \frac{N(-j\omega)}{D(-j\omega)} \\ &= \frac{N(j\omega)D(-j\omega) + N(-j\omega)D(j\omega)}{D(j\omega)D(-j\omega)} \\ &= \frac{F(\omega^2)}{D(j\omega)D(-j\omega)} \end{aligned} \qquad (3.39)$$

Since $Z(s)$ is positive real, $\Re[Z(j\omega)] > 0$ for all ω. Hence the numerator $F(\omega^2)$ in (3.39) has no real roots. From this fact it follows that $F(\omega^2)$ can be factored into the product

$$F(\omega^2) = f(j\omega)f(-j\omega)$$

where

$$f(j\omega) = f_1(j\omega)^{n-1} + f_2(j\omega)^{n-2} + \cdots + f_n$$

with all the f_i being *real* numbers. This means that we can write

$$Z(j\omega) + Z(-j\omega) = H(-j\omega)H(j\omega) = \frac{f(-j\omega)}{D(-j\omega)} \frac{f(j\omega)}{D(j\omega)} \qquad (3.40)$$

Kalman's trick: consider the identity

$$(-sI - A')P + P(sI - A) = -A'P - PA \qquad (3.41)$$

If $Z(s)$ is positive real, A must have its eigenvalues in left half plane and not on the imaginary axis, hence the Liapunov equation (3.34) has a solution P for any qq'; in fact P is positive definite. (See Chapter 2.) So we can write (3.41) as

$$(-sI - A')P + P(sI - A) = qq' \tag{3.42}$$

Multiply both sides of (3.42) on the left by $b'(-sI - A')^{-1}$ and on the right by $(sI - A)^{-1}b$ to obtain

$$b'P(sI - A)^{-1}b + b'(-sI - A')^{-1}Pb = b'(-sI - A')^{-1}qq'(sI - A)^{-1}b \tag{3.43}$$

Now (3.43) can be written

$$G(s) + G(-s) = b'(-sI - A')^{-1}qq'(sI - A)^{-1}b \tag{3.44}$$

where

$$G(s) = b'P(sI - A)^{-1}b$$

Hence

$$G(j\omega) + G(-j\omega) = b'(-j\omega I - A')^{-1}qq'(j\omega I - A)^{-1}b = |q'(j\omega I - A)^{-1}b|^2 \tag{3.45}$$

which implies that $G(j\omega) + G(-j\omega)$ can be factored the same way as $Z(j\omega) + Z(-j\omega)$. In fact, we can identify $Z(j\omega)$ with $G(j\omega)$, i.e.,

$$b'P(sI - A)^{-1}b = c'(sI - A)^{-1}b \tag{3.46}$$

from which we infer that

$$Pb = c$$

which is the condition we sought to prove.

To summarize the construction leading to the proof of the Kalman-Yacubovich lemma:

- Construct $Z(j\omega) + Z(-j\omega)$.

- Factor the result into the product of functions $H(j\omega)$ and $H(-j\omega)$ such that the corresponding $H(s)$ is minimum phase, i.e., that its poles and zeros lie in the left half of the s plane.

- Realize $H(s)$ in the controllable canonical form. Let q be the vector of coefficients of the numerator of $H(s)$.

- Solve the Liapunov equation

$$PA + A'P = -qq'$$

for P. Then

$$c = Pb$$

Example 3.7   Constructing Liapunov Function for Positive-Real Function

The impedance seen by the current source in Example 3.1 is readily determined to be

$$Z(s) = \frac{s^2 + 4s + 3}{s^3 + 5s^2 + 6s + 1}$$

and it can be verified that it is a positive-real function. (See Problem 3.6). Substituting $s = j\omega$ into this transfer function gives

$$Z(j\omega) = \frac{(j\omega)^2 + 4j\omega + 3}{(j\omega)^3 + 5(j\omega)^2 + 6j\omega + 1}$$

Hence

$$Z(j\omega) + Z(-j\omega) = \frac{(j\omega)^2 + 4j\omega + 3}{(j\omega)^3 + 5(j\omega)^2 + 6j\omega + 1}$$
$$+ \frac{(-j\omega)^2 + 4(-j\omega) + 3}{(-j\omega)^3 + 5(-j\omega)^2 + 6(-j\omega) + 1}$$

After a bit of calculation we find

$$Z(j\omega) + Z(-j\omega) =$$
$$\frac{2(\omega^4 + 8\omega^2 + 3)}{[(j\omega)^3 + 5(j\omega)^2 + 6j\omega + 1][(-j\omega)^3 + 5(-j\omega)^2 + 6(-j\omega) + 1]}$$

Note that the numerator $N(\omega^2)$ is a polynomial in ω^2 with positive coefficients. In fact it can be factored as follows:

$$N(\omega^2) = 2(\omega^2 + 4 + \sqrt{13})(\omega^2 + 4 - \sqrt{13})$$
$$= \sqrt{2}\left((j\omega + \sqrt{4 + \sqrt{13}})\left(j\omega + \sqrt{4 - \sqrt{13}}\right)\right)$$
$$\times \sqrt{2}\left((-j\omega + \sqrt{4 + \sqrt{13}})\left(-j\omega + \sqrt{4 - \sqrt{13}}\right)\right)$$

Identifying the first pair of factors as the numerator of $G(j\omega)$ we obtain

$$H(s) = \frac{\sqrt{2}(s + \sqrt{4 + \sqrt{13}})(s + \sqrt{4 - \sqrt{13}})}{s^3 + 5s^2 + 6s + 1} = \frac{1.4142s^2 + 4.7883s + 2.4495}{s^3 + 5s^2 + 6s + 1}$$

The vector q is obtained from the numerator of $H(s)$:

$$q = \begin{bmatrix} 2.4495 \\ 4.7883 \\ 1.4142 \end{bmatrix}$$

The controllable companion form development of the impedance $Z(s)$ yields the following matrices:

$$A = \begin{bmatrix} 0 & 1 & 0 \\ 0 & 0 & 1 \\ -1 & -6 & -5 \end{bmatrix} \quad b = \begin{bmatrix} 0 \\ 0 \\ 1 \end{bmatrix} \quad c' = \begin{bmatrix} 3 & 4 & 1 \end{bmatrix}$$

Using this A matrix and the vector q found above, the Liapunov equation (3.34) is solved numerically to yield:

$$P = \begin{bmatrix} 10.2710 & 12.5359 & 3.0000 \\ 12.5359 & 16.2283 & 4.0000 \\ 3.0000 & 4.0000 & 1.0000 \end{bmatrix}$$

The last column of P, which is Pb for the vector b given above, is observed to be the same as the observation vector c. Thus the condition $c = Pb$ is satisfied.

The relationship between the *"spectral factorization"* introduced by Kalman to relate frequency domain characteristics of postive real functions and their corresponding Liapunov functions has had very broad implications in modern control theory as discussed in Note 3.1.

The application of the Kalman-Yacubovich lemma to the stability of a nonlinear system comes almost as an anticlimax: Consider the system in the standard form of Figure 3.18 with the external excitation set to zero. To evaluate stability of the system we choose a candidate Liapunov function

$$V = x'Px \tag{3.47}$$

the time derivative of which is

$$\dot{V} = \dot{x}'Px + x'P\dot{x} \tag{3.48}$$

$$= (x'A' + ub')Px + x'P(Ax + bu) \tag{3.49}$$

$$= x(A'P + PA)x + 2ub'Px \tag{3.50}$$

If the linear part of the system is positive real, however, the results of the Kalman-Yacubovich lemma can be used. In particular $b'P = c'$ and $c'x = y$. Also note that $u = -\phi(y)$. Thus (3.50) becomes

$$\dot{V} = -x'qq'x - 2y\phi(y) = -(q'x)^2 - 2y\phi(y) \tag{3.51}$$

The first term is negative semi-definite. The second term is negative unless $\phi(y) = 0$ for some $y \neq 0$. Thus, if the nonlinear element has the characteristic of a strict nonlinear resistor (i.e., $y\phi(y) > 0$), the system is globally asymptotically stable. To ensure that even the case in which the element has zero resistance for $y \neq 0$, the more general Liapunov function (3.34) is needed.

The foregoing is an example of the property of *absolute stability*. It does not rely on the specific nonlinearity; rather it applies to every nonlinearity having characteristics that lie in the first and third quadrants. Unfortunately, however, it applies only to systems in which the linear part is positive real. As we remarked earlier, very few linear systems enjoy this property. Hence this absolute stability result does not apply to very many systems. As we will see in the next section, however, there are ways of obtaining results that are more broadly useful.

Before leaving this topic, a few observations regarding this derivation are appropriate:

- As are all results based on Liapunov's Second Method, this result is conservative. It does not imply that *only* positive-real plants can be absolutely stable in a closed-loop system with a dissipative nonlinearity. It is surely possible for plants that are not positive-real to be absolutely stable. Moreover, the derivative $\dot{V}$ in (3.51), being the sum of two negative quantities, is "very" negative. For some nonlinearities it might be possible to demonstrate that $\dot{V}$ is negative even when $\phi(\ )$ is not a perfect nonlinear resistor.

- The result can be generalized to multivariable systems. In electrical network theory the positive-real concept is extended to multiport networks [11], and the possibilities of extending the Kalman-Yacubovich lemma are also apparent.

- The notion of positive real functions has an important role in proof of stability of several adaptive control algorithms, as will be discussed in Chapter 10.

3.2.3   The Circle Criterion

Very few processes in practice have transfer functions that are positive real. Thus the Kalman-Yacubovich lemma, which assures the absolute stability of a system to which it applies, is rarely directly applicable. But it serves as the starting point for stability conditions that are more broadly applicable.

One of the generalizations of the lemma is the *circle criterion*. To understand how this criterion evolves, note that although the Nyquist diagram rarely remains to the right of the imaginary axis, in many cases it does remain to the right of a vertical line in the left half plane, as shown in Figure 3.20. In such cases a positive real function can be formed from the transfer function by shifting to the right: If

$$G(j\omega) \geq -a$$

with $a > 0$, then

$$Z(j\omega) = G(j\omega) + a \geq 0$$

So $Z(j\omega)$ is a positive real function. What does this imply about the stability of the loop containing $G(s)$? To see the answer to this question look at Figure 3.21(a) which is just another representation of the standard loop of Figure 3.1, because the two feedback paths containing the gain a cancel each other. In Figure 3.21(b) the local loop containing the nonlinearity $\phi(\ )$ and a is reduced to a single nonlinear element ψ. We conclude that the system is absolutely stable for the nonlinear element $\psi(\ )$ anywhere in the first and third quadrant. This can be related to the original nonlinearity $\phi(\ )$. By simple algebra we see that

$$\psi(x) = \frac{\phi(x)}{x - a\phi(x)}$$

from which we conclude that if $\psi(\ )$ lies in the first and third quadrants, then

$$0 \leq \phi(x) \leq x/a$$

This result can be stated as a stability theorem: a sufficient condition for the stability of a system with a sector-bounded nonlinearity: $0 \leq \phi(x) \leq x/a$ is that the Nyquist diagram lie to the right of the line $s = -a - j\omega$.

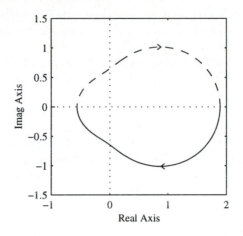

Figure 3.20 Nyquist diagram of plant lies to the right of $s = -a$.

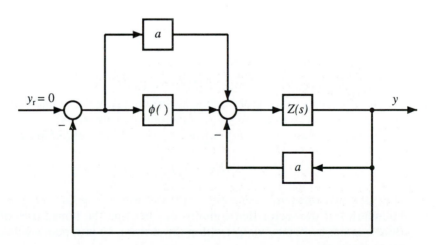

Figure 3.21 Loop transformation for development of stability theorem.

This result is useful but more restrictive than desirable, because the minimum real part (i.e., the most negative value) of $G(j\omega)$ for a typical plant is attained at some value other than on the negative real axis. It would be beneficial to have a criterion that relates to the behavior of the Nyquist diagram near the negative real axis, since the describing function of zero memory nonlinearities lie on this line. The desired result is the circle criterion, which is obtained from the result of the previous section by a somewhat more complicated set of loop transformations.

Suppose that the nonlinearity in question is sector bounded as shown in Figure 3.14. By a sequence of loop transformations we can convert $\phi(\ )$ to another nonlinearity $\psi(\ )$ that occupies the entire first and third quadrants. In order to leave the dynamics unchanged, however, the transfer function $G(s)$ must be transformed to another transfer function $Z(s)$. If $Z(s)$ is positive-real, what does this imply about $G(s)$?

The original system in standard form is shown repeated in Figure 3.22(a). Since the stability we are investigating does not depend on the input, it is omitted and replaced by a gain of -1 in Figure 3.22(b). Also shown is the addition of two paths each with a gain of K_1. The effect of these two paths cancel and hence the system remains unchanged. Finally two more paths, each with a gain of $1/(K_2 - K_1)$ are added. Again, the effect of these paths cancel leaving the overall dynamics unchanged. (Note that the feedforward path for ϕ is a feedback path for $G(s)$ and vice versa.)

The block containing the nonlinearity $\phi(\ )$ with its feedforward and feedback paths defines another nonlinearity $\psi(\ )$, the characteristics of which are defined by

$$u = -K_1 z + \phi(z) \tag{3.52}$$

$$e = \frac{K_2 z - \phi(z)}{K_1 - K_2} \tag{3.53}$$

Similarly the block containing $G(s)$ and its feedforward and feedback paths has the transfer function

$$Z(s) = \frac{G(s)}{1 + K_1 G(s)} + \frac{1}{K_2 - K_1} \tag{3.54}$$

Now suppose $\phi(\ )$ is sector-bounded between $K_1 x$ and $K_2 x$ and that $Z(s)$ is positive real. What does this imply about the functions $\psi(\ )$ and $G(s)$?

To see what happens to $\psi(\ )$, replace $\phi(x)$ in (3.52) and (3.53) by Kz. Then

$$u = (K_2 - K_1)\frac{K - K_1}{K_2 - K}e = \psi(e)$$

It is seen that when $K = K_1$, $\phi(e) = 0$ and when $K = K_2$, $\phi(e) = \infty$. Thus the boundaries of the sector that contains $\phi(\ )$ become the boundaries of the first and third quadrants. Points on $\phi(e)$ within the sector map into points within the first and third quadrants.

As regards the relation between $Z(s)$ and $G(s)$, note that (3.54) defines a bilinear transformation. Such transformations are known to transform circles and lines into circles and lines. Thus the imaginary axis, defined by the line

$$Z(s) = 0 + j\omega$$

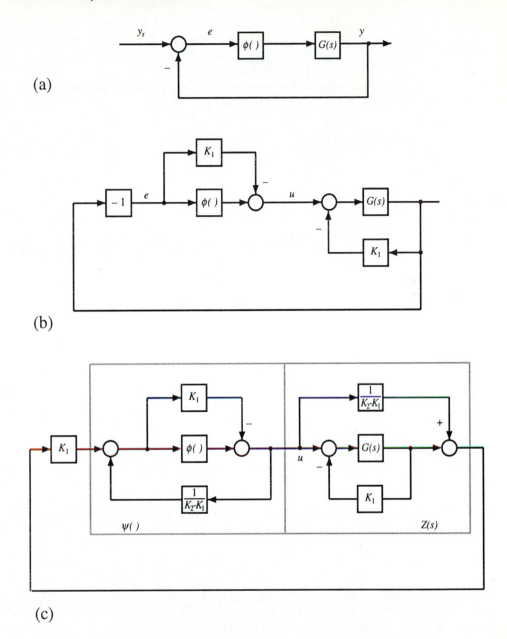

Figure 3.22 Loop transformations for development of circle criterion. (a) Standard form. (b) Reference input removed and K_1 paths added. (c) Two more paths added.

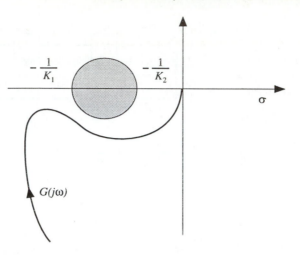

Figure 3.23   The circle criterion.

maps into a circle. To find the location of the circle, solve (3.54) for $G(s)$:

$$G(s) = \frac{(K_2 - K_1)Z(s) - 1}{K_2 - K_1(K_2 - K_1)Z(s)}$$

Hence, $Z(s) = 0$ corresponds to $G(s) = -1/K_2$ and $Z(s) = \infty$ corresponds to $G(s) = -1/K_1$. The imaginary axis of the $Z(s)$ plane thus maps into a circle whose diameter is defined by the points $-1/K_2, -1/K_1$ as shown in Figure 3.23. It is readily established that the points in the left half of the $Z(s)$ plane map into the interior of this "*critical circle.*" From this we can conclude that if the Nyquist diagram for $G(s)$ stays outside the critical circle, the system is asymptotically stable. This can be summarized by the following:

Theorem 3.1 (Circle Criterion) *If nonlinearity $\phi()$ in the standard representation is sector bounded in the sector defined by*

$$K_1 \leq \frac{\phi(x)}{x} \leq K_2$$

and the Nyquist diagram $G(j\omega)$ does not enter the critical circle with the diameter defined by

$$s = -1/K_1 \quad and \quad -1/K_2$$

then the system is globally asymptotically stable.

The circle criterion can be contrasted with the stability prediction of the describing function: For a sector-bounded nonlinearity the describing function is a subset of the line segment $[-1/K_1, -1/K_2]$ which is the diameter of the critical circle. Hence it would predict that if the Nyquist diagram does not cross the diameter of the critical circle, the system will be stable. Obviously it is possible for the Nyquist diagram to enter and leave the critical circle without crossing its diameter. (See Figure 3.24) In this case the describing function predicts that the system will be stable, whereas the circle criterion leaves the issue of stability in doubt (since failure to satisfy the circle criterion does not *ipso facto* imply instability). It is possible, of course, in a case depicted in Figure 3.24 for the prediction of the describing function to be be correct.

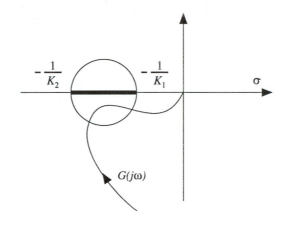

Figure 3.24   Describing function analysis and circle criterion may give conflicting predictions about stability.

3.2.4   Popov's Criterion and the Generalized Circle Criterion

As a practical matter the describing function is somewhat too lenient in predicting that systems for which the Nyquist diagram does not cross the "forbidden line segment" $\mathcal{D} = [-1/K_2, -1/K_1]$ will not be oscillatory; in contrast, the circle criterion is too severe, suggesting the possibility of oscillation if the Nyquist diagram crosses into the circle of which $\mathcal{D}$ is the diameter. The generalized circle criterion brings the two methods closer together.

The generalized circle criterion may be stated as follows:

Theorem 3.2 (Generalized Circle Criterion) *If nonlinearity $\phi(\ )$ in the standard representation is sector-bounded in the sector defined by*

$$K_1 \leq \frac{\phi(x)}{x} \leq K_2$$

and the Nyquist diagram $G(j\omega)$ does not enter a circle having a chord defined by

$$s = -1/K_1 \quad and \quad -1/K_2$$

then the system is globally asymptotically stable.

Since the diameter of a circle is a special case of a chord, the generalized circle criterion is more permissive that the ordinary circle criterion. The situations illustrated in Figure 3.25 correspond to asymptotically stable systems, but would not pass the test of the circle criterion.

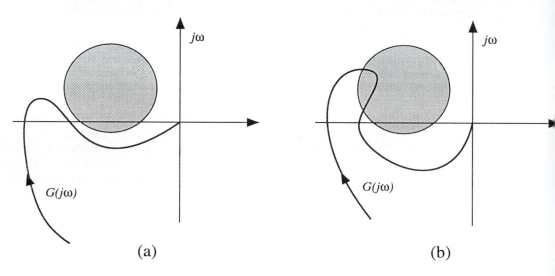

(a)   (b)

Figure 3.25   Generalized circle criterion. (a) System asymptotically stable. (b) Stability not certain.

It does not appear that the generalized circle criterion can be derived directly by means of loop transformations of positive-real functions. It can be derived, however, from another graphical criterion known as Popov's criterion, which can be stated as follows:

Theorem 3.3 (Popov's Criterion) *If the nonlinearity $\phi(\ )$ in the standard representation is sector-bounded in the sector*

$$0 \leq \frac{\phi(x)}{x} \leq K$$

and the Nyquist diagram satisfies the following condition:

$$\Re[(1 + j\omega q)G(j\omega)] + \frac{1}{K} \geq \delta > 0$$

for some real constant q and all real $\omega \geq 0$, then the system is globally asymptotically stable.

Proofs of the Popov criterion can be found in a number of texts on nonlinear systems, such as [6] and [12] as well as [1].

The generalized circle criterion can be derived by means of loop transformations of Popov's criterion. (See Problem 3.7.)

Remember that the describing function, the circle criteria, and the Popov criterion are useful as qualitative design guidelines. Simulation remains the only way to determine how a system really behaves.

3.3 NOTES

Note 3.1 Spectral factorization

The notion of spectral factorization first appeared in system theory as a means of solving the Wiener-Hopf equation that arose in Wiener's theory of linear filtering [13]. The steady-state Kalman filter is another solution of the linear, least squares filtering problem. Hence you might suspect a connection between the Kalman filter and spectral factorization. Moreover, the steady-state, linear, quadratic (LQ) control problem is the dual of the Kalman filtering problem. So you would also expect some connection between spectral factorization and the LQ control problem. If you are familiar with the latter, especially with Kalman's proof of the robustness of the LQ control [14], you will recognize Kalman's hand in the proof of the Kalman-Yacubovich lemma. (See especially (3.42) and (3.43).)

Spectral factorization creates a link between optimum control and Wiener-Hopf theory and suggests the use of the latter for formulating and solving optimum control problems. This link was exploited by Youla and Bongiorno in a series of papers on Wiener-Hopf control system design [15].

3.4 PROBLEMS

Problem 3.1 Effect of friction on performance of servomotor

A permanent-magnet d-c servomotor with an inertial load and negligible armature inductance is governed by the following equations:

$$\dot{\theta} = \omega \tag{3.55}$$

$$\dot{\omega} = -\frac{K^2}{RJ}\omega - \alpha(\omega) + \frac{K}{RJ}u \tag{3.56}$$

where θ is the shaft position, u is the input voltage, K is torque-to-current ratio, R is the armature resistance, and J is the combined inertia of the armature and the load. The function $\alpha(\omega)$ denotes the angular acceleration due to friction.

Numerical values of the above parameters are as follows:

$$\frac{K^2}{RJ} = 25., \quad \frac{K^2}{RJ} = 5.$$

When the servomotor is used in a closed-loop control system with position feedback, the system is stable for all positive values of loop gain.

Use the analysis technique outlined in Example 3.1 and the Describing Function Method and the circle criterion to determine the possible effect of friction on the system, assuming the following friction models:

(a) Coulomb friction

$$\alpha(\omega) = a\,\text{sgn}(\omega)$$

with (i) $a = 1$, and (ii) $a = 100$.

(b) Friction with "stiction"

$$\alpha(\omega) = (a_1 e^{-a_2|\omega|} + a_3)\,\text{sgn}(\omega)$$

where $a_1 = 90$, $a_2 = 0.1$, and $a_3 = 10$

Problem 3.2   Effect of friction on servomotor (Continued)

Suppose the armature inductance L of the servomotor in Problem 3.1 is not entirely negligible. (Say $L/R = 0.01$). Would the presence of the inductance have any significant effect on the stability analysis of Problem 3.1?

Problem 3.3   Temperature control with saturating linear control

Consider the temperature control system of Example 3.4 , except that a saturating linear control replaces the deadzone in implementing the feedback control. Discuss the effect of this replacement on the stability and transient response, when

(a) The reference input is zero.

(b) The reference input is nonzero.

On the basis of the results of this study, under what circumstances would you recommend using the saturating linear control in place of the deadzone, considering the additional expense of the former.

Problem 3.4   Describing function of sector bounded nonlinearity

Show that the describing function of a sector-bounded nonlinearity lies between a and b.

Problem 3.5   Positive real functions

(a) Show that a positive real function is an asymptotically stable, minimum phase function.

(b) Show that if $Z(s)$ is a positive-real function, then $Y(s) = 1/Z(s)$ is also a positive-real function.

(c) Show that the difference in degrees of the numerator and the denominator of a positive real function cannot differ in magnitude by more than 1.

Problem 3.6   Verification of positive-real function

Verify that the following function

$$Z(s) = \frac{s^2 + 4s + 3}{s^3 + 5s^2 + 6s + 1}$$

is positive-real.

Problem 3.7   Generalized circle criterion

Derive the generalized circle criterion by means of loop transformations of Popov's criterion.

3.5   REFERENCES

[1] J.C. Hsu and A.U. Meyer, *Modern Control Principles and Applications*, McGraw-Hill, New York, 1968.

[2] D. Atherton, *Nonlinear Control Engineering*, 2nd ed., Van Nostrand Reinhold, New York, 1982.

[3] M.A. Aizerman, "On a Problem Concerning Stability in the Large of Dynamical Systems," Uspekhi Mat. Nauk (USSR), Vol. 4, pp. 187-188, 1949.

[4] R.E. Fitts, "Two Counterexamples to Aizerman's Conjecture," IEEE Trans. on Automatic Control, vol AC-11, no. 3, pp. 553-556, July 1966.

[5] R.E. Kalman, "Physical and Mathematical Mechanisms of Instability in Nonlinear Control Systems," Transactions, ASME, Vol. 79, pp. 553-566, 1957.

[6] M. Vidyasagar, *Nonlinear Systems Analysis,* Prentice-Hall, Inc., Englewood Cliffs, NJ, 1978.

[7] R.R. Mohler, *Nonlinear Systems, Vol. 1. Dynamics and Control,* Prentice-Hall, Inc., Englewood Cliffs, NJ, 1991.

[8] J.-J. E. Slotine and W. Li, *Applied Nonlinear Control*, Prentice-Hall, Inc., Englewood Cliffs, NJ, 1991.

[9] R.E. Kalman, "Lyapunov Functions for the Problem of Lur'e in Automatic Control," Proc. National Academy of Sciences, USA, Vol. 48, pp. 201-205, 1963.

[10] S. Lefschetz, *Stability of Nonlinear Control Systems,* Academic Press, New York, 1965.

[11] W. Cauer, *Synthesis of Linear Communication Networks,* McGraw-Hill Book Co., New York, 1958.

[12] H.K. Khalil, *Nonlinear Systems*, Macmillan Publishing Co., New York, 1992.

[13] H.W. Bode and C.E. Shannon, "A Simplified Derivation of Linear Least Square Smoothing and Prediction Theory," Proc. IRE, Vol. 38, April 1959, pp. 417-425.

[14] R.E. Kalman, "When is a Linear Control System Optimal?" Trans. ASME (J. Basic Engineering) Vol. 86D, No.1, March 1964, pp. 51-60.

[15] D.C. Youla, H.A. Jabr, and J.J. Bongiorno, Jr. "Modern Wiener-Hopf Design of Optimal Controllers, Part II, The Multivariable Case, " IEEE Trans. on Automatic Control, Vol. AC-21, pp. 319- 338, 1976.

Chapter 4

CONTROLLING NONLINEAR SYSTEMS

Real systems are nonlinear; nevertheless they can usually be controlled, as countless existing control systems testify. While a nonlinear system is usually more bothersome than a linear system, often the nonlinearity is only a minor nuisance. Sometimes it is a design challenge. Only rarely does the presence of nonlinearity preclude a control system from being designed. And sometimes the presence of nonlinearity may make a system easier to control.

If the system is only mildly nonlinear, the simplest approach might be to ignore the nonlinearity in designing the controller (i.e., to omit the nonlinearity in the design model), but to include its effect in evaluating the system performance. The inherent robustness of the control law designed for the approximating linear system is relied upon to carry it over to the nonlinear system. This approach can be used when the performance is evaluated by simulation or by one of the analytical techniques described in Chapter 3. The importance of including the nonlinear effects in the truth model cannot be overemphasized. It is much better to devote a small effort to uncovering potential problems before a system is built than to spend much more effort to correcting a problem afterwards.

Even when the system is significantly nonlinear, a suitable control law can often be designed by extension of the methods developed for linear systems. When applied to nonlinear systems, however, these methods are generally not rigorous; hence the validity of the resulting design must be verified separately, either by use of the analytical methods discussed in Chapters 2 and 3, or by simulation.

Sometimes the dynamics of a system are strongly nonlinear when expressed in the

state variables most natural for the problem (such as those that can be directly measured), but can be made more nearly linear by a nonlinear transformation to a new
set of variables. In such cases an obvious approach would be to design the control law
for the new set of variables and then transform it back to the original set.

In every physical system there is a finite limit on the magnitude of the control
that can be provided by the actuator. The actuator is said to "saturate" or to have
"limited authority." A control law designed to avoid reaching the saturation limit is
usually overdesigned and inefficient, since the actuator has a physical capability that is
not used, like the engine in a heavily overpowered automobile. A well-proportioned
control system is one in which the actuator is selected so that its full capabilities are
sometimes needed. The actuator should be expected to saturate in extreme tasks.
This suggests that a well-designed closed-loop system should be *expected* to be nonlinear, if only through the saturation of the actuator. If the actuator only saturates
rarely, however, it is usually possible to ignore the saturation in the design and to assess its effect only in the performance evaluation. On the other hand, if the authority
of the actuator is extremely limited and it saturates frequently, a design method that
explicitly accounts for the saturation, such as the bang-bang technique discussed in
Chapter 6, would be more appropriate.

This chapter and the next are intended to present a few methods that can be used
to design control algorithms for plants in which the nonlinearity is not negligibly small
and hence must be accounted for explicitly in the design. The task of designing a
controller for a nonlinear plant is generally more challenging than for a linear plant.
But it is rarely hopeless, as the examples in these two chapters should demonstrate.

4.1 EXTENDED SEPARATION PRINCIPLE

One of the main benefits of state space methods for design of linear control systems
is the separation principle, which permits the control system to be designed in two
stages. First, a "full-state" feedback control law is designed, and then an observer
is provided to estimate all the states or those that cannot be measured directly. A
consequence of the separation principle is that the control system designed by this
procedure preserves the essential properties of the full-state feedback control and of
the observer. In linear systems, the poles of the closed-loop system are the poles of
the full-state feedback system and the poles of the observer.

In general, the separation principle does not rigorously apply to nonlinear systems:
There is no assurance that the control algorithm obtained by combining a nonlinear,
full-state feedback law (or, for that matter, a linear full-state feedback law), with an
observer, i.e., a dynamic system for estimating the unmeasured state variables, will result in a closed-loop nonlinear system with satisfactory performance. Pragmatically,
however, the separation principle can be extended to nonlinear systems and will often
prove to be an effective design method. This will be demonstrated in Chapter 5 where
the concept of a nonlinear observer is addressed. Indeed, it is possible to demonstrate
local asymptotic stability for a class of nonlinear designs that are based on the separa-

tion principle. But the practical way of predicting the behavior of a nonlinear control system is by simulation.

For the separation principle to hold, it is necessary that the model of the plant used in the observer be an accurate representation of the true dynamics of the process. If a simplified design model is used or if the plant parameters do not agree exactly with those of the model, the separation principle does not strictly hold. But if the model is a reasonable approximation to the plant, and robustness requirements are considered, the performance of the controller can be expected to be satisfactory.

There are methods, of course, that do not employ the separation principle but rather seek to achieve the entire control algorithm design in one step. The very oldest and the most modern design methods are among these. The most basic design method is to select a controller having a fixed structure (e.g., "PID"—"proportional-integral-derivative") with adjustable parameters. The design consists of "tuning" the adjustable parameters to "optimize" performance. (The quotation marks are used around the word *optimize* to suggest that the optimization is often more knob-turning than mathematical optimization.)

Control system designs of fixed structure are as likely to be effective for nonlinear systems as they are for linear systems. Since the number of variable parameters in the structure is likely to be small, it should be fairly easy to determine whether a simple design can provide the required performance by means of a simulation study. If the required performance can be achieved, your work is done. The methods of this book are not needed for this class of applications. In the most interesting applications (if not the largest in number), however, a control algorithm of simple fixed structure will fall short of providing the required performance. For such applications, design techniques are needed that are capable of systematically providing control algorithms that can provide the necessary performance.

The newest design methods, such as the H_∞ method, also do not make use of the separation principle. It is premature to predict whether the H_∞ design methods of the late 1980s, and their kin, will be successfully extended to nonlinear systems. Since they are based on frequency-domain analysis, they are rigorously applicable only to linear, time-invariant systems. But since they are aimed at achieving a high degree of robustness (insensitivity to parameter change) they might prove applicable to systems in which the nonlinearity can be represented as an uncertain parameter in a linear system.

Most factors considered, the design approach now the most capable of providing a useful design for a nonlinear system of moderate complexity is the extended separation principle. Following this approach, the present chapter is concerned with the design of control laws under the assumption that all the state variables can be measured. The next chapter deals with the design of observers for estimating the state in nonlinear dynamic control law designs.

4.2 LINEARIZATION ABOUT A SET POINT

Many control systems are required to maintain the state of the dynamic process at a specified constant value. In the process control industry this value is usually called the *set point*, or *operating point*. In aircraft flight control the set point is often referred to as the *trim* condition. Sometimes it is referred to simply as the *reference point*. The control system in these applications can be visualized as consisting of two distinct subsystems, the first to generate the control signal needed to keep the process at the set point once it gets there, and the second to generate the control needed to return the process to the set point if it should deviate from it. In an aircraft, for example, a trim condition might be represented as a specified combination of altitude and speed. To maintain this trim condition would require an appropriate setting of the throttle and control surfaces. Deviations from the trim condition, detected by the flight control instruments, would generate the control signals—changes in throttle and control surface deflections—to return the aircraft to the trim condition.

For a mathematical representation of the method of linearization about a set point, consider the system

$$\dot{x} = f(x, u) \tag{4.1}$$

$$y = g(x) \tag{4.2}$$

Suppose that $\bar{y}$ is the desired output, which is assumed to be constant. Suppose also that $\bar{x}$ and $\bar{u}$ are values of the state and control, respectively, that produce this constant output. (Note that $\bar{x}$ and $\bar{u}$ are not necessarily unique; if they are not unique, it may be possible to make them unique by the minimization of an appropriate auxiliary function, but there is no requirement to do so.) Finally, assume that $\bar{x}$ is an equilibrium state. Then (4.1) and (4.2) become

$$0 = f(\bar{x}, \bar{u}) \tag{4.3}$$

$$\bar{y} = g(\bar{x}) \tag{4.4}$$

The determination of $\bar{x}$ and $\bar{u}$ is an algebraic problem. If the problem is physically meaningful, a solution (possibly not unique, as noted above) exists. If the functions $f(\cdot, \cdot)$ and $g(\cdot)$ are simple, it may be possible to obtain these solutions analytically; otherwise it may be necessary to implement an on-line numerical method to solve (4.3) for u. In either case, the solutions can be denoted as

$$\bar{u} = U(\bar{y}) \tag{4.5}$$

$$\bar{x} = X(\bar{y}) \tag{4.6}$$

The function $U(\cdot)$ is sometimes called a *feedforward* control law, since it determines the setting of the control on the basis of the desired output; the function $X(\cdot)$ can be called the *set point generator*, since it determines the reference state or set point that produces the desired output.

As already noted, it often happens that not all the state variables are specified by the reference input $\bar{y}$. The number of reference input variables that can be independently specified cannot exceed the number of independent control variables, or *control degrees of freedom.* [1]

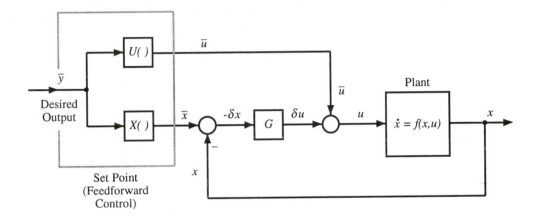

Figure 4.1 Structure of control system designed by linearization about a set point.

To control deviations of the actual state x from the set point $\bar{x}$, another subsystem is provided. This subsystem may be called the *regulator* or the *feedback controller,* since it generates the corrective controls based on the actually measured performance. Figure 4.1 shows the structure of this system.

The regulator operates on the difference δx between the actual state x and the set point $\bar{x}$ to produce a corrective control δu. The corrective control is summed with the feedforward control $\bar{u}$ to produce the actual control u input to the plant.

Note that the reference state $\bar{x}$ is shown as a possible input to the regulator: The parameters of the process for which the regulator is designed depend upon the reference input and the regulator may be designed to account for this dependence.

In some applications, the set point may be not a constant but rather a slowly varying function of time. If the transient response of the regulator is fast enough relative to

---

[1]The expression "degrees of freedom" is overused. This can be a source of confusion. In classical dynamics the number of degrees of freedom is usually thought of as the number of independent modes of energy storage. The dimension of the state space of a system with n dynamic degrees of freedom is $2n$. In systems comprising a variety of physical devices, it is often not possible to identify the physical modes of energy storage; moreover the state-space model may have an odd number of state variables, which would preclude identifying the state variables with physical degrees of freedom. To add to the confusion, the number of independent state variables is sometimes called the number of degrees of freedom.

the rate of variation of the set point, the control law may well prove to be adequate for
"tracking" the varying set point. If the rate of variation of the set point is comparable
to the transient response of the regulator, however, it may be necessary to explicitly
consider the set point's variation.

To design the regulator, note that the total state x is the sum of the reference state
$\bar{x}$ and a perturbation δx:

$$x = \bar{x} + \delta x \tag{4.7}$$

Similarly,

$$u = \bar{u} + \delta u \tag{4.8}$$

Then, since $\dot{\bar{x}} = 0$, (4.7) becomes

$$\dot{\delta x} = A(\bar{x})\delta x + B(\bar{x})\delta u + e(\bar{x}, \bar{u}, \delta x, \delta u) \tag{4.9}$$

where

$$A = \begin{bmatrix} \dfrac{\partial f_1}{\partial x_1} & \cdots & \dfrac{\partial f_1}{\partial x_k} \\ & \cdots & \\ \dfrac{\partial f_k}{\partial x_1} & \cdots & \dfrac{\partial f_k}{\partial x_k} \end{bmatrix}$$

$$\tag{4.10}$$

$$B = \begin{bmatrix} \dfrac{\partial f_1}{\partial u_1} & \cdots & \dfrac{\partial f_1}{\partial u_l} \\ & \cdots & \\ \dfrac{\partial f_k}{\partial u_1} & \cdots & \dfrac{\partial f_k}{\partial u_l} \end{bmatrix}$$

in which the partial derivatives are evaluated at the set point. The function $e(\ )$ in (4.9)
represents the nonlinear terms remaining after expanding the right hand side of the
dynamics about the set point.

The following reasoning is the basis of the regulator design: If the regulator works
effectively, the perturbation δx will be small; if the perturbation is small, the nonlinear
term $e(\ )$ in (4.9) will be small and can be regarded as a disturbance; since the regulator
is designed to counteract the effects of disturbances, the presence of $e(\ )$ should cause
no problem.

This reasoning is, of course, circular and cannot be justified rigorously. But nevertheless it usually works. Needless to say, it may not always work, so it is necessary to
test the design that emerges for stability and performance—analytically, where possible, by Liapunov's Second Method, for example, and by simulation.

The following is an example in which the method works well.

Example 4.1   Robot Arm

A "single-link" robot arm is nothing more than a pendulum, as shown in Figure 4.2. Its
motion, in the absence of friction, is governed by

$$\ddot{\theta} + \Omega^2 \sin\theta = u \tag{4.11}$$

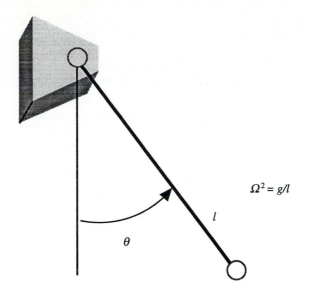

$$\Omega^2 = g/l$$

Figure 4.2   Robot arm.

where u is the specific torque (i.e., the torque divided by the moment of inertia).
    To bring the arm to a fixed angle $\bar{\theta}$ it is necessary to supply a constant torque

$$\bar{u} = \Omega^2 \sin \bar{\theta} \qquad (4.12)$$

To obtain the perturbation equation, note that $\dot{\bar{\theta}} = \ddot{\bar{\theta}} = 0$. Hence

$$\delta\ddot{\theta} = -\Omega^2 \sin(\bar{\theta} + \delta\theta) + \bar{u} + \delta u$$
$$\doteq -(\Omega^2 \cos \bar{\theta})\delta\theta + \delta u \qquad (4.13)$$

Assuming that the angle θ and its derivative $\dot{\theta} = \delta\dot{\theta}$ are measurable, a corrective
control in the linear form

$$\delta u = -G_1\delta\theta - G_2\dot{\theta} \qquad (4.14)$$

can be used. The gains G_1 and G_2 can be obtained by any of the methods for the design
of linear systems.
    Note, however, that the dynamics matrix corresponding to (4.13) depends on the set
point $\bar{\theta}$:

$$A = \begin{bmatrix} 0 & 1 \\ -p & 0 \end{bmatrix} \qquad (4.15)$$

where

$$p = \Omega^2 \cos \bar{\theta}$$

This means that the gains in (4.14) may also depend upon the set point. In particular,
if the gains are selected to minimize a quadratic integral in $\delta\theta$ and δu, the gains turn out

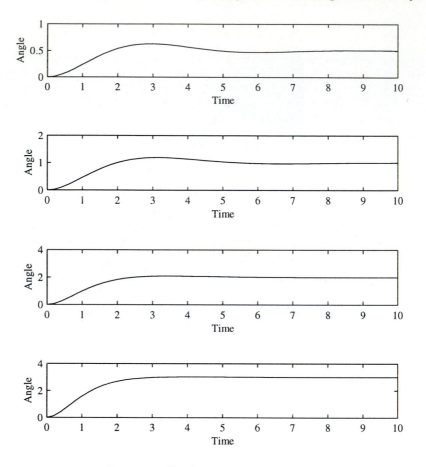

Figure 4.3   Transient response of robot arm.

to be given by

$$G_1 = \sqrt{p^2 + c^2} - p$$
$$G_2 = \sqrt{2g_1}$$

which are functions of $\bar{\theta}$.

The total control law is obtained by combining the feedforward control law (4.12) with the regulator (4.14).

The effectiveness of this control law is shown by the transient responses of Figure 4.3, for $c = 1.0$ and for $\Omega = 4$. It is seen that the responses are entirely satisfactory for set points ranging from 0.5 radians to 3.0 radians, the latter being a condition in which the arm is nearly vertically upward.

Example 4.2   Magnetic Levitation

From first principles (see [1], for example) it can be determined that the force on a ferrous block in a magnetic circuit shown in Figure 4.16 is approximately proportional to the square of the current and inversely proportional to the square of the air gap. The force is such as to reduce the size of the air gap. (Since the force would become infinite as the air gap approaches zero, this relation must obviously be limited to air gaps of reasonable size.) Assuming that the only forces acting on the block are its weight and the magnetic force, the differential equation governing the motion is

$$\ddot{x} = -k\frac{u^2}{x^2} + g \tag{4.16}$$

where x is the displacement of the block from the magnet body (i.e., the air gap), measured downward; u is the current, which will be regarded as the control variable in this example; g is the acceleration of gravity; and k is a constant which depends on the physical units of the problem.

In state-space form (4.16) can be written as

$$\dot{x}_1 = x_2 \tag{4.17}$$
$$\dot{x}_2 = -k(u^2/x_1^2) + g \tag{4.18}$$

To maintain a steady-state air gap of $\bar{x}$, a steady-state current $\bar{u}$ is required to keep the acceleration zero. From (4.16), the steady-state current is given by

$$\bar{u} = \pm\sqrt{g/k}\;\bar{x} \tag{4.19}$$

Note that only the square of the current matters; the sign can be chosen for convenience in implementation. Let's take the top sign, making

$$\bar{u} = \sqrt{g/k}\;\bar{x}$$

To design a control law by linearizing about the operating point $(x_1 = \bar{x}, x_2 = 0)$ let

$$x_1 = \bar{x} + \delta x_1$$
$$u = \bar{u} + \delta u$$

Calculating the requisite partial derivatives we find the perturbation equations

$$\delta\dot{x}_1 = x_2 \tag{4.20}$$
$$\dot{x}_2 = (2g/\bar{x})\delta x_1 - (2g/\bar{u})\delta u \tag{4.21}$$

Note that the coefficient of δx_1 is *positive* which implies that the set point is an *unstable* equilibrium state. This is a well-known property of all magnetic levitation systems: Active feedback control is essential to maintain the required equilibrium state.

A suitable linear, full-state feedback control law is of the form

$$\delta u = -G_1\delta x_1 - G_2 x_2 \tag{4.22}$$

Using this control law in (4.21) gives the closed-loop equation

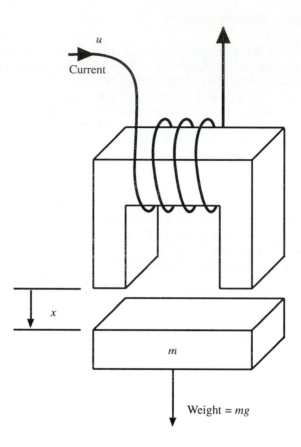

u

Current

x

m

Weight = mg

Figure 4.4   Magnetic levitation control.

$$\dot{x}_2 = 2g \left(\frac{1}{\bar{x}} + \frac{G_1}{\bar{u}} \right) \delta x_1 + \frac{2g}{\bar{u}} x_2 \tag{4.23}$$

For asymptotic stability about the equilibrium state, it is necessary that the coefficients of both δx_1 and x_2 in (4.23) be negative, i.e.,

$$G_1 > \frac{2g}{\bar{x}} \tag{4.24}$$

$$G_2 < 0 \quad (\text{for } \bar{u} > 0) \tag{4.25}$$

$$\tag{4.26}$$

These give a fairly wide latitude for the choice of feedback gains. Note, however, that G_2 must be present to provide damping; hence feedback of velocity is required. If the velocity cannot be measured directly it must be estimated using an observer, as discussed in Chapter 5.

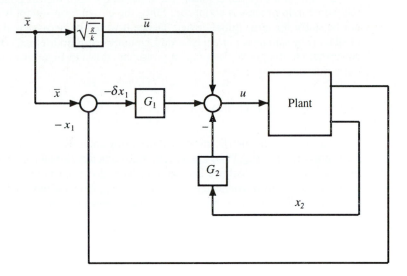

Figure 4.5   Control for magnetic levitation.

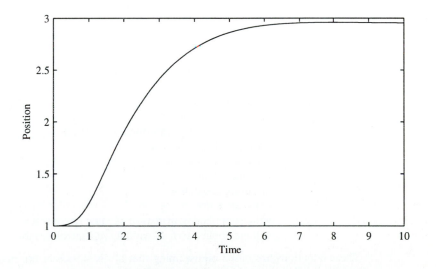

Figure 4.6   Transient response of magnetic levitation system
with control law designed by linearization about a set point.

The complete control law, including the set point control, is obtained by combining (4.20) with (4.22) as shown in Figure 4.5.

The control law (4.22) is linear in the deviation from the set point, and as a result the closed-loop process is nonlinear. Since the closed-loop system is nonlinear, it may not be stable for large initial errors. Hence it would be necessary to check the region of stability by simulation or analysis via Liapunov's Second Method. (See Problem 4.5.) Nevertheless, acceptable transient response, as shown in Figure 4.6, is obtained over a fairly wide operating range.

Example 4.3   Separately Excited D-C Motor Drive

Large d-c motors, such as those used for lifting and traction, typically delivering 10 or more horsepower, usually have externally excited fields. (Small motors usually have permanent magnet fields that need no external excitation.) Such motors are often operated with the field winding connected either in parallel (shunt) with the armature winding or with the two windings in series. If the two windings have external terminals, however, they may be separately controlled to provide superior performance to that which could be obtained with the windings either in parallel or in series.

The differential equations that govern the behavior of a separately excited d-c motor are

$$J\frac{d\omega}{dt} = -B\omega + Ki_f i_a - T_L \tag{4.27}$$

$$L_a\frac{di_a}{dt} = -R_a i_a - Ki_f \omega + e_a \tag{4.28}$$

$$L_f\frac{di_f}{dt} = -R_f i_f + e_f \tag{4.29}$$

where

ω	:	motor angular velocity
i_a	:	armature current
i_f	:	field current
J	:	inertia of motor
B	:	damping of motor
K	:	electromechanical transduction constant
T_L	:	torque delivered to load
L_a	:	armature inductance
L_f	:	field inductance
R_a	:	armature resistance
R_f	:	field resistance
e_a	:	voltage applied to armature winding (a control variable)
e_f	:	voltage applied to field winding (a control variable)

These equations are based on assuming that all the losses in the motor can be accounted for by the mechanical damping and the electrical resistance of the armature and the field, and that the behavior of the motor is linear. The dynamics of a real motor are somewhat more complicated, but (4.27) – (4.29) capture the essential dynamic characteristics and are thus acceptable as a design model.

Since there are two control variables (the voltages applied to the armature and to the field), it is possible to control two static quantities, or as an alternative, to control one quantity and to optimize some aspect of performance. For the sake of the example, suppose that the load torque is a constant. (This situation could arise in lifting a fixed mass by an elevator or a crane.) Suppose, moreover, that the two excitation voltages are to be adjusted to minimize the resistive power loss of the motor:

$$p = R_a i_a^2 + R_f i_f^2$$

It is easy to show that the minimum is achieved with

$$R_a i_a^2 = R_f i_f^2$$

which implies that the two currents are related by

$$i_a = \sqrt{R_f/R_a}\, i_f \tag{4.30}$$

The control law is to be designed to provide a specified load torque T_L and to maintain the currents in the ratio implied by (4.30). Note that the series connection of the armature and the field will usually not satisfy (4.30), since $R_a \ll R_f$ in most cases.

Suppose that the set point is defined by a fixed load torque $\bar{T}_L$ and a desired constant speed $\bar{\omega}$. From (4.27) with $\dot{\omega} = 0$ and (4.30) we obtain the relation for the set point armature current:

$$K i_a{}^2 \sqrt{R_f/R_a} = \bar{T}_L + B\bar{\omega}$$

or,

$$\bar{i}_a = \frac{(R_f/R_a)^{1/4}}{K^{1/2}}(\bar{T}_L + B\bar{\omega})^{1/2} \tag{4.31}$$

Also, again using (4.30),

$$\bar{i}_f = \frac{(R_a/R_f)^{1/4}}{K^{1/2}}(\bar{T}_L + B\bar{\omega})^{1/2} \tag{4.32}$$

The set-point voltages required to produce these set-point currents are obtained using (4.28) and (4.29) with the derivatives set to zero:

$$\bar{e}_a = R_a \bar{i}_a + K\bar{\omega}\bar{i}_f \tag{4.33}$$

$$\bar{e}_f = R_f \bar{i}_f \tag{4.34}$$

The dynamic equations (4.27) through (4.29) can be written in terms of the set point and perturbations from it. In particular, let

$$\omega = \bar{\omega} + \delta\omega$$
$$i_a = \bar{i}_a + \delta i_a$$
$$i_f = \bar{i}_f + \delta i_f$$
$$e_a = \bar{e}_a + \delta e_a$$
$$e_f = \bar{e}_f + \delta e_f$$

Substitution of these expressions into (4.27) – (4.29) and straightforward algebraic calculations gives

$$\frac{d}{dt}\begin{bmatrix} \delta\omega \\ \delta i_a \\ \delta i_f \end{bmatrix} = \begin{bmatrix} -B/J & K\bar{i}_f/J & K\bar{i}_a/J \\ -K\bar{i}_f/L_a & -R_a/L_a & -K\bar{\omega}/L_a \\ 0 & 0 & -R_f/L_f \end{bmatrix}\begin{bmatrix} \delta\omega \\ \delta i_a \\ \delta i_f \end{bmatrix}$$

$$+ \begin{bmatrix} 0 & 0 \\ 1/L_a & 0 \\ 0 & 1/L_f \end{bmatrix}\begin{bmatrix} \delta e_a \\ \delta e_f \end{bmatrix} + \begin{bmatrix} K\delta i_f \delta i_a \\ K\delta i_f \delta\omega \\ 0 \end{bmatrix} \qquad (4.35)$$

It should be noted that this matrix differential equation is equivalent to (4.27)–(4.29); it remains nonlinear because it retains the products of perturbations as the last vector on the right-hand side. On the assumption that the products of perturbations are negligible, however, the control system design can be based on linear theory.

A block-diagram representation of the entire system, assuming full state-variable feedback, is shown in Figure 4.7. The linear control law would have the form

$$\begin{bmatrix} \delta e_a \\ \delta e_f \end{bmatrix} = -\begin{bmatrix} g_{11} & g_{12} & g_{13} \\ g_{21} & g_{22} & g_{23} \end{bmatrix}\begin{bmatrix} \delta\omega \\ \delta i_a \\ \delta i_f \end{bmatrix} \qquad (4.36)$$

with the gains g_{ij} chosen by any appropriate method.

The closed-loop dynamics matrix is given by

$$A_c = A - BG = \begin{bmatrix} -B/J & K\bar{i}_f/J & K\bar{i}_a/J \\ -(K\bar{i}_f + g_{11})/L_a & -(R_a + g_{12})/L_a & -(K\bar{\omega} + g_{13})/L_a \\ -g_{21}/L_f & -g_{22}/L_f & -(R_f + g_{23})/L_f \end{bmatrix}$$

$$(4.37)$$

The armature and the field loops are decoupled by setting g_{21} and g_{22} to zero. This simplifies the design, so let's do it. In this case the armature loop is second order with a characteristic polynomial

$$\Delta_a(s) = s^2 + \left(\frac{B}{J} + \frac{R_a + g_{12}}{L_a}\right)s + \frac{B(R_a + g_{12}) + (K\bar{i}_f + g_{11})K\bar{i}_f}{JL_a}$$

Instead of making the gains g_{ij} be constant, they can be made functions of the set point to keep the transient response invariant to a change in set point. (Naturally, this makes the implementation more complex and almost surely would necessitate digital implementation.) In particular, the coefficients of $\Delta_a(s)$ are independent of the set point when

$$g_{12} = \text{constant}$$

$$g_{11} = -K\bar{i}_f + \frac{JL_a\Omega^2 - BR_a - g_{12}}{K\bar{i}_f}$$

where Ω is the desired closed-loop natural frequency.

The field loop characteristic polynomial is first-order:

$$\Delta_f(s) = s + \left(\frac{R_f + g_{22}}{L_f}\right)$$

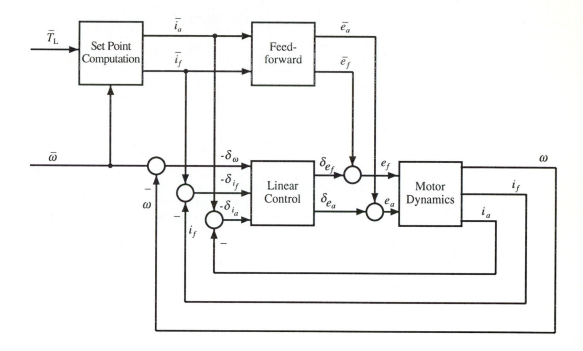

Figure 4.7   Block diagram of full-state feedback design of d-c motor drive system based on linearization about set point.

To illustrate performance achievable, consider numerical data for a typical motor as suggested by Hung and Park [2]. When connected as a shunt motor with $e_f = e_a = 115$ v, and with an output power of 1.5 Kw at a speed of 1750 rpm = 183 rad/s, the armature current i_a is 13 a and the field current i_f is 2.1 a. From these operating conditions, the following parameters can be calculated or estimated:

$$
\begin{array}{rcl}
T_L & : & 8.19\ \text{nm} \\
K & : & 0.30\ \text{nm/a}^2 \\
R_a & : & 8.8\ \Omega \\
R_f & : & 55\ \Omega \\
L_a & : & 1.0\ \text{H} \\
L_f & : & 10.0\ \text{H} \\
J & : & 0.1\ \text{Kg m}^2 \\
B & : & 0.01\ \text{Kg m}^2\ \text{sec}^{-1}
\end{array}
$$

The dynamic response shown by Hung and Park suggests a closed-loop characteristic polynomial for the armature loop of $(s + 25)^2 + (25)^2$ and a field loop characteristic polynomial of $s + 10$. Using these closed-loop parameters in the above expressions for the gains gives

$$g_{11} = -0.3\bar{i}_f + (277./\bar{i}_f)$$
$$g_{12} = 41.1$$
$$g_{23} = 4.5$$

Transient responses using these gains are shown in Figure 4.8 with the motor starting at rest, and with two reference speeds: 50 r/s and 150 r/s. For added realism, it is assumed that the supply voltage is 200v; hence the armature and field voltages are assumed to saturate at this voltage level. The field current is assumed to start at 1.0 amp prior to the start of the transient. For the reference speed of 50 r/s, the motor reaches this speed in about 0.7 sec, with the armature voltage at its saturation level for about 0.5 sec. For the larger reference speed of 150 r/s, the armature voltage is saturated for about 1.8 sec, but the desired speed is achieved shortly after the voltage goes out of saturation.

It is of interest to note that the motor starts going backward, but this does not appear to be objectionable. The transient response in all other respects seems excellent.

The design models in the foregoing examples, although nonlinear, are still simple enough to permit calculation of the nonlinear control law analytically. In most applications this will not be possible; it will be necessary to solve for the gain matrix G numerically, using an appropriate algorithm. To implement the control law by means of linear, quadratic control theory it may be necessary to solve a matrix Riccati equation "on-line" with the set point $\bar{x}$ as an input to the calculation, or "off-line" with the results stored in a look-up table, for example. Either method is feasible in principle, and can be practical, considering recent advances in computer hardware. But, since a linear, fixed gain, control law may be robust enough to be valid over the normal range of operation of the system, these methods may be excessively complicated.

4.3   EXTENDED LINEARIZATION

Another method of designing control systems for nonlinear processes is to explicitly consider the nonlinear dynamics, but to treat them as if they were linear. We can call this design method "extended linearization." In particular, suppose the dynamic process to be controlled has the nonlinear state-space representation

$$\dot{x} = A(x)x + B(x)u \tag{4.38}$$

where $A(x)$ and $B(x)$ are general nonlinear functions of the state x. Any method that could be used to design a control law when A and B are constant matrices is a candidate for use in the design of a control law for (4.38). Thus, for example, the linear control law

$$u = -Gx \tag{4.39}$$

where $G = -R^{-1}B'M$ with M being the solution to the algebraic matrix Riccati equation

$$MA + A'M - BR^{-1}B'M + Q = 0 \tag{4.40}$$

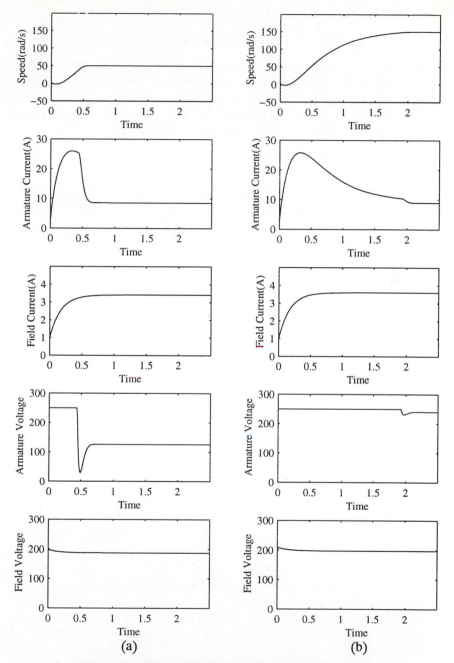

Figure 4.8   Transient responses of control for separately excited
d-c motor. (a) Reference speed: 50 r/s. (b) Reference speed: 150
r/s.

would become the nonlinear control law

$$u = -G(x)x \tag{4.41}$$

with the nonlinear function $G(x)$ being evaluated as in the above equations, but with $A(x)$ and $B(x)$ used in (4.40) as if they were constant matrices. There is no general principle that can be invoked to ensure that this design approach will be successful, but it often is. It is of course necessary to verify the validity of the design by use of one of the appropriate analytical methods or by simulation.

Example 4.4   Robot Arm (Continued)

Instead of using linearization about a set point to design a control law for the robot arm of Example 4.1 we can consider extended linearization by writing the dynamics as

$$\dot{x} = A(x)x + Bu$$

where

$$x = \begin{bmatrix} \theta \\ \dot{\theta} \end{bmatrix}$$

and

$$A(x) = \begin{bmatrix} 0 & 1 \\ p(x) & 0 \end{bmatrix}$$

with

$$p(x) = \frac{\Omega^2 \sin \theta}{\theta}$$

Since the process has the appearance of being linear, a "linear" control law

$$u = -G_1(\theta - \bar{\theta}) - G_2\dot{\theta}$$

can be contemplated. The gains G_1 and G_2 would depend upon the parameter $p(x)$.

Completion of the design and comparison of performance with the previous design is the subject of Problem 4.1.

4.4   FEEDBACK LINEARIZATION

Although linearization about a set point and extended linearization are valid methods, there is no assurance that they will result in stable control laws. It is necessary to evaluate the performance and verify stability for each individual design. The limitations of the previous methods have served to motivate investigations of design methods that can assure globally stable performance. One currently popular method is "feedback linearization," the basic idea of which is to represent the control u as the sum of two terms, one of which cancels out the nonlinearities and the other controls the resulting linear system.

To understand the basic idea, consider the nonlinear system represented by a single kth order differential equation

$$y^{(k)} + f_1(x)y^{(k-1)} + \cdots + f_{k-1}(x)y^{(1)} + f_k(x)y = u \tag{4.42}$$

where

$$y^{(i)} := \frac{d^i y}{dt^i}$$
$$x := [y, y^{(1)}, \ldots, y^{(k-1)}] \tag{4.43}$$

Define a new control variable

$$v := u - f_1(x)y^{(k-1)} - \cdots - f_{k-1}(x)y^{(1)} - f_k(x)y \tag{4.44}$$

then the original system (4.42) becomes simply

$$y^{(k)} = v \tag{4.45}$$

which is nothing more than a chain of k integrators driven by the new control v. The design of a control law for (4.45) couldn't be easier. For example, the linear control law

$$v = -a_1 y^{(k-1)} - \cdots - a_{k-1}y^{(1)} - a_k y \tag{4.46}$$

gives the closed-loop dynamics

$$y^{(k)} + a_1 y^{(k-1)} + \cdots + a_{k-1}y^{(1)} + a_k y = 0 \tag{4.47}$$

The coefficients a_i in (4.47) can be chosen to give the closed-loop system (4.11) any characteristic equation you want. After having found v, you can determine the actual control u from (4.44).

The notion of using part of the control to cancel the nonlinearities in the system can be generalized: The basic idea is to transform the state and control variables of the plant to another set of state and control variables, such that the dynamics of the transformed system are linear. Then design a linear control law for the transformed system. Finally, invert the transformations to get the control law for the original state and control variables. This idea can be expressed in mathematical terms as follows:

Suppose the original plant is defined by the nonlinear dynamic system

$$\dot{x} = f(x, u) \tag{4.48}$$

and we can find a nonsingular (i.e., invertible) nonlinear transformation T from $[x, u]$ to $[z, v]$, i.e.,

$$\begin{bmatrix} z \\ v \end{bmatrix} = T \begin{bmatrix} x \\ u \end{bmatrix} \tag{4.49}$$

with

$$\begin{bmatrix} x \\ u \end{bmatrix} = T^{-1} \begin{bmatrix} z \\ v \end{bmatrix} \tag{4.50}$$

such that in the transformed system the dynamics are linear in the state z and control v:

$$\dot{z} = Az + Bv \tag{4.51}$$

If we can do this, we can design a linear control law

$$v = -Gz \tag{4.52}$$

for the linear system (4.51). (This step is relatively easy: the gain matrix G can be determined by pole placement, for example, or linear quadratic optimization.)

Finally, we use the inverse transformation (4.50) to get the *nonlinear* control law in the original state and control variables.

A block diagram representation of the control law obtained by this general feedback linearization technique is given in Figure 4.9

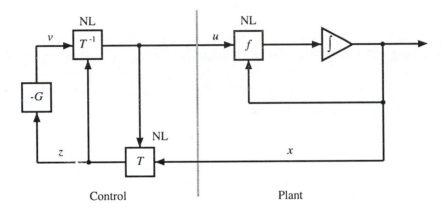

Figure 4.9   Generalized feedback linearization.

This method of control system design seems almost too good to be true. The mathematical operations are transparently simple, so the difficulty, if there is any, must lie elsewhere. Among the questions that may have occurred to you would be the following:

- The control law may not be realizable; the authority may not be adequate to cancel all the nonlinearity, or may entail an operation that cannot be performed in hardware. The feedback term $\bar{u}$ that cancels the nonlinearities may, for example, exceed the control limits— it may saturate the actuator. Or the required control law may entail computing the square root of a negative number.

- What assurance is there that the resulting control law is robust? Suppose the cancellation of the nonlinearities is imperfect, owing to imperfect knowledge of the plant dynamics.

- Under what conditions is it possible to represent a system with a single output y (which in general is represented by a set of k first-order differential equations) as a single differential equation of order k?

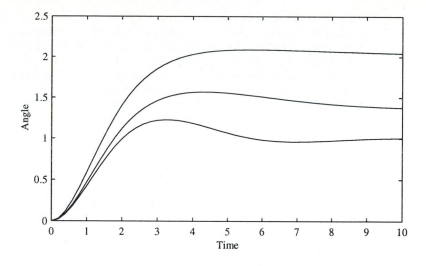

Figure 4.10 Robot arm transient response.

- Does this method apply to multivariable systems, i.e., to systems in which y is a vector of dimension $m > 1$?

The first two of these questions, which are concerned with practical issues, have hardly been addressed by the investigators who have developed the basic theory [3]. A great deal is known about the answers to the second two questions. Before discussing these, however, we illustrate the method with a few examples.

Example 4.5 Robot Arm (Continued)

For the dynamic process of Example 4.1, the linearizing control $\bar{u}$ in this case is simply

$$\bar{u} = \Omega^2 \sin(\delta\theta) \tag{4.53}$$

This converts the dynamics to a double integrator:

$$\ddot{\theta} = v \tag{4.54}$$

for which a suitable control law might be

$$v = -g_1\delta\theta - g_2\dot{\theta} \tag{4.55}$$

with

$$g_1 = c$$
$$g_2 = \sqrt{2c}$$

The complete control law is thus

$$u = \bar{u} + v$$
$$= \Omega^2 \sin \delta\theta - g_1\delta\theta - g_2\dot{\theta} \tag{4.56}$$

It is of interest to compare this control law with the one obtained in Example 4.1 by linearization about a set point. As regards performance, the transient response is that of a linear system; hence its appearance does not depend on the set point $\bar{\theta}$. For $c = 1.0$, the same value as used for the earlier design, the response is as shown in Figure 4.10. As regards implementation, it is noted that (4.56) provides a constant damping factor whereas (4.14) has a damping factor that depends on the set point. Another difference is that the control in (4.56) depends only on the position error $\delta\theta = \theta - \bar{\theta}$, whereas (4.14) depends on $\bar{\theta}$ through the feedforward term, as well as on the position error.

Example 4.6   Magnetic Levitation (Continued)

As an alternate to the design by linearization about an operating (set) point, a control law for the magnetic levitation system considered in Example 4.2 can be designed by feedback linearization. For the dynamics of (4.16) the obvious choice (but not the only one) for the linearizing control v is

$$v = -k(u^2/x_1^2) + g \tag{4.57}$$

Using this control in (4.17) gives

$$\dot{x}_2 = v \tag{4.58}$$

An appropriate control law for (4.16) and (4.17) would be

$$v = -G_1(x_1 - \bar{x}) - G_2 x_2 \tag{4.59}$$

with

$$G_1 = \Omega^2$$
$$G_2 = 2\zeta\Omega$$

where Ω is the desired closed-loop natural frequency and ζ is the desired damping factor. There is a fairly wide latitude in choosing these parameters, but they cannot be chosen arbitrarily, as discussed below.

The actual control law is obtained by solving (4.57) for u, after the substitution of (4.59) therein:

$$u = \pm\sqrt{\frac{g + G_1(x_1 - \bar{x}) + G_2 x_2}{k}}\, x_1 \tag{4.60}$$

The quantity under the radical may become negative if G_1 or G_2 are too large, relative to the gravitational acceleration g. If the radical were to become negative, the control law (4.60) would not be realizable. Hence this control law is valid only in the region for which

$$g + G_1(x_1 - \bar{x}) + G_2 x_2 > 0$$

Although we might be able to live with this restriction, there is a better way of dealing with the problem, namely, to use integral control. This is accomplished by defining the current u as a state variable rather than as the control variable. (This is also more consistent with the physics of the process: If the inductance of the magnet were considered, it would not be possible to control the current directly. See Problem 4.6.) Let the current be denoted by the state variable x_3. Then the plant dynamics previously given in Example 4.2 become

$$\dot{x}_1 = x_2$$
$$\dot{x}_2 = -k(x_3^2/x_1^2) + g \qquad (4.61)$$
$$\dot{x}_3 = \bar{u}$$

where $\bar{u}$ is now the control variable.

To apply feedback linearization to this plant we define part of the transformation T as follows:

$$z_1 := x_1$$
$$z_2 := x_2$$
$$z_3 := -k(x_3^2/x_1^2) + g$$

The inverse transformation is then given by

$$x_1 = z_1$$
$$x_2 = z_2 \qquad (4.62)$$
$$x_3 = z_1\sqrt{(g - z_3)/k}$$

(Note that the inverse transformation contains a square root, which could be a problem. In this case, however, as we will see, the inverse transformation is never explicitly used.)

The motivation for the choice of the transformation is that the transformed dynamics are now simply

$$\dot{z}_1 = z_2$$
$$\dot{z}_2 = z_3$$
$$\dot{z}_3 = v$$

where

$$v := -2k\left[\frac{x_3\bar{u}}{x_1^2} - \frac{x_3^2 x_2}{x_1^3}\right] \qquad (4.63)$$

completing the definition of the nonlinear transformation.

The differential equations (4.62) represent simply a chain of three integrators. The control law

$$v = -G_1(z_1 - x_r) - G_2 z_2 - G_3 z_3$$

will result in closed-loop dynamics of the transformed system having a characteristic equation:

$$s^3 + G_3 s^2 + G_2 s + G_1$$

and the gains G_i can be chosen to place the poles wherever desired.

Having determined the gains, (4.63) is used to obtain the nonlinear control law

$$\bar{u} = -\frac{x_1^2}{2kx_3}\left[-G_1(z_1 - x_r) - G_2 z_2 - G_3 z_3 + \frac{x_2 x_3}{x_1}\right]$$
$$= -\frac{x_1^2}{2kx_3}\left[-G_1(x_1 - x_r) - G_2 x_2 - G_3\left(-k\frac{x_3^2}{x_1^2} + g\right)\right] + \frac{x_2 x_3}{x_1}$$

Figure 4.11 shows the step response of this system with the gains set at the following values:

$$G_1 = 1., \qquad G_2 = 2.$$

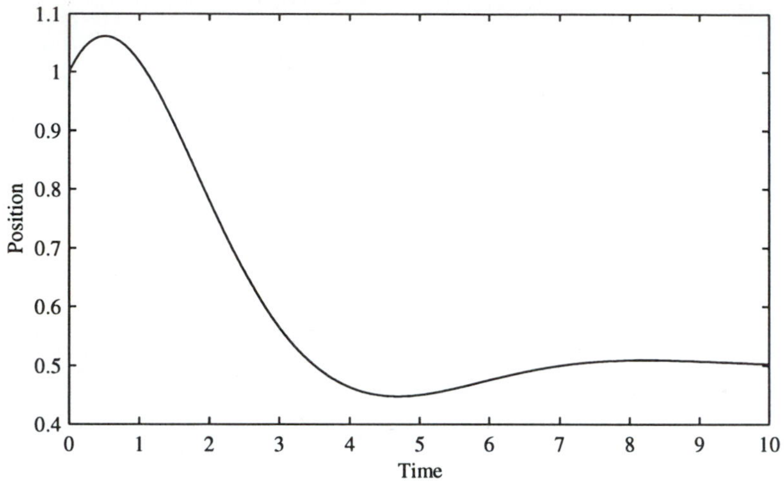

Figure 4.11   Transient responses of magnetic levitation system with control designed by feedback linearization.

Example 4.7   Separately Excited D-C Motor Drive (Continued)

Feedback linearization can also be applied to the separately-excited d-c motor drive system considered earlier. One way of doing this is to use ω and $\dot{\omega}$ as state variables. From (4.27) we obtain

$$\ddot{\omega} + \frac{B}{J}\dot{\omega} = \frac{K}{J}\left(i_f \frac{di_a}{dt} + i_a \frac{di_f}{dt}\right)$$
$$= \frac{K}{J}\left[-\left(\frac{R_a}{L_a} + \frac{R_f}{L_f}\right)i_a i_f + \frac{K}{L_a}i_f^2 \omega + i_f e_a + i_a e_f\right] \qquad (4.64)$$

In keeping with the feedback linearization approach, we define the control variable v as the entire right-hand side of (4.64):

$$v := \frac{K}{J}\left[-\left(\frac{R_a}{L_a} + \frac{R_f}{L_f}\right)i_a i_f + \frac{K}{L_a}i_f^2 \omega + i_f e_a + i_a e_f\right] \qquad (4.65)$$

We now have to design a control law for the plant

$$\ddot{\omega} + \frac{B}{J}\dot{\omega} = v$$
$$i_f + \frac{R_f}{L_f}i_f = e_a$$

which comprises two linear, uncoupled systems with control variables v and e_a.

A linear control law would be appropriate for v:

$$v = -G_1(\omega - \omega_r) - G_2\dot\omega \tag{4.66}$$

using (4.27) to relate $\dot\omega$ to the original state variables. Finally, solving (4.65) for e_a, we obtain the control law:

$$e_a = \frac{L_a}{\imath_f}\left(\frac{J}{K}v\frac{1}{L_f}\imath_a e_f\right) + L_a\left[\left(\frac{R_a}{L_a} + \frac{R_f}{L_f}\right)\imath_a + \frac{K}{L_a}\imath_f\omega\right] \tag{4.67}$$

with v given by (4.66).

To obtain the control law for the field current, we consider making the field current track the armature current by defining

$$\bar\imath_f := \sqrt{R_a/R_f}\,\imath_a$$

in conformity with (4.30). Then use the field control law

$$e_f = G_3(\bar\imath_f - \imath_f) \tag{4.68}$$

Transient responses for the control law given by (4.67) and (4.68) was simulated for the same parameters (including voltage saturation at 250 v) except that the gain for the field control was taken as 100. The transient responses are shown in Figure 4.12 for the same initial conditions as in Example 4.3 . The time histories for the motor speed and armature current and voltage are very similar to those of the earlier example. The field current and its corresponding voltage time-histories are rather different from those of the earlier example, because of the desire to optimize the field current. The performance of the control law, like that of the earlier control law for this plant, is quite acceptable.

Having seen examples of the use of feedback linearization, we turn now to address the questions raised at the beginning of this section: physical realizability, robustness, and theoretical existence of a transformation that can linearize the system.

Physical Realizability   In both of the examples, the control law that we obtained by feedback linearization entailed division by one or more of the state variables. Obviously, these control laws become invalid as the corresponding state variables approach zero. As a practical matter, these variables cannot even be close to zero without saturating the actuator. Hence, the practical implementation of any control law that entails division by a state variable or some other operation which can theoretically result in an infinite control must account for the physical saturation. This means that the physical control law and the theoretical control law will not be identical in some region of the state space. If the state of the system can fall into such a region, the resulting behavior may not be satisfactory and may even be unstable. The performance of a control law designed by feedback linearization must be evaluated in the presence of realistic limits on the control. The evaluation can be performed by simulation, or, preferably, by analysis (e.g., by using Liapunov's Second Method.)

The problem of saturating control is not limited to those control laws that might theoretically result in infinite control signals. The problem can be present even when

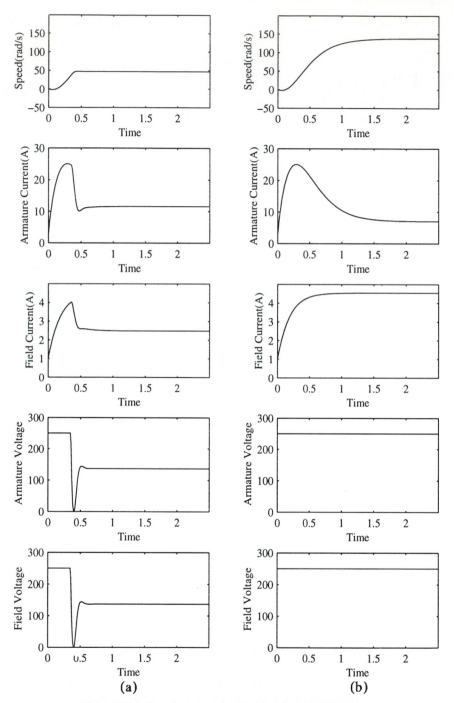

Figure 4.12   Transient response of separately excited d-c motor
with control law designed by feedback linearization. (a) Reference speed: 50 r/s. (b) Reference speed: 150 r/s.

the control law has no singularities, but calls for control signals that are too large for the actuator to deliver. When the control law is designed in consideration of the original state and control variables, which usually have physical meaning, it is usually possible to select the control gains by which serious saturation problems are avoided. But the state and control variables z and v of the transformed system may not have obvious physical significance; hence it may not be clear how to select the control gains for the transformed system.

Robustness   Another reason why a control law designed by feedback linearization may not live up to expectations is robustness: The true dynamics and those of the design model must inevitably be different. If the control law is sufficiently robust, small discrepancies between the true system and the system for which the control law was designed will be taken in stride. The basic problem, however, is whether the control law is sufficiently robust. Unlike a linear system in which the robustness can readily be evaluated, the robustness of a nonlinear system designed by feedback linearization is not easy to assess analytically. Simulation is one of the few tools currently available.

Existence of a Linearizing Transformation   Issues of control saturation and robustness are among the reasons why feedback linearization would not routinely be used with every nonlinear system to which it is applicable. Suppose, however, that notwithstanding its potential limitations, you want to try the method in your application. How do you know it can be used? How do you know that a linearizing transformation can be found?

In many applications, particularly when the dynamics are of fairly low order (less than 5, say) the transformation can be found directly by performing a few rather obvious manipulations of the original differential equations. But suppose that a long session of manipulation proves fruitless. Should you conclude that the problem is impossible, or should you continue your quest? This question can be answered for dynamic systems that are linear in the control variable:

$$\dot{x} = f(x) + G(x)u \tag{4.69}$$

and in which the nonlinear function $f(\cdot)$ and the matrix $G(\cdot)$ are *smooth*, i.e., possess derivatives of all order. The requirement of smoothness rules out systems with "hard" nonlinearities such as saturation. But, since it is always possible to approximate such hard nonlinearities with smooth ones, the restriction is not critical. Neither is the restriction that the dynamics must be linear in the control. Most systems are either already in this form or can be manipulated into this form by fairly obvious tricks, such as that which was used in Example 4.6 .

The results are couched in the terminology of *differential geometry*. In particular, the results require the use of *Lie brackets* and the notion of *involute sets of functions*. The background of differential geometry needed to fully appreciate the result is beyond the scope of this text; several recent texts and monographs are available for interested readers. See Note 4.1. The development that follows gives the basic theorems that will allow you to check whether a system can be linearized by feedback, and shows how they relate to the corresponding theorems for linear systems with which you are already familiar.

The *Lie bracket* of two nonlinear, vector-valued functions $f(x)$ and $g(x)$ is defined by

$$[f, g](x) := \frac{\partial g(x)}{\partial x} f(x) - \frac{\partial f(x)}{\partial x} g(x) \tag{4.70}$$

where $\partial g/\partial x$ denotes the Jacobian matrix of the vector $g(x)$ with respect to x. As the notation indicates, the Lie bracket of a pair of vector-valued functions of x is another vector-valued function of x.

Consider a set of vector-valued functions $f_1(x), f_2(x), \ldots f_k(x)$ where $x \in \mathbf{R}^n$, an n-dimensional space. At a given point x the set of vectors spans a subspace $\mathcal{S}$ of $\mathbf{R}^n$. This set is said to be *involutive* at the point x if and only if

$$[f_i, f_j](x) \in \mathcal{S}$$

for all vectors f_i, f_j in the set.

In principle you can test for involutivity by forming all the three column matrices

$$\begin{bmatrix} f_i & f_j & [f_i, f_j] \end{bmatrix}$$

and checking that they are not of rank 3. If any such matrix is rank 3, the set is not involutive. As you can imagine, the test can entail a lot of calculation.

The concepts of Lie brackets and involutivity enable us to state the theorem on feedback linearizability. First we introduce the notation

$$ad_f^0 g := g \tag{4.71}$$

$$ad_f^{i+1} g := [f, ad_f^i g] \tag{4.72}$$

We can then state the following:

Theorem 4.1 *Feedback Linearizability*
   The nonlinear kth order system

$$\dot{x} = f(x) + g(x)u \tag{4.73}$$

can be linearized by feedback at the point x_0 if and only if at the point x_0

   1. The matrix $\begin{bmatrix} g(x) & ad_f g(x) & \cdots & ad_f^{k-1} g(x) \end{bmatrix}$ has rank k, and

   2. The set of vectors $g(x), ad_f g(x), \ldots, ad_f^{k-2} g(x)$ is involutive.

Although this theorem looks complicated, and it is, in a linear system

$$\dot{x} = Ax + Bu$$

condition 2 is always satisfied and condition 1 reduces to the well-known controllability condition, namely, that the matrix

$$\mathbf{C} = \begin{bmatrix} B & AB & \cdots & A^{k-1}B \end{bmatrix}$$

be nonsingular. In fact, in order for a nonlinear system to to be linearizable at a point, say x_0 it is necessary (but not sufficient) for the linearized system

$$\dot{\xi} = A(x_0)\xi + B(x_0)u + h(x, u) \tag{4.74}$$

be controllable, where $\xi = x - x_0$,

$$A(x_0) = \left[\frac{\partial f(x)}{\partial x}\right], \quad B(x_0) = \left[\frac{\partial g(x)}{\partial x}\right]$$

and $h(0, u) = 0$.

The basic idea underlying the proof is to transform the given system (4.73) into a chain of integrators:

$$\dot{z}_1 = z_2, \quad \dot{z}_2 = z_3, \quad \ldots \dot{z}_{k-1} = z_k, \quad \dot{z}_k = v$$

The transformed system is called the *Brunovsky normal form*.

Theorem 4.1, which is proved in new textbooks on nonlinear control systems, such as [3], [4], actually provides a specific algorithm for constructing the linearizing transformation, but performing the required calculations is a job for software that performs symbolic calculations.

Conditions similar to those of Theorem 4.1 are available for multi-input processes (see [3]).

4.5   NOTES

Note 4.1   Differential geometry and feedback linearization

The idea underlying feedback linearization is fairly obvious. The method probably was used for many years in practical control system design without anyone having reported it. Concurrently, a number of mathematical investigations on the use of differential geometry in dynamic systems were being conducted without much notice by control system designers. The effort of G. Meyer [5] and his colleagues at the NASA Ames Research Center in the mid-1970s may have been the catalyst in bringing together the mathematicians and the control system designers. The work of Meyer et al. was distinguished from the earlier mathematical work in that it addressed the concrete problem of aircraft flight control and addressed some of the practical issues of such systems, including the inversion of nonlinear functions that are not expressible as differentiable analytical functions. (Interestingly, Meyer's computational techniques are not much discussed in the literature.) Another important characteristic of this work is that control systems designs based on this approach were implemented (in hardware! [6]) and found to operate very well.

The practical success of the NASA Ames approach stimulated substantial activity in the field and motivated the organization of a tutorial workshop on the subject at the 1985 IEEE Conference on Decision and Control (Ft. Lauderdale, FL) which introduced the theory to a broad audience, some of whom mounted their own efforts in the area.

Much of the current effort is concerned with theoretical issues and is based on the use of dynamic models with smooth dynamics, since it is based on the existence of partial derivatives of whatever order is needed in the theory. The theory seems to have outpaced its need in dealing with practical systems that fit the assumptions. On the other hand, a number of problems of practical systems, such as saturation, hard (i.e., discontinuous) nonlinearities, are not readily amenable to treatment by these methods.

4.6   PROBLEMS

Problem 4.1   Control of robot arm by extended linearization

Consider the robot arm of Example 4.1 . Determine control gains $G_1(x)$ and $G_2(x)$ that provide global asymptotic stability. Compare the performance (transient response) of this design with that obtained for Example 4.1 .

Problem 4.2   Robot arm with direct motor drive

Consider the motor driven robot arm of Problem 4.1 with the following numerical data:

$$\alpha = 5, \quad \beta = 1, \quad \Omega^2 = 64.$$

It is desired to control the system to hold any reference angle in the entire range $[0, 2\pi]$.

(a) Design a full-state feedback control law for the process using linearization about a nominal state.

(b) Design a full-state feedback control law for the process using feedback linearization.

In each case, simulate the performance of the system starting in the state [0,0] for the desired states:

$$\omega = 0, \quad \theta = \pi/4, \ \pi/2, \ 3\pi/4$$

Problem 4.3   Belt-driven robot arm

Consider the motor-driven robot arm of Problem 4.2, except that the arm and the motor are separated and connected by a resilient belt. The dynamics of the system are given by:

$$(J_L + ml^2)\ddot{\theta}_L = E(\theta_M - \theta_L) - mgl\sin\theta_L$$
$$J_M\ddot{\theta}_M = E(\theta_L - \theta_M) - (K^2/R)\theta_M + (K/R)e$$

where θ_M is the motor shaft angle, θ_L is the arm angle, E is the belt elasticity, J_L is the moment of inertia of the pulley attached to the arm, l is the length of the arm, and m is the mass of the payload.

Use the following numerical values:

$$
\begin{array}{lll}
J_M & : & 2 \times 10^{-4} \text{ Kg-m}^2 \\
J_L & : & 2 \times 10^{-4} \text{ Kg-m}^2 \\
m & : & 0.1 \text{ Kg} \\
l & : & 0.2 \text{ m} \\
R & : & 2\, \Omega \\
K & : & 0.04 \text{ N-m/amp} \\
E & : & 125 \text{ Kg/rad}
\end{array}
$$

It is required to move the moment arm from any angle to any other angle with negligible overshoot near the final position.

Design a full-state feedback control law for the system by linearization about the reference angle.

Simulate the performance for each of the reference angles of Problem 4.2.

Problem 4.4   Autonomous vehicle guidance

It is desired to navigate the autonomous vehicle of Example 2 such that it achieves and maintains a specified reference velocity $\bar{v}$. While doing this, the vehicle is to turn in a manner that it moves toward a specified straight line

$$ a_1 x + a_2 y + b = 0 $$

and stays on the line when it gets to it. Assuming all the state variables can be measured, devise a control law that achieves the desired control objectives.

Problem 4.5   Stability of magnetic levitation design

Use Liapunov's Second Method to investigate the stability of the control law design given in Example 4.2 .

Problem 4.6   Magnetic levitation with inductance considered

Consider the magnetic levitation system discussed in Example 4.2 , except that the control is not the current, but the voltage applied to the winding, and hence that the inductance $L(x)$ of the winding must be considered.

(a) Derive the differential equations of the system.

(b) Using the methods given in this chapter, but modified for the different dynamics, design control algorithms for the system.

Problem 4.7   Rapid thermal processing

The differential equation governing rapid thermal processing of semiconductor wafers is as follows:

$$ \dot{T} = A_1 T + A_2 T^4 + Bu $$

where $T = [T_1, T_2, T_3]$ is the vector of temperature at different points on the wafer, and A_1 is the heat conduction matrix. The nonlinear terms $T^4 := [T_1^4, T_2^4, T_3^4]$ are due to radiation.

With temperature measured in 10^2K and time in 100 sec, the matrices A_1, A_2, and B are given by

$$A_1 = \begin{bmatrix} -6. & 6. & 0. \\ 2. & -3. & 1. \\ 0. & 0.6 & -0.6 \end{bmatrix}, \quad A_2 = \text{diag}[1., 2., 3.] \times 10^{-4}, \quad B = \begin{bmatrix} 1. \\ .5 \\ .75 \end{bmatrix}$$

The performance requirements for the system are as follows:

1. The temperature of the wafer is to be uniform, i.e., $T_1 = T_2 = T_3$.
2. The temperature should track a reference temperature profile, which can be approximated by

$$T_{\text{ref}} = \begin{cases} 3. + 20t, & 0 < t < .1 \\ 5., & .1 < t < .2 \\ 5. - 20t, & .2 < t < .3 \end{cases}$$

(a) Design a control law for the process by linearizing about a nominal control.

(b) Design a control law for the process by feedback linearization.

Problem 4.8　Control of fermentation of sugar by linearization about a set point

The fermentation of sugar into ethanol (grain alcohol) by yeast can be approximated by the following differential equations [7]:

$$\dot{x}_1 = -x_1 + (1 - x_1)u$$
$$\dot{x}_2 = x_1 - x_2 u$$

where x_1 is the normalized sugar concentration, x_2 is the normalized ethanol concentration, and the control u is the normalized feedrate. The normalization is such that all variables are limited between 0 and 1.

(a) Show that equilibrium with constant control $u = \bar{u}$ there is a constraint between $\bar{x}_1$ and $\bar{x}_2$ and find the equilibrium control $\bar{u}$ in terms of a desired value of x_1.

(b) Design a control law by linearization about a set point.

(c) Simulate the performance of the control law for the following set points:

$$\bar{x}_1 = 0.25, \ 0.50, \ 0.75$$

Remember to take into account the constraint $0 \leq \bar{u} \leq 1$ in the simulation.

Problem 4.9　Control of fermentation of sugar by extended linearization

The dynamics of the fermentation process of Problem 4.8 can be written

$$\dot{x} = Ax + B(x)u$$

where

$$A = \begin{bmatrix} -1 & 0 \\ 1 & 0 \end{bmatrix}, \quad B(x) = \begin{bmatrix} 1 - x_1 \\ -x_2 \end{bmatrix}$$

Using these matrices, design a control law that maintains x_1 at a desired constant value. Compare the performance of this control law with that of Problem 4.8 for each of the specified set points.

Problem 4.10   Control of fermentation of sugar by feedback linearization

Design a control law for the fermentation process of Problem 4.8 by feedback linearization to maintain a constant value of x_1, and compare the performance with the control laws obtained in Problem 4.8 and Problem 4.9

Problem 4.11   Feedback linearization belt-driven robot arm

(a) Show that the belt-driven robot arm of Problem 4.3 can be linearized by feedback, using the following transformation:

$$z_1 = \theta_1$$
$$z_2 = \dot{\theta}_1$$
$$z_3 = \ddot{\theta}_1$$
$$z_4 = \dddot{\theta}_1$$

(b) Using the transformation thus obtained, design a control law for the process and compare the performance with that obtained in Problem 4.3.

4.7   REFERENCES

[1] H.G. Booker, *An Approach to Electrical Science*, McGraw- Hill, New York, 1959.

[2] S.C. Hong and M.H. Park, "Microprocessor-Based High-Efficiency Drive of a DC Motor," IEEE Trans. on Industrial Electronics, Vol.IE-34, No. 4, pp.433-440, November 1984.

[3] A. Isidori, *Nonlinear Control Systems: An Introduction,* 2nd Ed., Springer-Verlag, New York, 1989.

[4] M. Vidyasagar, *Nonlinear Systems Analysis* 2nd Ed., Prentice-Hall, Inc., Englewood Cliffs, NJ, 1993.

[5] G. Meyer and L. Cicolani, "A Formal Structure for Advanced Automatic Flight Control Systems," NASA Technical Note D-7940, 1975.

[6] G. Meyer and L. Cicolani, "Application of Nonlinear Systems Inverses to Automatic Flight Control Design—System Concepts and Flight Evaluations," In *Theory and Applications of Optimal Control in Aerospace Systems*, NATO AGAR-Dograph No.251, July 1981.

[7] J.P. Axelsson, *Modelling and Control of Fermentation Processes,* Doctoral Dissertation, Department of Automatic Control, Lund Institute of Technology, Lund, Sweden, March 1989.

Chapter 5

SATURATING AND DISCONTINUOUS CONTROL

Every physical control variable has a limited range: A control valve cannot be more than fully closed or more than fully open; the control voltage to a motor cannot exceed the supply voltage; the output of an electric heater cannot be negative nor more than a specified maximum amount. All practical control system designs must thus contend with control variables that are limited in amplitude, or, as it is said, subject to "saturation."

One design approach is to use an actuator with enough "muscle" to ensure that the saturation limits will never be reached during the normal course of operation of the system. But in some applications this brute-force approach may not be feasible. In an aircraft, for example, the control surfaces such as the rudder, the elevator, and the ailerons have limited ranges of motion (i.e., "control authority"). Rarely is it possible to deflect a control surface by more than 45 degrees—usually much less— without impairing its aerodynamic effectiveness. An efficient aircraft control system should be designed to operate with as small a control authority as feasible.

Even when the application permits use of an actuator that is powerful enough to avoid saturation, it is often not economical to do so, because the cost, size, and weight of the control systems may be dominated by the characteristics of the actuator. These characteristics depend on the maximum output (amplitude) that the device can produce. A control system is overdesigned if it has an actuator that is rarely called upon to produce its maximum output.

A typical saturating control has the characteristic shown in Figure 5.1. The limits U_{max} and U_{min} may or may not be symmetrical. In a valve or a heater, for example, U_{min} is zero.

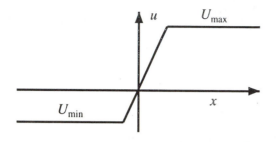

Figure 5.1   Characteristic of a saturating control.

The presence of a linear region between U_{max} and U_{min} makes it possible to "modulate" the control between these limits, i.e., to adjust it for a smooth variation. Control system designs are often based on the existence of a linear region, and many actuators are designed to provide it. But this modulation capability is expensive. It is much cheaper to make an actuator that has only two distinct settings: *on* and *off*, or *open* and *closed*. Many familiar control systems (thermostats, solenoid valves, etc.) use two-level actuators, so there is ample evidence that such systems can be designed to operate effectively. Control systems, properly designed, can come close to being optimal in the sense of minimizing a mathematically defined performance criterion, as discussed in Section 5.3.

The need to deal with saturating and discontinuous controls, however, may pose a challenge for the system designer. If the plant is "benign," i.e., if its dynamic behavior can be adequately approximated by a low-order set of differential equations with well-damped natural frequencies, acceptable performance can sometimes be achieved with a simple control law, such as

$$u = -U_{max}\, \text{sgn}(e)$$

where U_{max} is the saturation level, and e is the system error. This simple control algorithm, however, can often result in an unstable closed loop system, and a method of designing a more appropriate control law would be necessary.

Simulation of system performance with a discontinuous control can also be a problem. In the real world, no device can switch between two values instantaneously. There is always some dynamic effect that limits the maximum rate at which a variable can change. The precise physical mechanism that limits the rate of change, however, may not be understood or may be too complicated to justify accurate modeling. For design purposes it may be good enough to approximate the switching behavior by a discontinuous function. This approximation may cause problems in a computer simulation: The numerical integration algorithm may yield inaccurate results or run unacceptably slow. To avoid such problems, it is often expedient to include a narrow linear region in the simulation model of the discontinuous control. In doing so, you have to accept that the simulation may not faithfully represent the behavior of the process in the vicinity of the discontinuity.

5.1   SATURATING LINEAR CONTROL

Conceptually the simplest case to deal with is a linear dynamic process with a scalar saturating control:

$$\dot{x} = Ax + bU_{max}\text{sat}(u/U_{max}) \tag{5.1}$$

where $b = [b_1, b_2, \ldots, b_k]'$ is a column vector and

$$\text{sat}(x) = \begin{cases} 1 & x > 1 \\ x & |x| \leq 1 \\ -1 & x < -1 \end{cases} \tag{5.2}$$

In the linear region $|u| \leq U_{max}$ (5.1) becomes simply

$$\dot{x} = Ax + bu \tag{5.3}$$

If (5.3) is a controllable system, a linear control law

$$u = -g_1 x_2 - g_2 x_2 - \cdots - g_k x_k = -g'x \tag{5.4}$$

can be found that places the closed loop poles at any locations desired.

In the presence of saturation it is tempting to consider using the control law

$$u = U_{max} \, \text{sat}(-g'x/U_{max}) \tag{5.5}$$

which reduces to (5.4) for large values of U_{max}. If the saturation level U_{max} is truly large relative to the initial conditions, it seems reasonable that use of (5.5) as the control law would result in reasonably good performance: Occasional saturation during the transient interval might occur, but, as the error approaches zero, saturation would be less frequent, and as the steady-state approaches, the system would be expected to remain in the linear region. As remarked earlier, however, a good system design strives to keep U_{max} as small as possible relative to the range of possible initial states. Stated differently, it is desirable to achieve the largest possible set of initial states for which the system is asymptotically stable, given a specified value of U_{max}.

One way to design a control law that achieves asymptotic stability for a substantial range of initial conditions, even when the process is unstable, is to select the gain matrix

$$g = [g_1, g_2, \ldots, g_k] \tag{5.6}$$

in (5.5) to be the solution to a linear, quadratic regulator problem:

$$g = b'M/r \tag{5.7}$$

where M is the solution to the algebraic Riccati equation

$$MA + A'M - Mbb'M/r + Q \tag{5.8}$$

in which $r(> 0)$ is a scalar weighting parameter and Q is a positive-semidefinite matrix.

The region of the state space for which (5.5) with (5.7) is asymptotically stable can be estimated using Liapunov's Second Method. Following the approach of Bass and Gura [1] select the candidate Liapunov function for this situation as

$$V = x'Mx \tag{5.9}$$

where M is the solution to (5.8). The control matrix b limit can be scaled to make $U_{\text{max}} = 1$. With this scaling, the time-derivative of V is

$$\begin{aligned} \dot{V} &= \dot{x}'Mx + x'M\dot{x} \\ &= (x'A' - b'\text{sat}(g'x))Mx + x'M(Ax - b\,\text{sat}(g'x)) \\ &= x'(A'M + MA)x - (b'Mx + x'Mb)\,\text{sat}(g'x) \end{aligned} \tag{5.10}$$

(Note that $\text{sat}(g'x)$ is a scalar multiplier.)

By (5.7), $b'M = rg$ and, by (5.7) and (5.8), $MA + A'M = rgg' - x'Qx$. Thus (5.10) becomes

$$\dot{V} = -x'Qx - 2rg'x\,\text{sat}(g'x) + r(g'x)^2 \tag{5.11}$$

In the vicinity of the origin, i.e., for $|g'x|$ sufficiently small, (5.11) gives

$$\dot{V} = -x'Qx - r(g'x)^2 \tag{5.12}$$

which is, of course, negative definite. Hence, by Theorem 3.1, the origin is asymptotically stable. Since the gain matrix g was chosen to stabilize the system, this should come as no surprise. For a general (stabilizing) gain matrix, the asymptotic stability would be a strictly local result, i.e., we would not know how far from the origin the region of stability extends. In this case, however, the region of attraction can be conservatively estimated. As discussed in Chapter 2, such an estimate is a set within which $\dot{V} < 0$ bounded by the surface $V = \text{const}$. To achieve as large an estimate as possible, we should find the surface on which $\dot{V} = 0$ and find the $C = \max V$ on that surface. Then the region bounded by $V = C$ is an estimate of the region of attraction. From (5.9) and (5.11)

$$C = \max_x x'Mx \tag{5.13}$$

with

$$-x'Qx - 2rg'x\,\text{sat}(g'x) + r(g'x)^2 = 0 \tag{5.14}$$

Solution of (5.13) and (5.14) can be attempted by Lagrange multipliers, but it does not appear to be an easy problem.

A somewhat more conservative estimate of the region of attraction can be calculated more readily: Consider (5.14), which can be written

$$y^2 - 2y\,\text{sat}(y) + x'Qx/r = 0 \tag{5.15}$$

where

$$y = g'x$$

The solution to (5.15) must occur for $|y| \geq 1$, since for $|y| < 1$ we have $\dot{V} < 0$ by (5.12). Suppose $y \geq 1$. Then $\text{sat}(y) = 1$ and the solution for y of (5.15) is

$$y = 1 + \sqrt{1 + x'Qx/4r} > 2$$

Similarly, if $y < 0$, the solution of (5.15) is given by

$$y = -1 - \sqrt{1 + x'Qx/4r} < -2$$

Thus we see that the region

$$|g'x| \leq 2 \tag{5.16}$$

which is at least twice the size of the saturation region, is contained within the boundaries of the region for which $\dot{V} = 0$. It thus follows that the region of attraction includes the region for which

$$V(x) \leq \tilde{C} \tag{5.17}$$

where

$$\tilde{C} = \max_{|g'x|=2} x'Mx \tag{5.18}$$

By use of Lagrange multipliers, $\tilde{C}$ can be found:

$$\tilde{C} = \frac{4r}{b'Mb} \tag{5.19}$$

(See Problem 5.1.)

Example 5.1   Robot Arm with Linear, Saturating Control

Consider the control of a robot arm, linearized about its (unstable) upward position. The dynamics of the system are given by:

$$\dot{x}_1 = x_2 \tag{5.20}$$
$$\dot{x}_2 = \Omega^2 x_1 + u \tag{5.21}$$

The control law to minimize the quadratic performance criterion

$$V = \int_t^\infty (c^2 x_1^2 + u^2)\, d\tau \tag{5.22}$$

is given by

$$u = -g_1 x_1 - g_2 x_2$$

where, upon solving the corresponding algebraic Riccati equation, we find

$$g_1 = \Omega^2 + \sqrt{\Omega^4 + c^2}, \quad g_2 = \sqrt{2g_1}$$

and

$$M = \begin{bmatrix} m_1 & m_2 \\ m_2 & m_3 \end{bmatrix}$$

with

$$m_1 = \sqrt{2(\Omega^4 + c^2)(\sqrt{\Omega^4 + c^2} + \Omega^2)}, \quad m_2 = g_1, \quad m_3 = g_2$$

In accordance with (5.18) and (5.19) the region of attraction includes the states within the ellipse

$$\begin{bmatrix} x_1, & x_2 \end{bmatrix} M \begin{bmatrix} x_1 \\ x_2 \end{bmatrix} = \frac{4}{g_2^2} = \frac{2}{1 + \sqrt{1 + c^2}}$$

as shown in Figure 5.2

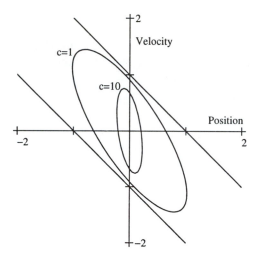

Figure 5.2   Estimate of region of attraction for unstable plant.

It must be emphasized that the estimate of the region of attraction is conservative; the actual region of attraction may be much larger.

As pointed out by Bass and Gura [1], control law (5.8), obtained by solving the algebraic Riccati equation (5.9) of a linear, quadratic regulator problem, has a remarkable property: asymptotic stability is assured even if the gain matrix g is multiplied by an arbitrary constant greater than 1/2. To see this suppose that the gain matrix g were multiplied by a constant α. Then (5.11) would become

$$\dot{V} = -x'Qx - 2r\,g'x\,\text{sat}(\alpha g'x) + r(g'x)^2 \tag{5.23}$$

In the linear region ($|g'x| < 1/\alpha$) the derivative of V is given by

$$\dot{V} = -x'Qx - r(1 - 2\alpha)(g'x)^2 \tag{5.24}$$

which is surely negative for $\alpha \geq 1/2$. (This is an alternate demonstration of the well-known $[1/2, \infty]$ gain margin property of linear, quadratic regulator designs [2].)

In the saturated region($|g'x| > 1/\alpha$) the derivative of V continues to be given by (5.11), and the argument preceding (5.16) remains valid. Hence we can conclude that the region for which $\dot{V} \leq 0$ is at least twice the region of saturation, i.e., $|g'x| \leq 2$ independent of the saturation. This property holds even when the size of the linear region shrinks to zero (i.e., when the control becomes discontinuous).

The ability of control laws designed by the linear, quadratic regulator (LQR) algorithm to provide asymptotic stability even for saturating and discontinuous control makes such control laws especially useful, particularly for high-order systems in which it is all but impossible to perform the calculations for a theoretically optimal control law. (See Section 5.3 for a further discussion.) The benefits of LQR designs notwithstanding, however, they are not a panacea for several reasons.

- An LQR design can guarantee asymptotic stability only for initial states in a finite region of the origin; global asymptotic stability cannot be assured. If the plant is unstable, global asymptotic stability cannot be achieved by any saturating control law. If the initial state is outside the "set of recoverable points," the size of which depends on the saturation level (the "control authority") no control law will be able to return the state to the origin. (This will be discussed in more detail in Section 5.3)

- All the benefits of an LQR design are achieved only with full- state feedback. If the state variables of the plant are not all accessible to measurement (the usual situation), an observer must be used to estimate some of them. In this case it may not be possible to ensure even local asymptotic stability. Nevertheless, a control system designed by the LQR algorithm, and which uses a suitable observer when necessary, often achieves outstanding performance.

5.2 THE SET OF RECOVERABLE STATES

When there is no bound on the amplitude of the control input to a linear controllable plant, its stability is not a major issue. A feedback control law can always be found for which the closed-loop system is globally asymptotically stable whether or not the plant is stable. It is always possible to find an input that can take the process from any state to any other state.

If the control amplitude is bounded, however, the situation is different: There is an essential distinction between a plant that is asymptotically stable and one that is not. If the plant is asymptotically stable, it will return to the origin from any initial state of its own accord, without a control input, and we would not expect saturation to be a problem. But if the plant is not asymptotically stable, it will not return to the origin without active control. Hence the nature of the control law is very much an issue.

A fundamental question arises: Can one always find a control law to stabilize a controllable, but unstable, plant when the control input is bounded? The answer, alas, is *no. In an unstable plant with a bounded control input there will always be states from which it is impossible to return to equilibrium.*

This fact is appreciated by designers of aircraft, who until recently had always insisted that the aircraft (i.e., the plant) be aerodynamically stable in the absence of control input, even if this requirement necessitated sacrificing performance or fuel efficiency. In recent years, some military aircraft designs have not adhered to this requirement and, for the sake of performance, have permitted the aircraft to be aerodynamically unstable, relying on the active control system to aerodynmaically-unstable aircrafr maintain stability. Since failure of the control system inevitably dooms the aircraft, no reasonable effort is spared to ensure the reliability of the control system: Triply or quadruply redundant components are used and the system architecture is designed to detect and compensate for component failures. Such measures make the overall system reliability incalculably high. Nevertheless, even perfect reliability of the control system does not ensure recovery from all states, because of the physical limits on the control authority. If, as a result of an exogenous disturbance such as wind shear or turbulence, the aircraft can get into a state from which it cannot be recovered, the design is flawed.

Unstable chemical processes and nuclear reactors share this problem. Not only is it necessary to ensure the reliable operation of the control system, but it is also necessary to establish that the process cannot wander into a state from which it cannot recover.

The hazards of controlling unstable plants with bounded control inputs make it important to determine or conservatively estimate the region of state space in which the plant cannot be stabilized by any bounded control, and to ascertain that the plant cannot be driven into this region by accident or by exogenous inputs.

To study the problem of finding the boundary of the set of recoverable states, consider the linear plant

$$\dot{x} = Ax + Bu$$

the solution of which can be expressed as

$$x(t) = e^{At}x_0 + \int_0^t e^{A(t-\tau)}Bu(\tau)d\tau \tag{5.25}$$

By definition, for the plant to return to equilibrium at the origin, we must have

$$\|x(t)\| \to 0$$

If the plant is asymptotically stable, this happens for $u = 0$ (although a nonzero control can help it along). On the other hand, if the plant is unstable, in order for $x(t)$ to approach zero it is necessary to use active control. To investigate the control requirements, multiply both sides of (5.25) by e^{-At} to obtain

$$e^{-At}x(t) = x_0 + \int_0^t e^{-A\tau}Bu(\tau)d\tau \tag{5.26}$$

Suppose that $x(t) = 0$ for some t. For this to happen, it is necessary for the control $u(t)$ to satisfy

$$\int_0^t e^{-A\tau}Bu(\tau)d\tau = -x_0 \tag{5.27}$$

for some control $u(\tau)$ within the constraint set and $\tau \leq t$. This is requirement is met in a controllable system when $u(t)$ is unconstrained, but may not be met when $u(t)$ can saturate and hence is constrained to lie in a bounded set $\mathcal{U}$.

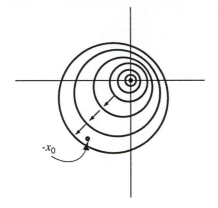

$-x_0$

Figure 5.3   In an asymptotically stable system the set $\mathcal{W}(t)$ expands to fill entire state space and eventually reaches $-x_0$.

Consider the set of points

$$\mathcal{W}(t) := \left\{ w(t) = \int_0^t e^{-A\tau} B u(\tau) d\tau \, \middle| \, u(\tau) \in \mathcal{U}, \tau \leq t \right\} \qquad (5.28)$$

where $\mathcal{U}$ defines the constraint set. Assuming that $\mathcal{U}$ is a *convex* set (i.e., that if $u_1, u_2 \in \mathcal{U}$, then $\alpha u_1 + (1 - \alpha)u_2 \in \mathcal{U}$ for $|\alpha| \leq 1$, which is the same as saying that the line segment joining u_1 to u_2 lies in $\mathcal{U}$), the fact that $w(t)$ in (5.28) is linear in $u(\tau)$ makes $\mathcal{W}(t)$ a convex set. Moreover, it is easily seen that

$$\mathcal{W}(t_2) \supseteq \mathcal{W}(t_1) \quad \text{for } t_2 \geq t_1$$

i.e., that the set $\mathcal{W}(t)$ expands with increasing time. Does it eventually expand to cover an arbitrarily large x_0? To answer this question note that e^{-At} appears in (5.28). If the plant is asymptotically stable and controllable, e^{-At} is a *growing* exponential (in all directions of the state space). Hence $\mathcal{W}(t)$ eventually expands to fill the entire state space. Thus (5.27) will be satisfied for some t, no matter how large $||x_0||$ is, as illustrated in Figure 5.3.

   If the plant is not asymptotically stable, however, e^{-At} will contain at least one exponential that does not grow with time, and the set $\mathcal{W}(t)$ will not expand infinitely in all directions; rather it will be bounded in at least one direction of the state space. If the initial state x_0 lies outside this boundary, no control in $\mathcal{U}$ will be capable of returning the plant to the origin.

Example 5.2   First-Order System

For a first-order plant

$$\dot{x} = ax + bu \quad (b > 0) \tag{5.29}$$

and $\mathcal{U}$ defined by $u(t) \leq 1$, the set $\mathcal{W}$ is the segment of the real line between

$$-\frac{1 - e^{-at}}{a}b \quad \text{and} \quad \frac{1 - e^{-at}}{a}b$$

If the plant is stable ($a < 0$), $e^{-at} \to \infty$ so the set $\mathcal{W}(t)$ eventually encompasses the entire real line. But if the plant is unstable, the set converges to $(-a/b, a/b)$; any initial state x_0 not contained in this interval cannot be brought back to the origin.

Example 5.3   Robot Arm

The robot arm considered in Example 5.1 is more instructive. From (5.20) and (5.21) we obtain the dynamics matrix of the system:

$$A = \begin{bmatrix} 0 & 1 \\ \Omega^2 & 0 \end{bmatrix} \tag{5.30}$$

which has a pair of eigenvalues on the real axis at $s = \pm\Omega$. The system is thus unstable. Hence the set of recoverable states is not the entire state space.

To find the set of recoverable states, it is convenient to transform (5.20) and (5.21) to a diagonal system

$$\dot{z}_1 = \Omega z_1 + u \tag{5.31}$$

$$\dot{z}_2 = -\Omega z_2 + u \tag{5.32}$$

using the transformation

$$\begin{bmatrix} z_1 \\ z_2 \end{bmatrix} = \begin{bmatrix} \Omega & 1 \\ -\Omega & 1 \end{bmatrix} \begin{bmatrix} x_1 \\ x_2 \end{bmatrix} \tag{5.33}$$

Suppose the control u saturates at ±1. In the transformed system z_2 is stable, hence it is possible to reach any state along the z_2 axis with bounded (saturating) control. But z_1 is unstable. Hence, by the analysis of the previous example, it is only possible to reach the points in the region $-1/\Omega \geq z_1 \geq 1/\Omega$.

Combining the results for the two axes, we infer that the set of recoverable points is the strip along the z axis bounded by the line $z_1 = \pm 1/\Omega$ as shown in Figure 5.4(a). Transforming these lines back to the original coordinate system using the inverse of the transformation

$$T^{-1} = \begin{bmatrix} 1/2\Omega & -1/2\Omega \\ 1/2 & 1/2 \end{bmatrix}$$

of (5.33) gives the boundaries of the set of recoverable points in the original state plane as

$$x_2 = -\Omega x_1 \pm \frac{1}{\Omega} \tag{5.34}$$

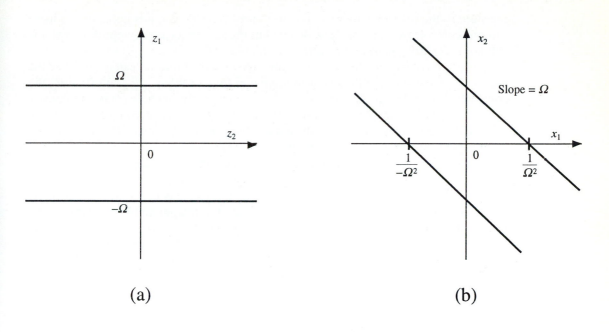

(a)   (b)

Figure 5.4   Sets of recoverable points for unstable robot arm.
(a) In transformed coordinates. (b) In original coordinates.

as shown in Figure 5.4(b).

It is of interest to note the points $z_1 = -z_2 = \pm\Omega$ are equilibrium points with constant control of $u = \pm1$. The boundaries of the set of recoverable points are the *unstable* eigenvectors through these equilibrium points. These equilibrium points map into the points

$$x_2 = 0, \quad x_1 = \pm1/\Omega^2$$

as shown in Figure 5.4(b).

The results of the previous two examples can be extended to a system with one unstable eigenvalue and any number of stable eigenvalues. The unstable eigenvalue λ gives rise to a strip of width inversely proportional to λ_i from the outside of which no state can return to the origin with bounded control.

The analysis is more complicated for a system with multiple (real or complex) unstable eigenvalues [3].

5.3   OPTIMUM DISCONTINUOUS CONTROL

If the state lies in the set of recoverable states, it is possible to find some control input in the bounded set that will return the state to the origin. But in order to return the state to the origin from every state in the recoverable set, it is necessary to use a suitable control law: A control law that makes the system locally asymptotically stable will not necessarily achieve the entire recoverable set.

Finding a control law that achieves the entire recoverable set can be approached indirectly by considering the classical *minimum time* problem, i.e., determining the control law that takes the system state to the origin in minimum time using only control signals lying within the specified bounds. A control law that accomplishes this will achieve the entire recoverable set, since states lying outside the recoverable set cannot be brought to the origin in *any* finite time.

The minimum-time control problem was at the core of the theory of optimal control, which occupied much of control systems research of the 1950s and 1960s. The general theory entails a special formulation of the calculus of variations known as the *maximum principle* that was developed by L.S. Pontryagin and his students in the USSR of this period. (A summary of the theory and its background is given in Note 5.1; a more thorough treatment is beyond the scope of this book.) For purposes of the present discussion, we can make use of one of the important results of this theory as it relates to a linear, time-invariant, controllable plant

$$\dot{x} = Ax + Bu$$

where each component of the control vector is subject to saturation:

$$|u_i| \leq 1$$

The maximum principle asserts that the minimum-time control law has the property that each control variable is always at either its upper or its lower bound. This result, often called the *"bang-bang" principle*, confirms intuition: If you want the process to get back to the origin as fast as possible, you have to use the largest available effort in the proper direction. (This has an incidental advantage with regard to implementation: An expensive linear power amplifier is not needed; your only need is a means for switching the control between its minimum and its maximum value.)

The direction in which the control must be applied is also determined by the maximum principle which asserts that

$$u_i(t) = [\text{sgn}(B'p(t))]_i$$

where $p(t)$ is the state of the so-called *adjoint* system

$$\dot{p} = -A'p$$

the solution of which is

$$p(t) = e^{-A't}p_0$$

The maximum principle thus says that all possible control signals (as functions of time) can be obtained by varying the initial vector p_0 of the adjoint system.

For a system of kth order with real (positive or negative) eigenvalues, it is not too difficult to show that the components of $B'p(t)$ change sign not more than $k-1$ times.

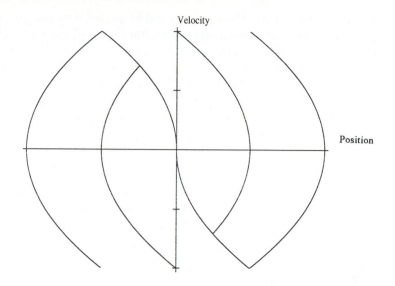

Figure 5.5   Trajectories to the origin with maximum control effort.

Accordingly, the optimal control law will not switch sign more than that number of times.

Consider the implication of this result for a single-input second-order system with real eigenvalues. According to the maximum principle, the state approaches the origin with the control set at either $+1$ or at -1. Each of these settings produces one and only one trajectory into the origin, as shown in Figure 5.5. If the initial state lies on either of these trajectories, the corresponding control will return the state to the origin with no sign changes. But if the initial state does not lie on these trajectories, the control will change sign once. The change will occur when the state reaches the point on the trajectory from which it can return to the origin without any further sign changes. Hence the two trajectories, which together separate the state space into two regions, constitute a *switching curve*. If the state starts on one side of the curve, the control will be (say) positive until the state reaches the switching curve and then will change sign; if the state starts on the other side of the switching curve, the control will be negative until it reaches the switching curve and then will change to positive.

This description can be embodied in a feedback control law. Suppose the equation of the switching curve is

$$x_2 - f(x_1) = 0$$

Then the feedback control law is simply

$$u = -\mathrm{sgn}(x_2 - f(x_1))$$

An efficient method of calculating the switching curve for second-order linear, time-invariant systems is based on the recognition that the curve is a trajectory to the origin with $u = \pm 1$. The trajectory can be obtained by integrating the differential *backwards in time* from the origin using either $u = +1$ or $u = -1$. The differential equation for the plant, run backward in time is

$$-\frac{dx}{dt} = Ax + Bu \tag{5.35}$$

Starting at the origin (i.e., with initial conditions of zero), with $u = U = \pm 1$, the solution to (5.35) is

$$x(\tau) = -\int_0^\tau e^{-At} B dt\, U \tag{5.36}$$

The resulting functions of τ are parametric representations of the switching curve. Each value of U (+1 and –1) provides a branch of the curve. To obtain a non-parametric expression for the curve, it is necessary to eliminate τ from the equations. The calculation can be messy, except in a few special cases considered in the following examples.

Example 5.4   Double Integrator

The differential equations for a double integrator are

$$\dot{x}_1 = x_2$$
$$\dot{x}_2 = u$$

to which there correspond the matrices

$$A = \begin{bmatrix} 0 & 1 \\ 0 & 0 \end{bmatrix}, \quad B = \begin{bmatrix} 0 \\ 1 \end{bmatrix}$$

from which, using (5.36) we find

$$\begin{bmatrix} x_1(\tau) \\ x_2(\tau) \end{bmatrix} = U \begin{bmatrix} \tau^2/2 \\ -\tau \end{bmatrix}, \quad U = \pm 1$$

From $x_2(\tau) = -U\tau$ we obtain the switching curve

$$x_1 = U x_2^2$$

In the second quadrant $U = +1$ defines a trajectory to the origin, as shown in Figure 5.6. In the fourth quadrant $U = -1$ defines another trajectory to the origin. The switch curve separates the entire state plane into two regions. Above the curve the control is negative and the trajectories move downward to the switching curve at which the sign of the control signal changes to positive; below the curve the control is positive and the trajectories move upward to the switching curve at which the sign of the control signal changes to negative.

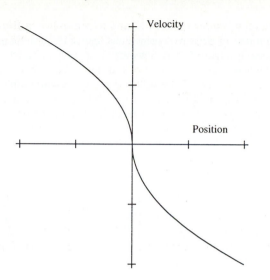

Figure 5.6 Switching curve for double integrator.

Example 5.5 Real Eigenvalues

If the system has two real eigenvalues $\lambda_1, \lambda_2 \neq \lambda_1$, the switching curve can be readily obtained after the system is diagonalized to

$$\dot{z}_1 = \lambda_1 z_1 + \bar{b}_1 u$$
$$\dot{z}_2 = \lambda_2 z_2 + \bar{b}_2 u$$

For $u = U = \pm 1$ we find, using (5.36), the locus of states that can be reached with constant control:

$$z_i(\tau) = -U \left(\frac{1 - e^{-\lambda_i \tau}}{\lambda_i} \right) \bar{b}_i, \quad i = 1, 2 \tag{5.37}$$

Note that $z_i(\tau)$ expands to cover the entire (negative) half of the real z_i axis if and only if the corresponding λ_i is negative, i.e., if the eigenvalue is stable. An unstable eigenvalue ($\lambda_i > 0$) will limit $z_i(\tau)$ to the interval $[0, b_1/\lambda_1]$. There is no switching curve outside this region, which hence demarcates the set of recoverable states.

Solving each of (5.37) for τ and equating the solutions gives

$$\log \left(1 + U \frac{\lambda_1 z_1}{b_1} \right)^{-1/\lambda_1} = \log \left(1 + U \frac{\lambda_2 z_2}{b_2} \right)^{-1/\lambda_2} \tag{5.38}$$

It is important to note that for positive eigenvalues each side of this expression is valid only for the range in which the logarithms have positive arguments. In the region of validity, we can express z_2 as a function of z_1 from (5.38):

$$1 + U \lambda_2 \bar{z}_2 = (1 + U \lambda_1 \bar{z}_1)^{\lambda_2/\lambda_1}], \quad U = \pm 1 \tag{5.39}$$

where

$$\bar{z}_i = \frac{z_i}{b_i}, \quad i = 1, 2$$

The switching curves for several combinations of λ_1 and λ_2 are shown in Figure 5.7. Figure 5.7(a) corresponds to a pair of negative eigenvalues, hence the switching curve extends to the entire state space. In Figure 5.7(b), however, λ_2 is positive, which implies that the set of recoverable states is a strip on both sides of the $\bar{z}_2$ axis. Since $\lambda_2 = 2$, the boundaries of the strip are $\bar{z}_2 = \pm 1/2$. The switching curve is asymptotic to these boundaries. In Figure 5.7(c) both eigenvalues are positive, so the set of recoverable states is bounded by the box: $\bar{z}_2 = \pm 1/2, \bar{z}_1 = \pm 1$; the entire set of recoverable states lies inside the box shown in Figure 5.7(c).

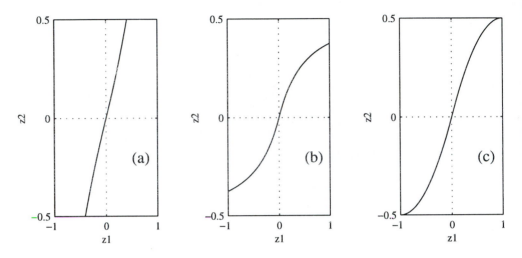

Figure 5.7   Switching curves for several sets of real eigenvalues. (a) $\lambda_1 = -1, \lambda_2 = -2$, (b) $\lambda_1 = -1, \lambda_2 = 2$, (c) $\lambda_1 = 1, \lambda_2 = 2$.

Example 5.6   Minimum Time Control of Robot Arm

The results of the previous example can be used to determine the switching curve for the robot arm considered in Example 5.3 . In this case

$$\lambda_1 = \Omega = -\lambda_2$$

and

$$b_1 = b_2 = 1$$

Moreover, from (5.33),

$$z_1 = x_2 + \Omega x_1$$
$$z_2 = x_2 - \Omega x_1$$

Hence, (5.39) becomes

$$1 - U\Omega(x_2 - \Omega x_1) = \frac{1}{1 + U\Omega(x_2 + \Omega x_1)}$$

After a bit of algebra this equation simplifies to

$$x_2^2 = \Omega^2 x_1^2 + 2U x_1$$

or

$$x_2 = \pm \sqrt{\Omega^2 x_1^2 + 2U x_1} \qquad (5.40)$$

Taking into account the proper sign, we can express the control law as

$$u = -\mathrm{sgn}(x_2 + U \sqrt{\Omega^2 x_1^2 + 2U x_1}) \qquad (5.41)$$

where

$$U = \mathrm{sgn}(x_1)$$

The switching curve (5.40) for $\Omega = 4$ is shown in Figure 5.8. Note that the switching curve is asymptotic to the boundaries of the set of recoverable states, as given by (5.34). Also shown are several trajectories to the origin.

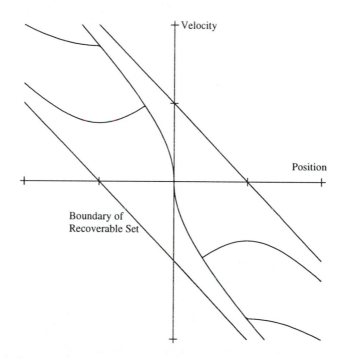

Figure 5.8   Switching curve and minimum time trajectories for linearized robot arm.

If the system has complex eigenvalues, the trajectories into the origin with constant control are spirals and do not completely define the switching curves. As discussed in several textbooks (e.g., [4], [5]), the optimum switching curves have a scalloped shape with discontinuous slopes as shown in Figure 5.9. The spiral trajectories to the origin constitute the inner segments of the switching curves. In a plant with unstable eigenvalues the switching curve terminates on the boundary of the set of recoverable points as shown in Figure 5.9(b).

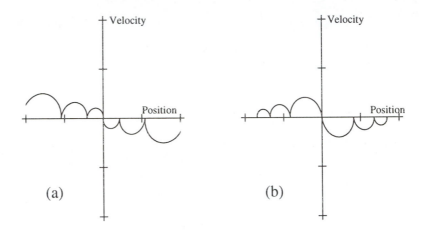

Figure 5.9   Switching curves for second-order systems with complex eigenvalues. (a) Stable plant. (b) Unstable plant.

Although minimum time of response is often desirable in a control system, it is not the only factor to be considered in the design. Other considerations include the following:

- If the order of the plant is greater than 3, especially if complex eigenvalues are present, the calculation of the switching surface is a very difficult task.

- Even if the switching surface can be calculated, it might be difficult to represent the surface in the computer used for real-time implementation of the control law.

- Implementation of the minimum-time control law requires that all the state variables be available for measurement. If they are not, it is not possible to exactly realize the minimum time control law, although use of an observer to estimate the states that cannot be measured directly (as discussed in Chapter 6) can give a very acceptable approximation.

- The switching surface is "tuned" to the parameters of the plant. If the plant parameters are uncertain or change during the operation of the control system, the control law will not be optimum.

- There are very few applications in which a response time slightly larger than the minimum would not be acceptable.

5.4 "CHATTERING"—THE "SLIDING MODE"

As indicated above, one of the reasons for not trying too hard to implement an optimum switching surface is that its shape depends on the parameters of the plant. If the parameters of the plant are uncertain, or change during the course of time, the switching surface will not be correct. What effect does this have on the behavior of the system?

Insight into what happens with an incorrect switching surface can be gained with the aid of Figure 5.10, which shows two possible erroneous switching curves for a second-order system.

If the actual switching curve is *below* the ideal switching curve, as shown in Figure 5.10(a), the switching would occur later than it would on the ideal switching curve, and the system would follow a trajectory parallel to the ideal switching curve until it encounters the next branch of the switching curve, also later than the corresponding branch of the ideal switching curve. This sequence of events continues indefinitely, as the trajectory works its way to the origin without reaching it in a finite time.

The situation is different if the actual switching curve is *above* the ideal switching curve, as shown in Figure 5.10(b). In this case, the switch would occur before it reaches the ideal switching curve. Upon the occurrence of the switch, the sign of the control is such that the state would move on a trajectory that returns it to the region from which it had just come. As soon as this happens, the control wants to switch again. In theory, after the trajectory reaches the switching curve for the first time, the control would switch at an infinite frequency ("chatter") while the state "slides" along the switching curve.

Such behavior is of course not possible with real hardware. What actually happens depends on the details of the hardware. One possibility, for example, is that there is a finite time-delay between the instant that a switching command is sent to the actuator and the instant that the command is executed. In that case, the trajectory will penetrate into the region on the other side of the switching curve before the switch occurs and will not return to the switching curve until the elapse of a finite time after which the process is repeated. Thus the real trajectory zigzags along the switching curve as it works its way to the origin. The extent of the excursion from the switching curve depends on the duration of the time delay: The longer the delay, the larger the excursion. More generally, there is always some dynamic process that intervenes between the (switched) output of the actuator and its input. The nature of the process determines the detailed behavior of the system in the vicinity of the switching surface. If the dynamics are fast, the excursions are small and the switching frequency is high; if the dynamics are slow, on the other hand, the excursions are relatively large and the switching frequency is relatively low.

To prevent the performance of the control system from depending on the vagaries of an uncertain (and often unmodeled) actuator, you might prefer to obtain smoother behavior by replacing "hard" switching function, sgn, in (5.36) by a saturation function or by a deadzone. The saturation function places a linear region on both sides of the switching curve; the deadzone turns off the control entirely, allowing the system to

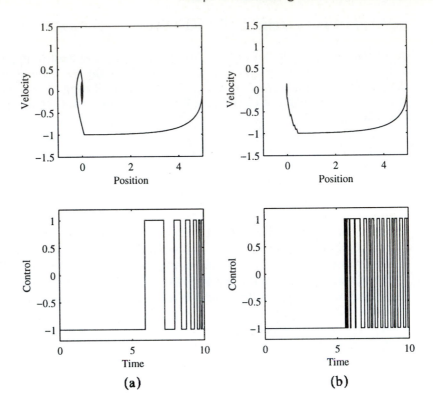

Figure 5.10   Trajectories when switching curve is incorrect. (a)
Late switching. (b) Early switching.

coast. In either case, the trajectory will still zigzag about the switching curve, but you
can adjust the amount of zigzag by the width of the linear region of the saturation
function or of the deadzone.

A saturating linear control in general will provide smoother behavior than a deadzone but requires an expensive actuator with the ability to produce a continuously
variable output level. A deadzone might give acceptable performance at much lower
cost.

The difficulty in calculating the optimum switching surface, and the dependence
of this surface on the plant parameters, suggests the desirability of approximating it by
a surface that can be determined (and implemented) more readily. The simplest and
most readily implemented surface is of course a plane. If you should decide to use a
"switching plane" instead of a more complicated surface, how would you determine
the orientation of the plane? Fortunately, there is a very effective method. Recall

from the discussion of Section 5.1 that the control law (for a single-input system)

$$u = -\text{sat}(\alpha g' x) \tag{5.42}$$

in which the gain matrix g is chosen to minimize a quadratic performance criterion

$$V = \int_t^\infty (x'Qx + ru^2)d\tau \tag{5.43}$$

has the property that it ensures local asymptotic stability for any value of α. In the limit as $\alpha \to \infty$ (5.42) becomes

$$u = -\text{sgn}(g'x) \tag{5.44}$$

Thus an appropriate linear switching surface is given by

$$g'x = 0$$

Any gain matrix that minimizes V in the absence of constraints on the control is suitable for use in (5.44). But is there a gain matrix that can be called "optimum"? In a sense there is. Since the control u is constrained to be less than 1 in amplitude, there is no need to further penalize it in the performance criterion (5.44). Hence the gain matrix obtained as $r \to 0$ would become optimum. As $r \to 0$, however, it is well known that the gain matrix g becomes infinite, which would preclude its use in a linear control law. The control law in (5.44), however, uses only the sign of $g'x$. Thus, only the relative magnitudes of the gains and not their absolute values are significant. You can use the limiting values of the elements in the gain matrix, as $r \to 0$, or any numbers proportional to these values, to define the desired switching plane.

Note that although the control law (5.44), with g chosen to minimize (5.43) will result in a closed-loop system that is asymptotically stable (in some region of the origin), it is not in general optimum. The control law that minimizes (5.43), subject to the constraint $|u| \le 1$ is generally more complicated: There is a region surrounding the origin in which the control is linear; surrounding this region is another in which the control law is nonlinear but in which the control is less in magnitude than 1; and surrounding the latter is still another region in which the control is saturated. Calculation of the boundaries of the regions and the control laws in each region is generally not simple, as the 1964 study by Johnson and Wonham [6] reveals. In most applications, however, it is scarcely important that the control law (5.44) is not optimum. What really matters is that the control law is easy to calculate and to implement and that the resulting performance is satisfactory.

Example 5.7   Robot Arm with Switched Linear Control

We return to the control problem considered in Example 5.6. It is instructive to calculate the linear control law and compare the performance with that of the optimal control law.

As seen in Example 5.1 , the linear control gains for the state weighting matrix

$$Q = \begin{bmatrix} c^2 & 0 \\ 0 & 0 \end{bmatrix}$$

are related by

$$g_2 = \sqrt{2g_1}$$

Thus, as both gains become infinite with decreasing control weighting, the switching curve approaches the vertical line $x_1 = 0$.

The simulated performance for an initial position error of $x_1 = 0.2$ and with $g_1 = 50.$, $g_2 = 10.$ is compared with the minimum-time control in Figure 5.11. (The "hard" limit, sgn($\cdot$) was replaced by a saturation function with a slope of 1000 in the linear region.)

It is seen that with linear switching the state of the system returns to zero nearly as rapidly as with the exact minimum time control law. The difference between the two is the behavior of the control signal. With optimal switching, the state returns to zero in a finite time, after which the control signal is turned off. With linear switching, the control chatters as the trajectory tries to follow the linear switching curve to the origin. Thus, if chattering control is acceptable, linear switching can be nearly as good as optimal switching and may be somewhat easier to implement.

Example 5.8 Temperature Control

Temperature control systems probably comprise the largest application of saturating and discontinuous control. In many cases, the only control law needed is to turn the heat on when the temperature is too low, and turn it off when it is too high. (An ideal discontinuous control will have trouble with the situation in which the temperature is exactly correct, but hysteresis or a similar mechanism eliminates the problem in practical cases. See Section 6.4)

This simple control law, however, may not always be satisfactory. If the plant is third-order or higher and the temperature sensor is physically separated from the source of heat, for example, this control law can result in a limit cycle. This situation is illustrated by the third-order system considered in Example 3.4 *et seq.*, in which the plant dynamics are given by

$$\dot{x}_1 = -x_1 + x_2 + u$$
$$\dot{x}_2 = x_1 - 2x_2 + x_3$$
$$\dot{x}_3 = x_2 - 2x_3$$

where x_i denote the temperature at three nodes in the system.

Suppose that the temperature is measured at node 3, i.e.,

$$y = x_3$$

Furthermore, suppose that the maximum current (heat) is one unit (in some suitable system of units). Using this maximum heat input, the steady-state temperatures are given

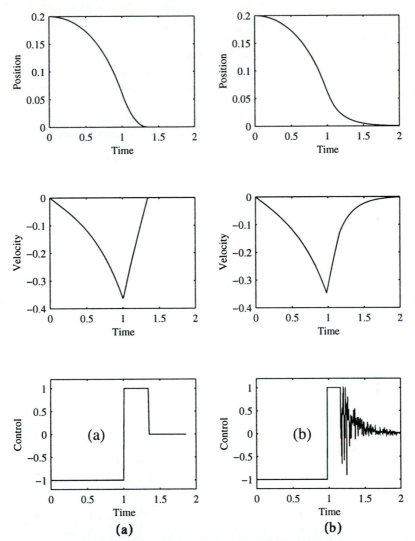

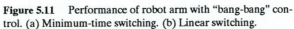

Figure 5.11 Performance of robot arm with "bang-bang" control. (a) Minimum-time switching. (b) Linear switching.

by the solution to

$$0 = -x_1 + x_2 + 1$$
$$0 = x_1 - 2x_2 + x_3$$
$$0 = x_2 - 2x_3$$

which is

$$x_1 = 3, \quad x_2 = 2, \quad x_3 = 1$$

Thus the largest temperature that can be achieved by node 3 is one unit (of temperature). A steady-state temperature of less than one unit can be achieved by a suitable control law.

The simplest control law is based simply on the difference e between the reference temperature $x_R(\leq 1)$ and the measured temperature $y = x_3$:

$$u = \begin{cases} 1 & e > 0 \\ 0 & e \leq 0 \end{cases} \qquad (5.45)$$

where

$$e = x_R - x_3 \qquad (5.46)$$

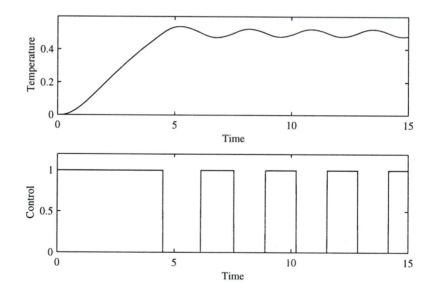

Figure 5.12   Limit cycle in response of temperature control system with output feedback.

As discussed in Example 3.4, this control law exhibits an undesirable limit cycle.

The theory in the foregoing section provides a design method that eliminates the limit cycle by the use of a full-state feedback control law with the gain matrix computed using

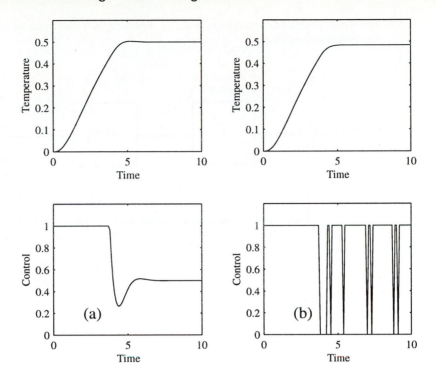

Figure 5.13 Transient response with full-state feedback has no limit cycle.

(5.7). There is a problem, however, with this control law: It does not reduce the steady-state error to zero. In order to reduce steady-state error to zero, it is necessary to include the reference input explicitly in the control law:

$$u = -Gx - G_0 x_R$$

where

$$G_0 = (CA_c^{-1}B)^{-1}CA_c^{-1}E$$

where

$$E = \begin{bmatrix} 0 & 1 & -2 \end{bmatrix}'$$

is the input matrix for the reference input and

$$A_c = A - BG$$

One set of gain matrices, for example, is

$$G = [2.6217, 6.0583, 10.6569], \quad G_0 = -20.9816$$

For a reference input of 0.5, the transient response at x_3 reaches the desired steady-state value in about 5 sec with essentially no overshoot (although overshoot is present in the temperatures analogous to x_1 and x_2.) The control is saturated for more than half the transient interval.

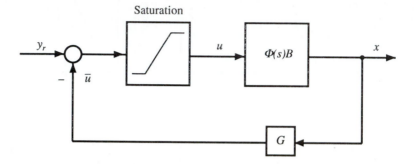

Figure 5.14   Block diagram for calculating full-state feedback transfer function.

The absence of a limit cycle can also be explained by the describing function method. With the aid of Figure 5.14, you can see that the transfer function between the input $\phi(u)$ to the linear part of the plant and the input e to the nonlinearity is given by

$$H(s) = G(sI - A)^{-1}B$$

Using the numerical values of this example, we find that

$$H(s) = \frac{2.6217s^2 + 16.5451s + 30.6386}{s^3 + 5s^2 + 6s + 1}$$

The Nyquist diagram for this transfer function is shown in Figure 5.15. It never crosses the negative real axis. Hence the Describing Function Method predicts that there will be no limit cycle for *any* static nonlinearity. This suggests that a deadzone (of any width) can be used without the risk of a limit cycle. (See Problem 5.7.)

The dynamic behavior using full-state feedback is surely preferable to the limit cycle that is present when only output feedback is used. But to implement this control law it is necessary to measure two more temperatures, x_1 and x_2. At the possible loss of some performance, the expense of instrumenting these additional measurements can be avoided through the use of an observer, as discussed in Chapter 6.

Note that plant is type zero. Hence, without feeding forward the reference temperature or by providing integral action, there will be a steady-state error in response to a constant temperature reference. (See Problem 5.8.)

As the above examples illustrate, the performance of a chattering control system as the trajectory slides into the origin can be quite satisfactory— very nearly optimal in many cases. Because of the ability of a chattering control to track a predetermined

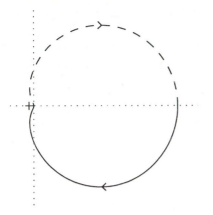

Figure 5.15 Nyquist diagram for temperature control with full-state feedback.

curve in state space, even with uncertain plant parameters, it has engendered considerable interest for a number of theoreticians. (See Note 5.2.) The somewhat pejorative (but accurate) connotation of the term "chattering" may explain why this type of control is often referred to as *"sliding mode" control*. Another term that is often used for this type of control, especially in eastern Europe, is *"variable structure,"* because the switch changes the feedback from negative to positive thus giving rise to two feedback configurations. Although the use of the term "variable structure" is technically correct, one is apt to confuse it with other usages. The appellations "chattering" or "sliding mode," which are more descriptive, are preferable.

The chattering behavior of this mode of control may have undesirable consequences that could preclude its use. Among these are the following:

- Because the actuator is always in operation, it draws power continuously. In addition to the direct cost of this power, it is usually dissipated in the form of heat, which may need to be removed from the system at an additional cost.

- The high operating frequency of the actuator may excite dynamic modes (of uncertain frequency) in the system that are not considered in the design model.

- In mechanical actuators, high-frequency operation may result in excessive wear and require frequent replacement.

- In mechanical actuators, the high actuator frequency may be manifest in the form of acoustical noise that may propagate and have deleterious effects elsewhere in the system.

Reasons such as these generally preclude the implementation of chattering control in practical control systems without a substantial linear region or deadzone. The size of the linear region or deadzone needed in a contemplated application can be investigated analytically by use of the describing function method, for example, or by simulation.

5.5 NOTES

Note 5.1   Optimum control: Pontryagin's Maximum Principle and dynamic programming

The successes of the Soviet Union in aerospace, culminating in the launch of Sputnik in October 1957, motivated a flurry of activity in the West to uncover developments in Soviet science and technology. Among the topics that captured the attention of Western scientists and engineers was the theory of optimum control, particularly in the formulation presented by the Soviet mathematician L.S. Pontryagin and his students. A number of investigators pored over partial translations of papers published in the USSR by these investigators; the publication in 1962 [7] of an English translation of the basic treatise aroused much excitement.

The basic problem addressed by the Soviet investigators was to minimize a performance criterion

$$V = \int_t^T L(x, u) d\tau \tag{5.47}$$

for the dynamic system

$$\dot{x} = f(x, u) \tag{5.48}$$

where x is the state of the system, u is the control input, t is the starting (i.e., the present) time and T is the stopping (terminal) time.

The problem thus stated has long been known in the calculus of variations as Bolza's problem. This problem had been studied since early in the twentieth century. Much of the work was done at the University of Chicago during the 1920s and 30s [8]. The ingredient that the Soviet investigators added to the problem is that the control variable is bounded, i.e.,

$$u \in \Omega \tag{5.49}$$

where Ω is a compact set. (This problem had in fact been addressed by the University of Chicago school [9], but, by Pontryagin's own admission, the results were not fully appreciated in the USSR.)

The solution advanced by Pontryagin et al. is embodied in the theorem known as Pontryagin's Maximum Principle, which can be summarized as follows:

Define the *"Hamiltonian"* function

$$H(x, p, u) := -L(x, u) + p' f(x, u) \tag{5.50}$$

where p is a vector of the same dimension as the state vector x which satisfies the differential equation

$$\dot{p} = -\frac{\partial f(x, u)}{\partial x} p \tag{5.51}$$

where

$$\frac{\partial f(x, u)}{\partial x} := \left[\frac{\partial f_i(x, u)}{\partial x_j} \right]$$

is the Jacobian matrix of the vector $f(\ )$ with respect to the state x.

(The vector p is called the *"adjoint"* or *"costate"* vector and has a role similar to the Lagrange multiplier vector in finite dimensional optimization problems.)

Let $\bar{u}(\tau)$ be the optimum control in the sense that the value of the performance integral V using $\bar{u}$ is smaller than it is for any other control signal $u(\tau)$. According to the maximum principle, the necessary conditions for control signal

$$\{\bar{u}(\tau) : \tau \in [t, T]\}$$

to be optimum are:

1. The Hamiltonian function attains its maximum value for $u = \bar{u}$
2. The maximum value of H is zero.

The first of these conditions is often written

$$\bar{u} = \text{argmax}_{u \in \Omega} H(x, p, u) \tag{5.52}$$

In terms of the Hamiltonian, the dynamic and adjoint equations can be written concisely:

$$\dot{x} = \frac{\partial H}{\partial p} \tag{5.53}$$

$$\dot{p} = -\frac{\partial H}{\partial x} \tag{5.54}$$

which are formally the same as the Hamiltonian equations of classical physics in which x and p represent the vectors of generalized coordinates and conjugate momenta.

Calculating the optimum control $\bar{u}$ as a function of x and p, using (5.52), and substituting the result into (5.53) and (5.54), yields a system of $2k$ ordinary differential equations for a dynamic system of order k. To solve these equations, $2k$ conditions are needed; k of these are the value of the state vector $x(t)$ at the starting time t. The other k conditions are specified at the *end* T of the control interval. These are known as the *transversality conditions* and depend on the conditions that must be satisfied by the state $x(T)$ at the terminal time. If $x(T)$ is free, then

$$p(T) = 0$$

More generally, if $x(T)$ is required to lie in a manifold defined by

$$\phi_i(x(T)) = 0, \quad i = 1, 2, \ldots, j < k$$

the costate $p(T)$ must be orthogonal to that manifold. This condition can be expressed by

$$p(T) = \left[\frac{\partial \phi}{\partial x}\right] \lambda$$

where λ is an arbitrary vector of dimension $k - j$.

The initial conditions and transversality conditions at different ends of the control interval create a "two-point boundary value (TPBV) problem."

Two features of this solution are notable:

1. The Maximum Principle only provides *necessary* conditions for an optimum control, analogous to the condition that the first derivative of a scalar function must be zero at its (local) minimum. Sufficient conditions are expressed in terms of the second variation of the Hamiltonian.

2. The solution of the two-point boundary problem gives the open-loop control and the corresponding optimum trajectory $\{\bar{x}(\tau) : \tau \in [t, T]\}$.

An open-loop optimum control signal and the corresponding trajectory from a specified starting state are useful in missile and space-vehicle launching and similar applications in which the initial state can be very accurately specified, but they do not constitute the feedback algorithm that is generally desired in control engineering. For closed-loop feedback control, you want an expression for the *present* optimum control $\bar{u}(t)$ as a function of the present state $x(t)$. The optimum control algorithm, however, is imbedded in the two-point boundary value (TPBV) problem: The initial value $u(t)$ of the control signal $\{u(\tau) : \tau \in [t, T]\}$ is the optimum control corresponding to the state $x(t)$. If you can solve the TPBV problem for any starting state $x(t)$ and time t, you have in effect determined the optimum control algorithm: Compute $u(t)$; apply it to the dynamic process for an infinitesimal instant to take you to a new state, say $x(t + dt)$; and repeat the calculation of the solution of the two-point boundary value problem.

The equivalence of the open-loop trajectory optimization problem and the feedback control problem was elucidated by the late American mathematician Richard Bellman as the Principle of Optimality which is the cornerstone of his theory of *"Dynamic Programming"* [10]. The principle of optimality states that the defining property of every optimum trajectory is that the trajectory from every state starting on the trajectory is also an optimum trajectory. The principal of optimality can be used to derive a partial differential equation, frequently referred to as the Hamilton-Jacobi-Bellman equation

$$\frac{\partial V}{\partial x} f(x, \bar{u}) + L(x, \bar{u}) = 0 \tag{5.55}$$

the solution of which determines the optimum control algorithm. Solution of this equation by the method of *"characteristics"* leads to the TPBV problem defined earlier; the adjoint vector p can be identified with the negative of the gradient of V:

$$p = -\frac{\partial V}{\partial x}$$

Much, if not most, of the research effort in the field of control theory during the 1960s and early 70s was concerned with optimum control— understanding and extending the fundamental work of Pontryagin and Bellman and their disciples. With the publication of textbooks by Athans and Falb in 1966 [4] and by Bryson and Ho [5] in 1969, the theory of optimum control entered the canon of control theory.

Research in the field fell into several categories, including trajectory optimization, closed-loop control methods, and extension of the general theory.

The U.S. space program motivated significant effort on the computation of optimum open-loop trajectories, i.e., efficient methods of solving the TPBV problem. By the mid-1970s, a number of algorithms and computer codes were available to solve the TPBV problems that arose in aerospace applications of optimum control. Interest in computational methods waned, however, as the Apollo program wound down.

The principle of optimality reveals a method of solving the closed-loop optimum control problem: Simply solve the TBPV problem instantly, starting at the present time and state; compute the corresponding control; and repeat the operation. The only practical limitation to this approach is the speed with which the TPBV problem can be solved.

Considering the advances in computer technology during the past two decades, this approach is not as far-fetched in the 1990s as it may have been in the 1970s. Then, however, methods were sought that avoided the necessity of solving the complete TPBV problem.

Extension of the basic theory proceeded in several directions, including state-variable constraints and the theory of differential games.

Note 5.2   Sliding mode control

Controlling a dynamic process by rapidly switching between two constant levels is not a new idea. Upon reflection, you can recognize that this is the method used to control temperature in virtually all heating, ventilating, and air-conditioning systems.

The development of a systematic theory for designing such systems seems to have been initiated in the USSR of the late 1950s. The technique has also gone by the name *"variable structure control."* By the mid- 1970s several books in Russian, e.g., [11] and in English translation, e.g., [12] had already appeared. Development of the theory has continued apace. An introductory account of the method can be found in [13].

Although many papers appear each year that contain simulations of performance using sliding-mode control, applications that have been actually implemented in hardware are seldom reported.

5.6   PROBLEMS

Problem 5.1   Boundary of region of attraction

Use the method of Lagrange multipliers to estimate the boundary of the region of attraction, as discussed in Section 5.1.

Problem 5.2   Set of recoverable points for complex eigenvalues

Determine the set of recoverable states for a second-order linear system with complex eigenvalues in the right half plane.

Problem 5.3   Set of recoverable states for robot arm with linear switching

Consider the robot arm control with linear switching, as discussed in Example 5.7 . Determine (by simulation, or by any other method) the set of recoverable points and compare with the maximum set of recoverable states achieved by the optimum (minimum-time) switch curve.

Problem 5.4   Robot arm with nonlinear dynamics

Investigate the behavior of the robot arm of Problem 5.3 when the linearized dynamics is replaced by the more accurate dynamics

$$\ddot{x}_1 = \Omega^2 \sin x_1 + u$$

where $u(x)$ is the linear switching law of Example 5.7 .

Problem 5.5   Limit cycle in robot arm with linear switching and deadzone

Investigate the behavior of the robot arm control of Example 5.7 except, instead of switching on the sign of $g'x$, a deadzone is inserted, so that the control is given by

$$u = -\text{dez}(\alpha g'x)$$

where

$$\text{dez}(x) = \left\{ \begin{array}{cc} 1, & x > 1 \\ 0, & |x| < 1 \\ -1, & x < -1 \end{array} \right.$$

(Note that α controls the width of the deadzone.)

Study the behavior for several values of α and g_2.

Problem 5.6   Motor-driven robot arm with saturating control

Investigate the performance of the control laws designed in Problem 5.2 when the control saturates at an amplitude $u_{max} = 100$.

Discuss the behavior when $u_{max} = 50$.

Problem 5.7   Stability of temperature control

Consider the temperature control problem of Example 5.8. By simulation or analysis (e.g., Liapunov's Second Method) determine whether the system is stable for a deadzone of arbitrary width.

Problem 5.8   Temperature control with nonzero reference input

Modify the control design of Example 5.8 to provide integral action and assess the stability of the design.

5.7 REFERENCES

[1] R.W. Bass and I. Gura, "High-Order System Design Via State-Space Considerations," Proc. Joint Automatic Control Conference, Troy, NY, pp. 311–318, June 1965.

[2] R.E. Kalman, "When Is a Linear Control System Optimal?" Trans. ASME (J. Basic Engineering), Vol. 86D, No. 1, pp. 51–60, March 1964.

[3] J.R. LeMay, "Recoverable and Reachable Zones for Control Systems with Linear Plants and Bounded Controller Outputs," Proceedings, 1964 Joint Automatic Control Conference, Stanford, CA, pp. 305–312, June 24- 26, 1964,

[4] M. Athans and P.L. Falb, *Optimal Control*, McGraw-Hill, New York, 1966.

[5] A.E. Bryson, Jr. and Y.-C. Ho, *Applied Optimal Control*, Blaisdell Publishing Co., Waltham, MA, 1969.

[6] C.D. Johnson and W.M. Wonham, "On a Problem of Letov in Optimal Control," Proceedings, 1964 Joint Automatic Control Conference, pp. 317-325, Stanford, CA, June 24-26, 1964.

[7] L.S. Pontryagin, V. Boltyanskii, R. Gamkrelidze, and E. Mischchenko, *The Mathematical Theory of Optimal Processes*, Interscience Publishers, Inc. New York, 1962.

[8] G.A. Bliss, *Lectures on the Calculus of Variations*, The University of Chigago Press, Chicago, IL, 1946.

[9] E.J. McShane, "On Multipliers for Lagrange Problems," Am. J. Math, Vol. 61, pp. 809–819, 1939.

[10] R. Bellman, *Dynamic Programming*, Princeton University Press, Princeton, NJ, 1957.

[11] V.I. Utkin, *Sliding Modes and their Application in Variable Structure Systems* (in Russian), Nauka, Moscow, 1974.

[12] U. Itkis, *Control Systems of Variable Structure,* John Wiley and Sons, New York, 1976.

[13] J.-J. Slotine and W. Li, *Applied Nonlinear Control*, Prentice-Hall, Inc., Englewood Cliffs, NJ, 1991.

Chapter 6

OBSERVERS FOR NONLINEAR SYSTEMS

The previous two chapters deal with methods for designing control laws for nonlinear systems under the assumption that all state variables are accessible for measurement. This assumption, alas, is rarely valid in practical applications: It is either impossible or too expensive to measure all the state variables. In such cases, you must design a control law either that does not need the unmeasurable state variables, or that provides acceptable estimates of the missing state variables using only the available measurements. Both of these approaches are feasible, but there is no systematic approach for the design of control algorithms for nonlinear systems using only available outputs. On the other hand, the separation principle— in which the controller is designed in two parts: a full-state feedback law and an observer—can be extended to nonlinear systems. This "extended separation principle," which, as we will see later in this chapter retains some of the properties of the separation principle of linear systems, motivates the study of nonlinear observers.

Informally, an observer can be described as a dynamic system designed to be driven by the output of another dynamic system (called the *plant*) and having the property that the state of the observer converges to the state of the plant.

A major application of nonlinear observers is to provide an estimate of the process state for use in implementation of a nonlinear feedback control law. In such an application, estimation of the state of the process under control is only the means to another objective—stable closed-loop control. If the ultimate objective is achieved, inaccuracy in the estimate of the state of the process is rarely objectionable.

In diverse applications, an estimate of the state of a dynamic process is needed for purposes other than closed-loop control. Such applications include navigation of aircraft and spacecraft, monitoring and diagnosis of failures in mechanical and biological systems, communications signal processing, and many others. If only a rough estimate of the state is acceptable, there is a great deal of latitude in the design of the observer; but if it is necessary to obtain an accurate estimate, and the required statistical information about the measurement noise and the excitation noise is available, a near-optimum estimate can be obtained by use of an "extended Kalman filter"("EKF")—a nonlinear observer in which the gain matrix is computed on-line by numerical integration of the variance equation. Even if the required statistical information is not available, the observer can be designed in the form of an extended Kalman filter, but with the values of the required statistical data obtained by educated guesswork. Although the observer may perform well, it cannot legitimately be called optimal.

The approach to the theory of observers for nonlinear processes used in this chapter follows the lines of that used in *Control System Design* [1]: It is postulated that a dynamic system exists that possesses the property that the state of the observer converges to the state of the observed process. The equations that must be satisfied by an observer having this property are derived. Finally, the dynamic of the observer are "optimized" in the statistical sense to result in the well-known extended Kalman filter.

6.1 NONLINEAR OBSERVERS

An observer for a plant, consisting of a dynamic system

$$\dot{x} = f(x, u) \tag{6.1}$$

with observations given by

$$y = g(x, u) \tag{6.2}$$

is another dynamic system, the state of which is denoted by $\hat{x}$, excited by the output y of the plant, having the property that the error

$$e = x - \hat{x} \tag{6.3}$$

converges to zero in the steady state.

One way of obtaining an observer is to imitate the procedure used in a linear system, namely to construct a model of the original system (6.1) and force it with the "residual":

$$r = y - \hat{y} = y - g(\hat{x}, u) \tag{6.4}$$

The equation of the observer thus becomes

$$\dot{\hat{x}} = f(\hat{x}, u) + \kappa(y - g(\hat{x}, u)) \tag{6.5}$$

where $\kappa(\ )$ is a suitably chosen nonlinear function. (How to choose this function will be discussed later.) A block-diagram representation of a general nonlinear observer is shown in Figure 6.1.

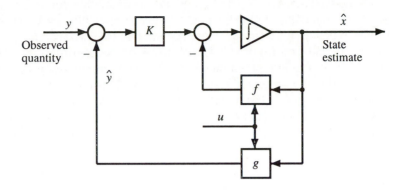

Figure 6.1   Structure of nonlinear observer.

The differential equation for the error e can be used to study its behavior. This equation is given by

$$\dot{e} = \dot{x} - \dot{\hat{x}}$$
$$= f(x, u) - f(\hat{x}, u) - \kappa(g(x, u) - g(\hat{x}, u))$$
$$= f(x, u) - f(x - e, u) + \kappa(g(x - e, u) - g(x, u)) \tag{6.6}$$

Suppose that by the proper choice of $\kappa(\ )$ the error equation (6.6) can be made asymptotically stable, so that an equilibrium state is reached for which

$$\dot{e} = 0$$

Then, in equilibrium, (6.6) becomes

$$0 = f(x, u) - f(x - e, u) + \kappa(g(x - e, u) - g(x, u)) \tag{6.7}$$

Since the right-hand side of (6.7) becomes zero when $e = 0$, *independent* of x and u, it is apparent that $e = 0$ is an equilibrium state of (6.6). This implies that if $\kappa(\ )$ can be chosen to achieve asymptotic stability, the estimation error e converges to zero.

It is very important to appreciate that the right-hand side of (6.7) becomes zero independent of x and u only when the nonlinear functions $f(\cdot, \cdot)$ and $g(\cdot, \cdot)$ used in the observer are exactly the same as in the equations (6.1) and (6.2) that define the plant dynamics and observations, respectively. Any discrepancy between the corresponding functions will generally prevent the right-hand side of (6.7) from vanishing and hence will lead to a steady-state estimation error. Since the mathematical model of a physical process is always an approximation, in practice the steady-state estimation error will generally not go to zero. But, by careful modeling, you can usually minimize the discrepancies between the f and g functions of the true plant and the model used in the observer. This can keep the steady-state estimation error acceptably small.

For the same reason that the model of the plant and the observation that is used in the observer must be accurate, it is important that the control u that goes into the plant is the very same control used in the observer. If the control to the plant is subject to saturation, for example, then the nonlinear function that models the saturation must be included in the observer. Failure to observe this precaution can cause difficulties, as will be illustrated in Example 6.1 .

Choosing the function $\kappa(\ )$ is particularly easy when the dynamics are linear in the state and in the observations; the plant is nonlinear only in the control:

$$f(x, u) = Ax + \phi(u) \tag{6.8}$$

$$g(x, u) = Cx \tag{6.9}$$

In this case $\kappa(\ )$ can simply be a gain matrix, and (6.6) becomes

$$\dot{e} = (A - KC)e \tag{6.10}$$

which is the same as for a linear system. To ensure convergence of the error to zero it is necessary only to pick K such that the eigenvalues of

$$\hat{A} = A - KC$$

lie to the left of the imaginary axis. This can be achieved in a variety of ways, as explained in texts on linear control systems, such as *Control System Design*[1].

Example 6.1   Temperature Control

We resume a temperature control application that we started exploring in Example 1.5. In Example 3.4, we examined the effect of saturation on the performance of the control system using only output feedback and observed that a limit cycle is present for all desired levels of reference y_r. The limit cycle can be eliminated through the use of full-state feedback, but this may be too high a price for eliminating the limit cycle. Can the limit cycle be eliminated by using an observer to estimate the state (or the unmeasured state variables)? We seek an answer by considering an observer-based control system design.

The plant, having no poles at the origin, is "type zero." In order to maintain a constant output temperature with zero steady-state error, the compensator must provide "integral action," i.e., to provide a constant output to the plant (which the plant needs to maintain a constant output temperature) although its own input, namely the system error, is zero.

We consider designing the observer to provide this integral action. This is accomplished by using the system error as a state variable and estimating the exogenous reference input. (The method is explained in greater detail for linear systems in [1].)

Since a performance goal is to achieve zero steady-state error for a constant reference input

$$x_3 = \bar{x}_3 = \text{const.}$$

we define the error

$$e := x_3 - \bar{x}_3$$

In terms of the error, the plant dynamics as given in Chapter 5 thus become

$$\dot{x}_1 = -x_1 + x_2 + \phi(u) \tag{6.11}$$

$$\dot{x}_2 = x_1 - 2x_2 + e + \bar{x}_3 \tag{6.12}$$

$$\dot{e} = x_2 - 2e - 2\bar{x}_3 \tag{6.13}$$

With the new state defined by

$$x := \begin{bmatrix} x_1 & x_2 & e \end{bmatrix}'$$

the plant dynamics take the form

$$\dot{x} = Ax + B\phi(u) + Ex_0$$

where

$$A = \begin{bmatrix} -1 & 0 & 0 \\ 1 & -2 & 1 \\ 0 & 1 & -2 \end{bmatrix}, \quad B = \begin{bmatrix} 1 \\ 0 \\ 0 \end{bmatrix}, \quad E = \begin{bmatrix} 0 \\ 1 \\ -2 \end{bmatrix}, \quad x_0 = \bar{x}_3$$

In the absence of saturation, the full-state feedback control law would have the form

$$u = -Gx - G_0 x_0 \tag{6.14}$$

An appropriate set of gains were determined in Chapter 5. (See Example 5.8.)

In this application it is assumed that the only quantity that can be measured is the desired temperature x_3. Thus an observer is used to estimate all the state variables. (A reduced-order observer could also be used to estimate only the state variables that are not directly measured. See Problem 6.1) Thus, instead of using the actual state

$$\mathbf{x} = \begin{bmatrix} x \\ x_0 \end{bmatrix}$$

we use an observer

$$\dot{\hat{\mathbf{x}}} = \mathbf{A}\hat{\mathbf{x}} + \mathbf{B}\phi(u) + \mathbf{K}(y - \mathbf{C}\hat{\mathbf{x}}) \tag{6.15}$$

where

$$\hat{\mathbf{x}} = \begin{bmatrix} \hat{x}_1 \\ \hat{x}_2 \\ \hat{e} \\ \hat{x}_0 \end{bmatrix}$$

$$\mathbf{A} = \begin{bmatrix} A & B \\ 0 & 0 \end{bmatrix}, \quad \mathbf{B} = \begin{bmatrix} B \\ 0 \end{bmatrix}, \quad \mathbf{C} = \begin{bmatrix} C & 0 \end{bmatrix} \tag{6.16}$$

The gain matrix $\mathbf{K}$ can be selected by pole placement or by solving the algebraic Riccati equation for the gain of the steady-state Kalman filter. By the latter method we find that an appropriate observer gain matrix is

$$\mathbf{K} = \begin{bmatrix} 2.2208 & -3.4411 & 9.1180 & -31.6288 \end{bmatrix}'$$

To illustrate performance, the control gain matrix that was used in Chapter 5 is also used here:

$$\mathbf{G} = \begin{bmatrix} 2.6217 & 6.0583 & 10.6569 & -20.9816 \end{bmatrix}$$

For the purpose of comparison, three situations will be considered:

- Linear compensator in system without saturation
- Linear compensator in system with saturation
- Nonlinear compensator in system with saturation

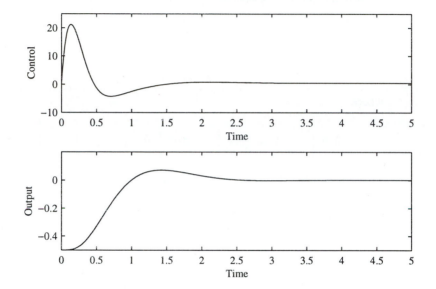

Figure 6.2   Response of temperature control system unsaturated linear control.

Linear System and Compensator   In the absence of saturation, i.e., with

$$\phi(u) = u$$

in (6.15) the transfer function of the compensator defined above is fourth-order, having the form

$$D(s) = \frac{b_1 s^3 + b_2 s^2 + b_3 s + b_4}{s(s^3 + a_2 s^2 + a_3)}$$

The pole at the origin provides the requisite integral action. Using the above gains we find the compensator transfer function:

$$D(s) = \frac{475.6s^3 + 4075.2s + 5539.5s + 1000.5}{s(s^3 + 16.74s + 133.61s + 457.49)}$$

The step response of the closed-loop system is shown in Figure 6.2(a). It has an overshoot of about 20% and a settling time of about 2 sec. With saturation absent, the control signal u that achieves this performance reaches a peak value of about 20 units, for a reference step of 0.5 units and reaches state value of 0.5. (It is readily established that the steady-state control is numerically equal to the desired reference value of x_3.)

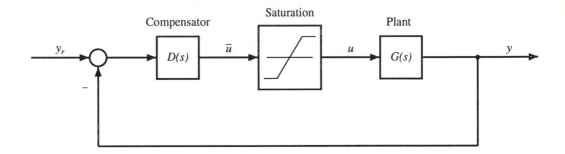

Figure 6.3 Block diagram of temperature control with linear compensator and saturation.

Linear Compensator in System with Saturation Suppose now that the control saturates at a level of 1.0. This level is sufficient to sustain the required steady-state level, but not large enough to support the transient that occurs in the linear system. Hence the output of the compensator will saturate the actuator (i.e., the heater). If the compensator $D(s)$ is used, but the control saturates, a system with the configuration of Figure 6.3 results. The resulting response of the system is shown in Figure 6.4(a). As you would expect, the response is slower than that of the unsaturated system, since the former receives a larger input during the early portion of the transient. Not only is the response slower, but it exhibits a substantially larger overshoot (about 40 %) than the linear system.

The response cannot be accelerated by increasing the gain in the linear region. As 6.4(b) shows, if the gain is increased by a factor of 100 (thus narrowing its width by that factor), the response is hardly changed, but a limit cycle (with a frequency of about 0.66 Hz) appears in the output. The amplitude of the limit cycle is probably tolerable in most applications, but the incessant switching of the heater may be objectionable. (A dead zone in place of the linear region may be helpful. See Problem 6.3.)

Nonlinear Compensator in System with Saturation Instead of using the linear compensator, we include the saturation nonlinearity in the observer (6.15). The resulting transient response is shown in Figure 6.4. As you can see, the response is very nearly time optimal. Moreover, increasing the gain in the linear region by a factor of 100 does not cause a limit cycle. As we saw in Section 5.2, there would be no limit cycle if full-state feedback were used. Thus, by including the saturation in the observer, we essentially achieve the behavior characteristic of full-state feedback.

The above example reveals the advantage of including the control saturation in the observer. Why is it achieved? By comparing the control signals in the two cases, you can see that the control remains at its upper limit too long when the saturation is

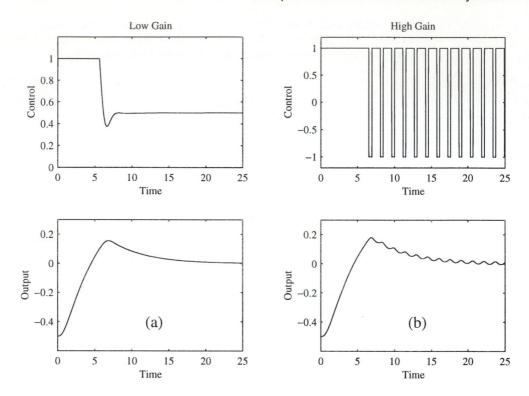

Figure 6.4   Response of temperature control system with saturation and linear compensator. (a) Slope of linear region = 1 (b) Slope of linear region =100.

omitted in the observer. The reason for this is the phenomenon known as *"integrator windup"*: The compensator, which has a pole at the origin, provides integral action. Imagine the transfer function of the compensator represented by an integrator in parallel with a second-order subsystem. The control signal to the integrator is oblivious to the fact that the input to the plant has saturated and hence continues "winding up" the integrator; the error signal changes sign when the temperature reaches the set point (at about 4.2 sec), but the control signal does not drop from its maximum value for about another second. When the saturation is included in the observer, on the other hand, the control signal drops from its maximum value even before the error changes sign, thus correctly taking into account the dynamics (i.e., the lag) of the process.

The foregoing example illustrates the unfavorable consequences of failure to include the plant nonlinearity in the observer. In that example the result is relatively benign: an excessive overshoot (and, as shown in Problem 6.2 and Problem 6.3, a possible limit cycle of low amplitude). In other situations, as the following example illustrates, the consequences may be much more serious.

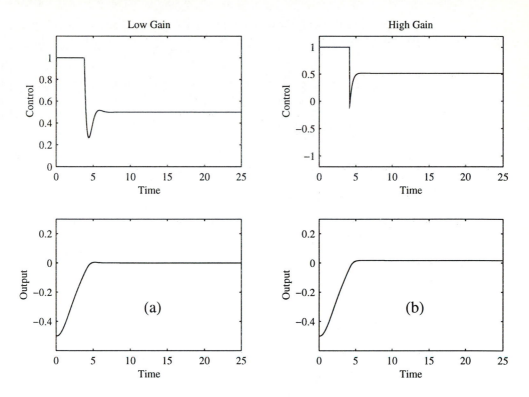

Figure 6.5 Response of temperature control system with saturation and nonlinear compensator. (a) Slope of linear region = 1 (b) Slope of linear region = 100.

Example 6.2 Robot Arm Control with Saturated Input

We return to the robot arm (inverted pendulum) considered in Chapter 5.

Since the plant is type zero, it is necessary to provide integral control action to track a step input. This action is achieved by adjoining a constant exogenous state $x_3 = x_0$ to be estimated.

For this example a linear control law is considered:

$$u = -G_1 x_1 - G_2 x_2 - G_3 x_3 \qquad (6.17)$$

A suitable gain matrix in this case is found to be:

$$G = \begin{bmatrix} 51.4401 & 10.1430 & 16. \end{bmatrix}$$

Note that the last gain G_3 is the gain needed to maintain zero steady-state error for a step input.

It is assumed that only the position and not the velocity is measured. Accordingly, the observer is given by

$$\dot{\hat{x}}_1 = \hat{x}_2 + K_1(y - \hat{x}_1) \tag{6.18}$$

$$\dot{\hat{x}}_2 = \Omega^2 x_1 + \hat{x}_3 + \bar{u} + K_2(y - \hat{x}_1) \tag{6.19}$$

$$\dot{\hat{x}}_3 = +K_3(y - \hat{x}_1) \tag{6.20}$$

An observer gain matrix for this example is

$$K = \begin{bmatrix} 12.8769 & 82.9075 & 10.0 \end{bmatrix}'$$

We consider the following cases:

- Linear compensator with linear plant
- Linear compensator with nonlinear plant
- Nonlinear compensator with nonlinear plant

Each of these cases are investigated with an initial position of $x_1 = 0.12$. The velocity and all the observer state variables are zero. Moreover, although the control is designed to achieve and maintain a nonzero reference position, in this case the reference is taken as zero: The goal in this example is to return the system to the vertical equilibrium state.

Linear Compensator With Linear Plant To serve as the basis of comparison with the other cases, we consider the case in which the control is not saturated, i.e.,

$$\bar{u} = u$$

in both (6.17) and (6.19).

The time response for this case is shown in Figure 6.6(a). It is observed that the state returns to the origin (with a small overshoot) in about 1 sec. To achieve this performance, however, the control has a peak value of about -7. With a saturation level of 1, this transient response cannot be obtained.

The state trajectory into the origin is shown in Figure 6.6(b). It should be noted that the state space has 5 dimensions; hence the trajectory shown is actually a projection of a curve in a space of this dimension onto the position-velocity subspace.

Linear Compensator With Nonlinear Plant In this case, we investigate the effect of using the linear compensator when the control is subject to saturation.

The control that is used in (6.19) of the observer is

$$\bar{u} = -G_1\hat{x}_1 - G_2\hat{x}_2 - G_3\hat{x}_3$$

while the control that is applied to the plant is

$$u = \text{sat}(\bar{u})$$

The transient response is shown in Figure 6.7. The response verges on being unstable. The trajectory in the position-velocity subspace (Figure 6.7)(b) shows this more vividly. Although the trajectory does ultimately return to the origin, it is very close to being an (unstable) limit cycle. In fact if the initial position error is raised to only slightly, the trajectory will be unstable. Hence, we conclude that this control system exhibits an unstable limit cycle with a period of about 5 sec. The describing function method, for example, would predict this limit cycle. (See Problem 6.4)

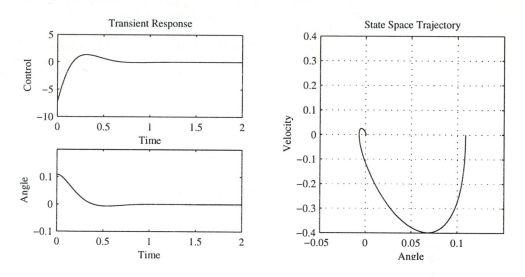

Figure 6.6 Transient response of unsaturated linear control system for robot arm.

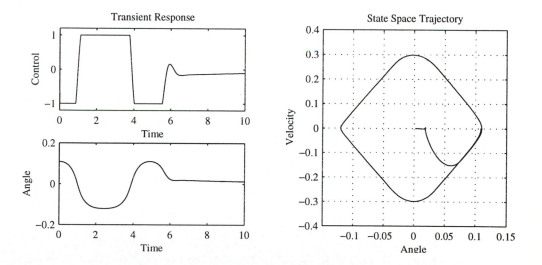

Figure 6.7 Transient response of linear control system for robot arm with saturated input.

In this case, the effect of the integrator windup due to not including the saturation in the observer is to produce an unstable limit cycle which substantially reduces the size of the set of recoverable points.

Nonlinear Compensator With Nonlinear Plant   In contrast to the previous case, the control used in (6.19) is

$$\bar{u} = \text{sat}(-G_1\hat{x}_1 - G_2\hat{x}_2 - G_3\hat{x}_3)$$

which is also the control input to the plant. The compensator is nonlinear, because the control saturation is included in the observer.

The transient response is shown in Figure 6.8. It is seen that the error is reduced to zero in about 2 sec, with essentially no overshoot. The response is close to being time-optimal.

Again, the advantage of including a saturation model in the observer is evident.

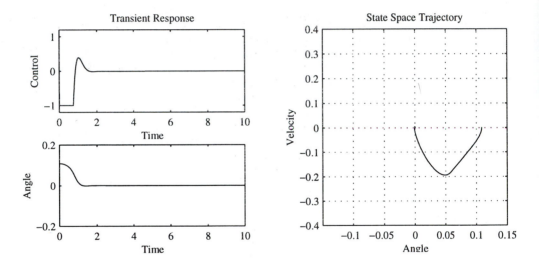

Figure 6.8   Transient response of robot arm with nonlinear control.

The function $\kappa(\ )$ in the observer must be selected to ensure asymptotic stability of the origin ($e = 0$ in (6.7)). By the theorem of Liapunov's First Method (Section 2.6), the origin is asymptotically stable if the Jacobian matrix of the dynamics, evaluated at the equilibrium state, corresponds to an asymptotically stable linear system. For the dynamics of the error (6.6) the Jacobian matrix with respect to the error e evaluated at $e = 0$ is given by

$$A_c(x) = (\partial f / \partial x) - K(\partial g / \partial x) \tag{6.21}$$

This is the nonlinear equivalent of the closed-loop observer equation of a linear system:

$$A_c = A - KC$$

where A and C are the plant dynamics and observation matrices, respectively. The problem of selecting the gain matrix for a nonlinear observer is analogous to that of a linear observer, but somewhat more complicated by the presence of the nonlinearities that make the Jacobian matrices in (6.21) dependent on the state x of the plant. Nevertheless, the techniques used for selecting the gain for a linear observer can typically be adapted for a nonlinear observer. Pole placement is one method; another is to make the observer an extended Kalman filter which, as explained in Section 5.5, entails on-line computation of the gains via the linearized variance equation.

Example 6.3 Magnetic Levitation

Full state control law for an idealized magnetic levitation system were considered in Chapter 4. As indicated there, a stable closed-loop system requires velocity as well as position feedback. The position of the levitated object is the controlled quantity; hence the reference position $\bar{x}$ and the position error $\delta x = x - \bar{x}$ are presumably measurable quantities. It may not be easy to measure the velocity, however. An observer to estimate the velocity might be preferable.

In this example we will consider a full-state observer: Both position and velocity will be estimated. In Example 6.5 we will consider a reduced-order observer, in which it will be assumed that only the velocity needs to be estimated.

To design the observer the complete *nonlinear* dynamics, not the linearized dynamics, are used:

$$\dot{x}_1 = x_2 \tag{6.22}$$
$$\dot{x}_2 = -k(u^2/x^2) + g \tag{6.23}$$

It is assumed that the observed quantity y is the position of the levitated body. Thus the observation residual is the difference between the observed position and the estimated position:

$$r = y - \hat{x}_1$$

Using these dynamics, the full-order observer is given by

$$\dot{\hat{x}}_1 = \hat{x}_2 + k_1(y - \hat{x}_1) \tag{6.24}$$
$$\dot{\hat{x}}_2 = -k(u^2/\hat{x}_1^2) + g + k_2(y - \hat{x}_1) \tag{6.25}$$

where $u = u(\hat{x})$ is the control law designed for full-state feedback but with the estimated state variable $\hat{x}_1$ used in place of the true state x_1. To set the observer gains k_1 and k_2 we determine the closed-loop dynamics matrix:

$$A_c(x) = (\partial f / \partial x) - K(\partial g / \partial x)$$
$$= \begin{bmatrix} -k_1 & 1 \\ 2k\bar{u}^2/\bar{x}_1^3 - k_2 & 0 \end{bmatrix} \tag{6.26}$$

The corresponding characteristic polynomial is

$$|sI - A_c| = s^2 + k_1 s + k_2 - 2k\bar{u}^2/\bar{x}_1^3 \qquad (6.27)$$

Thus the observer is asymptotically stable for

$$k_1 > 0 \qquad (6.28)$$
$$k_2 > 2k\bar{u}^2/\bar{x}_1^3 = 2g/\bar{x}_1 \qquad (6.29)$$

The above equations define the full-order observer for the case in which the gravitational acceleration is known. The levitation system can be turned into an instrument for measuring the acceleration of gravity, however, by assuming that g is a quantity to be estimated. (Actually the device will act as an accelerometer, measuring the specific force, i.e., the non-gravitational acceleration on a massive body upon which the instrument is mounted—see [1].) To accomplish this, the acceleration g is treated as a state variable to be estimated. Assuming that g is an (unknown) constant results in the additional equation

$$\dot{\hat{g}} = k_3(y - \hat{x}_1) \qquad (6.30)$$

which, when combined with (6.24) and (6.25), result in a third-order nonlinear observer. The block diagram representation of this observer is given in Figure 6.9.

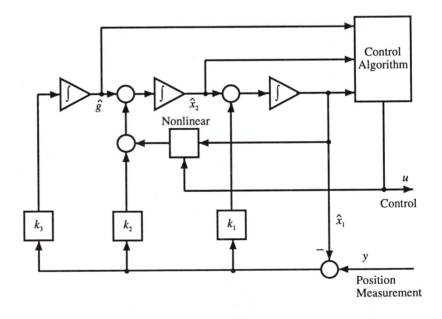

Figure 6.9   Observer for magnetic levitation with gravity estimation.

The Jacobian matrix for the observer is given by

$$A_c(x) = (\partial f/\partial x) - K(\partial g/\partial x)$$

$$= \begin{bmatrix} -k_1 & 1 & 0 \\ 2k\bar{u}^2/\bar{x}_1^3 - k_2 & 0 & 1 \\ -k_3 & 0 & 0 \end{bmatrix} \tag{6.31}$$

to which there corresponds the characteristic polynomial

$$|sI - A_c| = s^3 + k_1 s^2 + (k_2 - 2k\bar{u}^2/\bar{x}_1^3)s + k_3 \tag{6.32}$$

The range of variation of the gains k_1, k_2, and k_3 can be found with the aid of the Routh-Hurwitz criterion [1] (see Problem 6.9).

6.1.1   Using Zero-Crossing or Quantized Observations

The ability of an observer to utilize nonlinear observations is not limited to observations that exhibit only moderate nonlinearity; even highly nonlinear observations can be accommodated. Perhaps the most nonlinear observation is that of a zero-crossing detector, in which

$$y = \operatorname{sgn}(x) = \begin{cases} 1, & x > 0 \\ -1, & x < 0 \end{cases}$$

This is the extreme special case of a quantizer, in which the output is quantized to only two levels. If this extreme case can be successfully dealt with, it is surely possible to use the same technique with multi-level quantizers.

Suppose, for example, that the only observation is of the zero-crossing of x_1. The observer for this process is then given by:

$$\dot{\hat{x}} = f(\hat{x}, u) - K[y - \operatorname{sgn}(\hat{x}_1)] \tag{6.33}$$

as illustrated in Figure 6.10.

Provided that a gain matrix K can be found that stabilizes the observer at $e = 0$, the estimation error will be reduced to zero. Since the partial derivative of the nonlinear function with respect to the observation does not exist in this case, since the observation is discontinuous with respect to the state, you can't establish the stability of the observer by linearizing about the origin. You have to use some other method, such as Liapunov's Second Method, or determine the appropriate range of K by simulation.

Some insight into how the observer operates can be gained by considering that both y and $\operatorname{sgn}(x_1)$ are signals that take on the values of ± 1; their difference, which is the residual that appears in (6.33), is either 0 or ± 1. Suppose the observer is working well; most of the time y and $\operatorname{sgn}(x_1)$ will have the same sign and the residual will be zero. The residual will be nonzero for the short time interval in which y and $\operatorname{sgn}(x_1)$ have different signs. The residual will thus consist of a train of narrow pulses, each of height ± 1 and of width proportional to the phase difference between y and $\operatorname{sgn}(x_1)$, as shown in Figure 6.11. The effect of each pulse is to nudge the state of the observer to agree with the state of the plant.

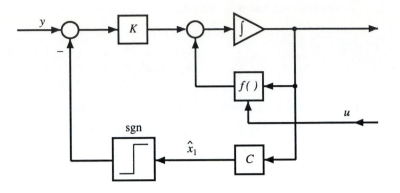

Figure 6.10   Observer using zero-crossing data.

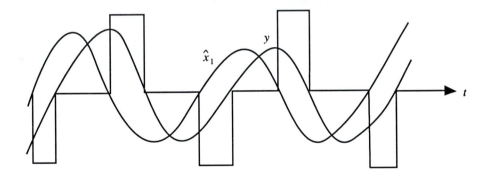

Figure 6.11   Residual for zero-crossing observations consists of pulses.

The same idea extends to quantized observations, since a quantizer can be regarded as a multiple-level crossing detector.

Of course, if there are other observations in addition to the zero-crossing observation, they can be combined with the latter and, with an appropriate choice of gains, can provide enhanced performance.

Example 6.4   Estimating the State of an Oscillating System

The use of zero-crossing measurements is illustrated by an application in which it is required to estimate the motion of a lightly damped mechanical oscillator, consisting of a mass, a spring, and a damping mechanism, the motion of which is sustained by means of a known signal. The dynamic behavior of the system is governed by the second-order differential equation

$$\ddot{x} + 2\zeta\Omega\dot{x} + \Omega^2 x = u \tag{6.34}$$

where u is a known signal.

It is required to obtain a continuous estimate of the position x of the oscillating mass, using only measurements of the zero-crossings:

$$y = \text{sgn}(x) \tag{6.35}$$

To apply the above observer design technique, (6.34) is written in state space form:

$$\dot{x}_1 = x_2 \tag{6.36}$$
$$\dot{x}_2 = -\Omega^2 x_1 - 2\zeta\Omega x_2 + u \tag{6.37}$$

with the observation given by

$$y_1 = \text{sgn}(x_1) \tag{6.38}$$

In accordance with (6.33), the observer is given by

$$\dot{\hat{x}}_1 = \hat{x}_2 + k_1 r_1 \tag{6.39}$$
$$\dot{\hat{x}}_2 = -\Omega^2 \hat{x}_1 - 2\zeta\Omega\hat{x}_2 + u + k_2 r_1 \tag{6.40}$$

where

$$r_1 = y_1 - \text{sgn}(\hat{x}_1) \tag{6.41}$$

is the residual.

The performance that such an observer can deliver is investigated by simulation, using the following numerical data:

$$\Omega = 1.2, \quad \zeta = 0.477, \quad k_1 = k_2 = 0.0002$$

The input to the process was simulated by white noise, which is assumed to be known to the observer. (See Problem 6.7.) The observations were simulated by assuming that the velocity observation is as given by (6.42) and that the position observation is given by

$$y_1 = \text{sgn}(x_1 + n)$$

where n is a first-order Markov process, obtained by passing white noise through a first-order low-pass filter.

The performance of the observer is shown in Figure 6.12 with observation noise (a) absent and (b) present. To the scale of the figure, the difference between the actual and the estimated position is not discernable. With observation noise absent, the error actually goes to zero after a short transient period. (Note the behavior of the residual.) With moderate observation noise, the error does not go to zero, but it is less than 0.1 percent of the magnitude the actual position.

As you might expect, the performance of the observer relies on a good model of the process, since it is the model that allows you to predict the zero crossings and make appropriate corrections. If the model used in the observer is not matched to the actual process,

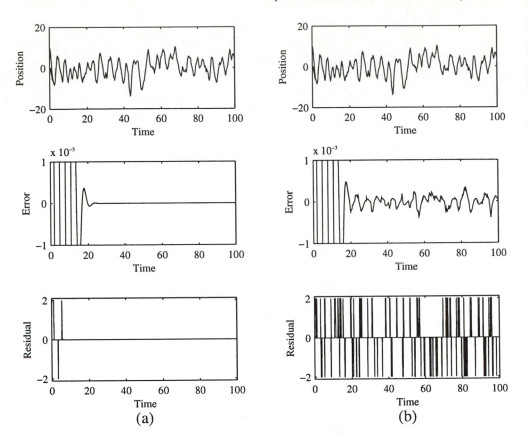

Figure 6.12   Performance of zero-crossing observer in estimating motion of an oscillating object. (a) Noise absent from measurement. (b) Noise present in observation.

performance will deteriorate. The effect of a mismatch is shown in Figure 6.13, for the same conditions as for Figure 6.12, except that the parameters used in the observer were:

$$\Omega = 1.12, \quad \zeta = 0.447$$

representing about a 10 percent error in the parameter values. The difference between the actual and estimated position is now barely discernable. The error level is about 10 percent of the actual position. It is interesting that the performance appears to be relatively independent of the observation noise, although the activity in the residual is greater with noise present in the observations.

In addition to the measurement of the position zero-crossings, it may be possible to measure the velocity of the oscillating object. In principle, if it were possible to measure the velocity with complete accuracy and the parameters ζ and Ω were known with complete accuracy, it would be possible to estimate the position of the body using only the

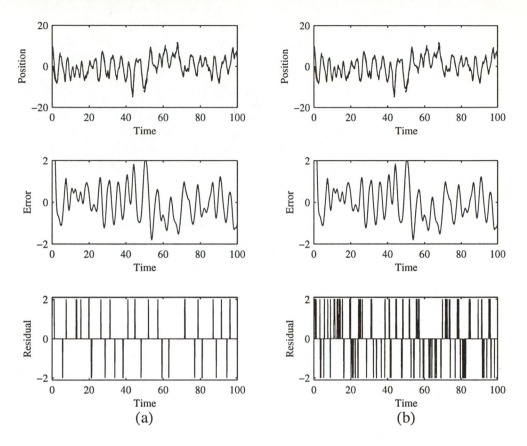

Figure 6.13 Performance of zero-crossing observer in estimating the motion of an oscillating object when observer is not matched to plant. (a) Noise absent from measurement. (b) Noise present in observation.

velocity measurement. In practice, however, neither of these conditions hold. Moreover, there is typically a bias in the measurement of velocity. Thus the velocity observation is given by

$$y_2 = x_2 + x_3 + w_2 \tag{6.42}$$

where x_3 represents the bias, modeled by the differential equation

$$\dot{x}_3 = 0 \tag{6.43}$$

and v represents the observation noise.

To use the velocity measurement it is necessary to estimate the bias x_3 as well as x_1 and x_2. The equations for the observer are thus given by

$$\dot{\hat{x}}_1 = \hat{x}_2 + k_{11}r_1 + k_{12}r_2 \tag{6.44}$$

$$\dot{\hat{x}}_2 = -\Omega^2\hat{x}_1 - 2\zeta\Omega\hat{x}_2 + u + k_{21}r_1 + k_{22}r_2 \qquad (6.45)$$

$$\dot{\hat{x}}_3 = k_{31}r_1 + k_{32}r_2 \qquad (6.46)$$

where r_2 is the residual in the velocity measurement:

$$r_2 = y_2 - \hat{x}_2 - \hat{x}_3 \qquad (6.47)$$

A block-diagram representation of the observer implementation is given in Figure 6.14. With the paths associated with the velocity measurement omitted, this block diagram also shows the configuration of the observer with only the position zero-crossing measurement.

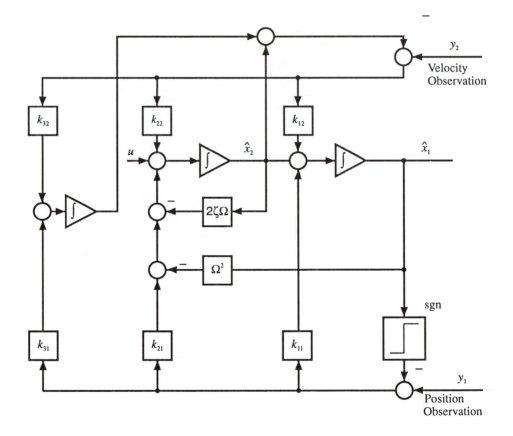

Figure 6.14   Observer for estimating the motion of an oscillating object using measurement of position zero-crossings and noisy, biased velocity measurements.

Both observers described above are based on having the model of the process and knowing the input to the process. An alternate observer that requires neither a knowledge of the process dynamics nor the input applied thereto is to regard the equation that defines the velocity measurement as the process dynamics:

$$\dot{x}_1 = y_2 - x_3 - w_1 \tag{6.48}$$

where the observation y_2 is treated not as an observation, but as a known quantity. This equation and (6.43) define the process dynamics. The observer resulting from this dynamic model is simply

$$\dot{\hat{x}}_1 = y_2 - \hat{x}_3 + \bar{k}_1 r_1 \tag{6.49}$$

$$\dot{\hat{x}}_3 = \bar{k}_3 r_1 \tag{6.50}$$

as shown in the block diagram of Figure 6.15. An observer based on this principle can perform quite well. (See Problem 6.8.)

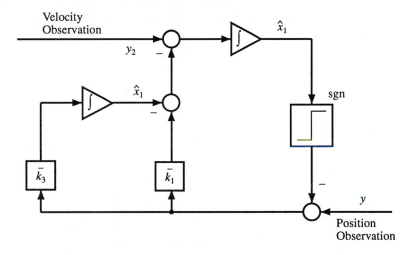

Figure 6.15 Block diagram of observer which uses velocity measurement in place of model of process dynamics.

6.2 REDUCED-ORDER OBSERVERS

The observer described in the previous section has the same dynamic order as the plant they observe, irrespective of the number of observations. In the absence of noise, however, the observations themselves can be used to determine some of the state variables, thereby reducing the number of state variables that must be included in the observer. If the number of (independent) observations is small relative to the order of the plant, the difference between a reduced-order observer and a full-order observer may be computationally insignificant. But when there are many observations relative to the order of the plant, the reduced-order observer may reduce computational requirements significantly.

More important than saving in computation, however, is the superior performance that can usually be achieved by use of a reduced-order observer. Intuitively, if a control law based on full-state feedback is optimal, the closer you come to full state feedback, the better. Hence, if some of the state variables can be measured directly, they should be used to implement the control law; it should not be necessary to estimate those state variables that can be measured. This intuitive reasoning is often borne out in practice. It usually turns out that a control law based on a reduced-order observer can be more robust than one using a full-order observer. A rigorous justification of this statement can be provided for a linear system [2]. In a nonlinear system it is more difficult to justify this statement, but experience appears to confirm it.

To develop the equations for a reduced-order observer, it is convenient to assume that the state variables are arranged in two groups: The first group comprises the state variables that are observed directly, and the second comprises those that are not observed at all. A further simplification is that the observed quantities are themselves the state variables in the first group. In equations, this all means that

$$x = \left[\begin{array}{c} x_1 \\ x_2 \end{array} \right] \tag{6.51}$$

where the observation is

$$y = x_1 \tag{6.52}$$

This is scarcely less general than the case

$$y = g(x_1, u)$$

provided that this expression can be solved for x_1 as a function of y and u:

$$x_1 = \psi(y, u) = \bar{y}$$

since we can use $\bar{y}$ as the observation. The completely general case in which y contains more state variables than its dimension can probably be handled in a manner similar to that used for linear systems [1]. The derivation is very tortuous in linear systems and is likely to be even more so in nonlinear systems, and is hence omitted.

Corresponding to the partitioning of the state vector x as in (6.51), the dynamic equations are written:

$$\dot{x}_1 = f_1(x_1, x_2, u) \tag{6.53}$$
$$\dot{x}_2 = f_2(x_1, x_2, u) \tag{6.54}$$

The nonlinear, reduced-order observer is assumed to have the same structure as the corresponding linear observer [1]: For the estimate of the substate x_1 we use the observation itself:

$$\hat{x}_1 = y \tag{6.55}$$

while the substate x_2 is estimated using an observer of the form

$$\hat{x}_2 = Ky + z \tag{6.56}$$

where z is the state of a dynamic system of the same order as the dimension of the subvector x_2 and is given by

$$\dot{z} = \phi(y, \hat{x}_2, u) \tag{6.57}$$

A block-diagram representation of the observer having the structure of (6.55–6.57) is given in Figure 6.16.

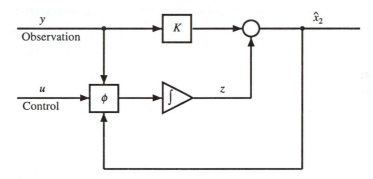

Figure 6.16   Reduced-order nonlinear observer.

The object of the observer design is the determination of the gain matrix K and the nonlinear function ϕ. As for the full-order observer, these are to be selected such that:

- The steady-state error in estimating x_2 converges to zero, independent of x and u. (The error in estimating x_1 is already zero when $\hat{x}_1 = y$.)

- The observer is asymptotically stable.

As in the case of the full-order observer, we proceed by deriving the differential equation for the estimation error

$$e = x_2 - \hat{x}_2 \tag{6.58}$$

Using (6.54), (6.56), and (6.53), we get

$$\begin{aligned}
\dot{e} &= \dot{x}_2 - \dot{\hat{x}}_2 \\
&= f_2(y, x_2, u) - K f_1(y, x_2, u) - \phi(y, x_2 - e, u) \tag{6.59}
\end{aligned}$$

In order for the right-hand side of (6.59) to vanish when $e = 0$, it is necessary that the function $\phi(\cdot, \cdot, \cdot)$ satisfy

$$\phi(y, x_2, u) = f_2(y, x_2, u) - K f_1(y, x_2, u) \tag{6.60}$$

for all values of y, x_2, and u.

To achieve asymptotic stability, the linearized system

$$\dot{e} = A(x_2)e \tag{6.61}$$

with

$$A(x_2) = (\partial\phi/\partial x_2) = (\partial f_2/\partial x_2) - K(\partial f_1/\partial x_2) \tag{6.62}$$

As in the case of the full-order observer, there are several techniques for selecting an appropriate gain matrix.

Example 6.5   Magnetic Levitation (Continued)

Assuming that the state variable x_1 can be directly observed, a reduced-order observer can be designed to estimate x_2 if g is treated as a known quantity, or to estimate x_2 and g. For the latter, following (6.56) and (6.57), the equations governing the observer are

$$\hat{x}_1 = y \ = \ x_1$$
$$\hat{x}_2 = z_1 + k_1 y$$
$$\hat{g} = z_2 + k_2 y$$

where

$$\dot{z}_1 = \phi_1 = -ku^2/\hat{x}_1^2 + \hat{g} - k_1\hat{x}_2$$
$$\dot{z}_2 = \phi_2 = -k_2\hat{x}_2$$

These equations are depicted in Figure 6.17.

The Jacobian matrix for this system is

$$A = \begin{bmatrix} -k_1 & 1 \\ -k_2 & 0 \end{bmatrix}$$

which, because x_1 is regarded as a known quantity and not a variable to be estimated, does not contain $\hat{x}_1$. The dynamics of the observer are thus linear and time-invariant (see Figure 6.17). Thus the reduced-order observer, in addition to having a simpler structure than the full-order observer of Example 6.3, is also dynamically simpler; its asymptotic stability is assured if the gains k_1 and k_2 are positive.

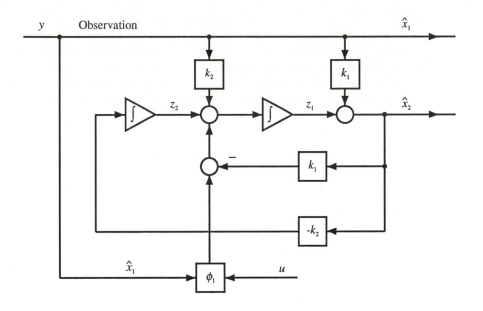

Figure 6.17   Reduced-order observer for magnetic levitation control.

6.3 EXTENDED SEPARATION PRINCIPLE

When an observer having the structure described above is used to estimate the state of a linear system, and the estimate is used in place of the actual state, the poles of the closed-loop system comprise the poles of the observer and the poles that would be present if full-state feedback were implemented. This is the "separation principle," which gives you the opportunity of designing two separate subsystems and then putting them together without worrying about the stability of the resulting closed-loop system. (But remember that this result holds only when the model of the plant used in implementing the observer is a faithful model of the physical plant.)

The separation principle of linear systems can be extended to nonlinear systems. Consider, in particular, the nonlinear system

$$\dot{x} = f(x, u) \tag{6.63}$$

for which a control law

$$u = \gamma(x) \tag{6.64}$$

has been designed. Use of (6.64) in (6.63) gives the closed-loop dynamics

$$\dot{x} = f(x, \gamma(x)) = F(x) \tag{6.65}$$

Assume that the closed-loop dynamics of the system with full-state feedback, as represented by $F(x)$ has been designed—by whatever method might be appropriate— to achieve satisfactory behavior. How will the process behave when the state $\hat{x}$ of an observer is used in place of the true process state x, i.e., when $u = \gamma(\hat{x})$? As earlier, let

$$\hat{x} = x - e$$

where e is the error in estimating the state. Then (6.64) becomes

$$u = \gamma(\hat{x}) = \gamma(x - e)$$

Then (6.65) becomes

$$\dot{x} = f(x, \gamma(x - e)) \tag{6.66}$$

which, together with (6.6) which now becomes

$$\dot{e} = f(x, \gamma(\hat{x})) - f(x - e, \gamma(\hat{x})) + K[g(x - e, \gamma(\hat{x})) - g(x, \gamma(\hat{x}))] \tag{6.67}$$

define the closed-loop dynamics.

There is not much that can be done with (6.66) and (6.67) for when f, g, and γ are general functions. But suppose these functions are sufficiently smooth to permit the use of Taylor's Theorem, i.e.,

$$f(x, \gamma(x - e)) = f(x, \gamma(\hat{x})) - (\partial f/\partial \gamma)(\partial \gamma/\partial e)e + O(e^2) \tag{6.68}$$
$$f(x - e, \gamma(\hat{x})) = f(x, \gamma(\hat{x})) - (\partial f/\partial x)e + O(e^2) \tag{6.69}$$
$$g(x - e, \gamma(\hat{x})) = g(x, \gamma(\hat{x})) - (\partial g/\partial x)e + O(e^2) \tag{6.70}$$

where $O(e^2)$ represents terms that go to zero as $\|e\|^2$. Then (6.66) and (6.67) become

$$\dot{x} = F(x) - (\partial f/\partial \gamma)(\partial \gamma/\partial e)e + O(e^2) \tag{6.71}$$
$$\dot{e} = [\partial f/\partial x + K(\partial g/\partial x)]e + O(e^2) \tag{6.72}$$

With the terms of $O(e^2)$ omitted, a block diagram representation of (6.71) and (6.72) is shown in Figure 6.18. Note that the equation for the error has no input from the state estimation. The error thus converges asymptotically to zero as in the linear case. Since the error is the input to the full-state feedback control system, if the latter is asymptotically stable, the effect of the estimation error vanishes.

6.4   EXTENDED KALMAN FILTER

In the foregoing examples the observer gains were selected merely to maintain ob- server stability and to provide reasonable dynamic response. This method of selecting

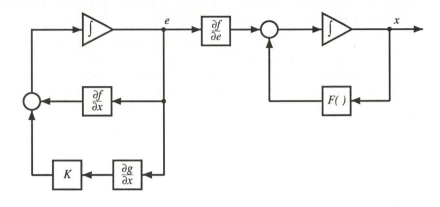

Figure 6.18 Illustration of extended separation principle.

the observer gains is usually adequate. There are two cases, however, when it is necessary to use a more systematic method. The first case is that in which there is no obvious method for selecting the observer gains. This situation usually arises when more than one quantity is observed, and hence when there are more terms in the observer gain matrix than are needed to establish the observer poles. The second case arises when it is important that the state estimate produced by the observer be as accurate as possible, and where the statistical properties of the noise in the measurements can be determined. For these cases it is often appropriate to make the observer an extended Kalman filter, i.e., to calculate the gain matrix on-line from the concurrent solution of the variance equation of the Kalman filter.

The estimation theory developed by R.E. Kalman is concerned with a linear process:

$$\dot{x} = Ax + Bu + Gv \tag{6.73}$$

$$y = Cx + w \tag{6.74}$$

where x is the process state, y is the observation, u is a known input, and v and w are white noise processes having spectral density matrices V and W, respectively.

The fundamental result of Kalman filter theory is that the optimal observer, in the sense of minimizing the expectation of the estimation error (and also in several other senses), has the form of a linear observer:

$$\dot{\hat{x}} = A\hat{x} + Bu + K(y - C\hat{x}) \tag{6.75}$$

in which the gain matrix K is given by

$$K = PC'W^{-1} \tag{6.76}$$

(when u and v are uncorrelated), with the covariance matrix P of the estimation error being the solution to the "variance equation"

$$\dot{P} = AP + PA' - PC'W^{-1}CW + GVG' \tag{6.77}$$

(When v and w are correlated, the relationship for the gain matrix involves the cross-spectral density matrix X as discussed in [1], for example.)

Few of the many applications for which Kalman filters have been used have met the linearity requirements of the theory. But nevertheless the theory has been successfully, if not rigorously, applied. This is done by using a nonlinear observer of the form of (6.7), but with the gain matrix K therein being computed, along with the state estimate, using (6.76) and (6.77). The matrices A and C in these equations are the Jacobian matrices of the dynamics and observations:

$$A = [\partial f / \partial x]_{x=\hat{x}} \tag{6.78}$$
$$C = [\partial g / \partial x]_{x=\hat{x}} \tag{6.79}$$

for the nonlinear process defined by

$$\dot{x} = f(x, u) + Gv \tag{6.80}$$
$$y = g(x, u) + w \tag{6.81}$$

—the white noise processes v and w being regarded as additive.

In the linear case, the error covariance matrix and through it the gain matrix K do not depend on the estimated state. In principle, these matrices can be computed before the filter is implemented and stored in the filter's memory. In the nonlinear case, however, the matrices A and C that are used in computing P and K depend on the state estimate. Hence, in the extended Kalman filter, the observer and the Kalman filter gain matrix computation are coupled. This means that the equations for both the variance equation and the observer must be implemented on-line as shown schematically in Figure 6.19.

The requirement for on-line computation of the extended Kalman filter gain matrix can be a computational burden. Even considering that the covariance equation P is symmetric, there still are $k(k-1)/2$ scalar differential equations in (6.77) that must be integrated numerically in addition to the k scalar observer equations for a kth-order dynamic process. In a tenth-order process, for example, a total of 55 equations must be integrated. It doesn't take a process of much higher order to overwhelm even a supercomputer. Moreover, the matrix Riccati equation (6.77) for P is notorious for being poorly behaved. Unless special measures are taken, the numerical solution to (6.77) is likely to lose its positive- definite character as the theory requires. If this happens, the resulting state estimate $\hat{x}$ will probably be useless.

Fortunately, it is rarely necessary to be a stickler for accuracy in the implementation of (6.77). In the first place, the entire theory of the extended Kalman filter is only approximate. Moreover, the numerical values of the spectral density matrices V and W that appear in (6.77) are hardly ever known to better than an order of magnitude.

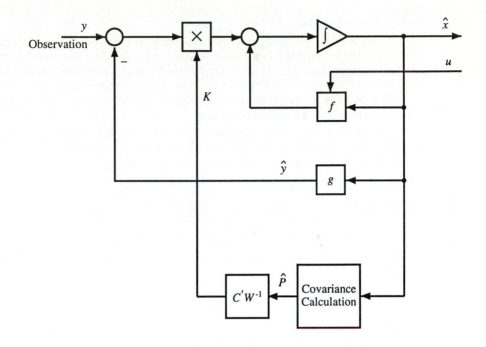

Figure 6.19   Schematic of extended Kalman filter, showing coupling between state estimation and covariance computation.

Hence any computational method that gives a reasonable approximation to P and K is usually acceptable. You might consider the following approximation techniques:

- Regard P as being piecewise-constant and compute it relatively infrequently, using the discrete-time version of the Kalman filter. (See Chapter 9.)

- Simulate the observer and (6.77) off-line; examine the results and use whatever approximations seem suitable. It may be possible, for example, to approximate some of the gains by constants. The effect of the approximations must be evaluated by further simulation.

- Use a simpler model to represent the process in (6.77) than is used as the process model in the observer. But be careful not to use an overly simple model in the implementation of the process dynamics.

6.5   NOTES

Note 6.1   Extended Kalman Filter

The Kalman filter is arguably the most important development in control theory of the second half of the twentieth century. Not only has this theory influenced much of the analytical methodology of modern control theory, but it is also the theoretical development that has been most widely used in practice. Aerospace applications are particularly notable and include trajectory estimation of satellites and missiles, space vehicle guidance and navigation, and aircraft navigation.

Because most practical processes entail nonlinear dynamics, most applications employ the extended Kalman filter outlined in Section 6.4. An account of the background, theory, and a number of applications is contained in [3].

6.6   PROBLEMS

Problem 6.1   Reduced-order observer for temperature control

Consider the temperature control problem of Example 6.1 . Design a reduced-order observer for the process and compare the performance of the closed-loop control system with that obtained by using the full-order observer as shown in Example 6.1 .

Problem 6.2   Temperature control with saturation

Using the Describing Function Method described in Chapter 3, analyze the limit cycle (estimate its amplitude and frequency) of the temperature control problem of Example 6.1 . Compare the results of the describing function analysis with the simulation results.

Problem 6.3   Temperature control with deadzone

Repeat the analysis of Problem 6.2, except using a deadzone rather than a saturation nonlinearity.

Problem 6.4   Limit cycle in linear control of robot arm

Consider the robot arm of Example 5.3 with saturating input and a linear compensator.

(a) Find the transfer function of the compensator using the data given in the example.

(b) Using this transfer function and that of the plant, plot the Nyquist diagram of the forward loop.

(c) Determine how well the describing function method predicts the existence, amplitude, and frequency of the limit cycle.

Problem 6.5   Motor-driven robot arm

Design a full-state observer for the motor-driven robot arm for which full-state feedback control laws were designed in Problem 5.6. Simulate the performance for each of the full-state control laws considered in Chapter 5.

Investigate the effect of saturation at the level $u_{max} = 100$.

(a) If the saturation is omitted from the observer

(b) If the saturation is included in the observer

Problem 6.6 Motor-driven robot arm

Design a reduced-order observer for the motor-driven robot arm for which a full state observer was designed in Problem 6.5. Assume that the angle θ can be measured with negligible error. Simulate the performance for each of the full-state control laws considered in Chapter 5.

Investigate the effect of saturation at the level $u_{max} = 100$.

(a) If the saturation is omitted from the observer

(b) If the saturation is included in the observer

Problem 6.7 Observer for oscillating system using zero crossings

Set up a simulation of the oscillating system of Example 6.4 and verify the results presented. Investigate the effects of different levels of noise.

Problem 6.8 Reduced-order observer using velocity as known input

Investigate the performance of the reduced-order observer defined by (6.48), (6.49), and (6.50) for estimating the position of an oscillating object. Use the same numerical data as used in Problem 6.7.

Problem 6.9 Magnetic levitation

Find the range of gains for which the observer in Example 6.3 is asymptotically stable.

Problem 6.10 Navigation of autonomous vehicle

Consider the navigation system for the autonomous vehicle of Example 1.2. Design an observer to estimate the state variables, using the following sensors:

1. Shaft encoders, which measure the total angular rotation of each of the wheels of the vehicle.

2. A rate gyro, which measures the angular velocity of the platform about a vertical axis.

3. A range sensor, which measures the range (distance) of a point on the moving platform to a fixed reference point.

(a) Is the state observable with the following combinations of sensors?

(i) [1 and 2], (ii) [1 and 3], (iii) [2 and 3]

(b) Design an observer using all three sensors. Select appropriate gains and simulate the performance of the observer.

6.7 REFERENCES

[1] B. Friedland, *Control System Design: An Introduction to State Space Methods*, McGraw-Hill, New York, 1966.

[2] B. Friedland, "On the Properties of Reduced-Order Kalman Filters," IEEE Trans. on Automatic Control, Vol. 34, No. 3, pp. 321–324, March 1989.

[3] H.W. Sorenson (ed.) *Kalman Filtering: Theory and Applications* IEEE Press, New York, 1985.

Chapter 7

PARASITIC EFFECTS

Often present in real-world control systems are effects that are troublesome to account for when the systems are designed. Rarely disabling, but debilitating if if not dealt with effectively, these effects can be called "parasitics" Such effects can include:

- Nonlinearity
- Noise
- Friction
- Backlash
- Hysteresis
- Unmodeled resonances
- Limit stops
- Time delay

These effects are present to some extent in most systems. Whether they are significant and, if so, what to do about them is often the question. The most optimistic course is simply to ignore them and hope for the best. Often good fortune will be on your side. But Murphy's Law says you ignore them at your peril.

A more prudent approach is to include the parasitic effects that you think might be troublesome in the simulation model (i.e., the "truth model") of the plant and run the simulation enough times to convince yourself that they are really negligible. If, to the contrary, you find that they are not negligible and can adversely affect system performance, you need to do something about them. But what?

The most appropriate remedies for deleterious effects of parasitics are usually tailored to the specific parasitics, but there are several general measures that can be helpful.

In designs based on the extended separation principle, as presented in Chapters 5 and 6, it is beneficial to include a model of the parasitic effect in the observer of the plant. The estimate of the plant state produced by an observer that includes the parasitic effects is likely to be closer to the true plant state than an estimate by an observer that omits them. Hence the control law design based on knowledge of the true state is more likely to perform as designed.

Another technique that is often effective is to alter the design to reduce the effective loop gain of the compensator, especially at high frequencies where the effects of parasitics are often predominant. Reduction in loop gain generally entails a sacrifice in performance. But this might be an acceptable price to pay for avoiding other problems due to the parasitic effects.

Techniques specifically intended to enhance the robustness of the design are also likely to be effective but may entail use of a more complicated control algorithm.

The remainder of this chapter addresses some common types of parasitic phenomena and illustrates how some of the techniques described above can be used to deal with them.

7.1 NONLINEARITY

As we have seen in Chapter 4, many processes are inherently nonlinear in the sense that their dynamics include products of state variables, "hard" nonlinearities (e.g., saturation), and similar operations that require the use of design methods specifically developed for such nonlinear systems as have been discussed in the last three chapters.

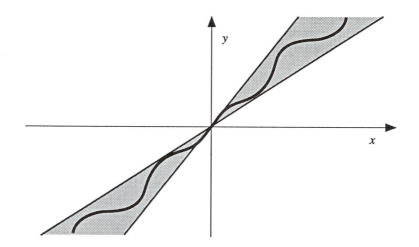

Figure 7.1 Nonlinearity typical of imperfect operation of any physical device.

Another type of nonlinearity frequently encountered in practical control systems arises because the physical world does not generally conform to ideal models. The difference between the ideal model and the real device often has the form of a nonlinear function. No real sensor, for example, is perfect. The quality of the sensor (and its cost) may be characterized by a number of parameters, including "linearity" and the range over which the specified linearity is maintained. A manufacturer's warranty that the linearity of a device is ±10 percent, ensures that the output of device will not deviate from the output of an ideal sensor by more than 10 percent over the range for which the instrument is certified. This means that if the ideal output y of the device is a constant times the input:

$$y = cx$$

the 10-percent linearity specification guarantees that the output of the sensor will lie in the sector defined by

$$0.9cx < y < 1.1cx$$

as shown in Figure 7.1. While no representation is made about its behavior within the sector, a typical instrument can be expected to have a fairly smooth nonlinear characteristic: a "droop" or a rise at the end of the operating range. Since there is usually no assurance that the characteristic is smooth, however, you must convince yourself that the design will be satisfactory for all variations that conform to the specifications. In particular, a nonlinear characteristic that meets a particular linearity specification may have a slope change in the operating range.

The effect of the static nonlinearities of the type described in this section can be assessed effectively by use of the circle criterion discussed in Chapter 3. To apply the circle criterion, start by computing the loop transfer function around the nonlinearity, i.e., from its output, through the system, back to the input. Then plot the Nyquist diagram corresponding to the transfer function. Assuming that the design is asymptotically stable (and the nominal gain of the nonlinearity is unity), the Nyquist diagram will not cover (encircle) the point $-1 + j0$. For a static nonlinearity specified to lie in the sector bounded by $[1 - \epsilon, 1 + \epsilon]$, draw the circle that passes through the points $-1/(1 - \epsilon)$, $-1/(1 + \epsilon)$. (For small ϵ, this circle is approximated by one of radius ϵ centered at -1.) If the Nyquist diagram does not pass through the circle, the design will remain asymptotically stable in the presence of the nonlinearity. Since the circle criterion is conservative, the system may be stable even if the Nyquist diagram does pass through the circle. But it's imprudent to be optimistic: better to seek a device with less nonlinearity or to design the compensator that yields a Nyquist diagram that avoids the critical point $-1 + j0$ by a larger margin.

The use of analog-to-digital converters for digital-computer control introduces a similar form of static nonlinearity: *quantization*. (See Figure 7.2.) The nonlinearity of quantization differs from a sector-bounded nonlinearity, however, in an important way. Whereas the latter passes through the origin, the output of a quantizer can range over $[-q, 0, q]$, where q is the quantization level, when the input is zero, depending on the detailed implementation of the quantizer and random effects (noise on the input of the magnitude of the quantization level, for example). The only sector that is guaranteed to bound the quantization nonlinearity is the entire first and third quadrant. To assure stability for this sector, the circle criterion would require the Nyquist diagram to remain to the right of the imaginary axis, an unrealistically severe requirement. The effect of quantization can be assessed by simulation. The possibility of limit cycles can be studied by the Describing Function Method discussed in Chapter 3.

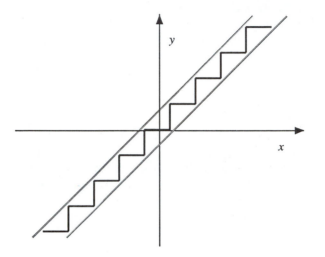

Figure 7.2   Nonlinearity due to quantization.

7.2   NOISE

Many signals occur in physical processes that cannot be predicted in advance. Informally, such signals can be regarded as noise. They may occur either as exogenous inputs to the plant or as extraneous signals in the outputs of sensors.

Theoretical treatments of control systems are generally based on the assumption that the statistical characteristics of the noise signals are known. For convenience, they are often assumed to be Gaussian noise with given spectral density functions. But in practice the effort and expense necessary to gather sufficient empirical data to verify the statistical assumptions about the noise is usually not warranted.

Moreover, "noise" in practice is any spurious signal, regardless of its statistical characteristics. For example, one of the most common types of noise is the result of the a-c power used to operate the system or other systems in its vicinity. The power supply frequency (60 Hz in the U.S., 50 Hz elsewhere) and its harmonics can be a source of noise the effects of which can be minimized, but not entirely elimininated, by careful electrical design. This frequency is well above the operating bandwidth of many control systems and thus may have a negligible effect on their performance. But in high bandwidth systems the effect of power system harmonics can be significant.

Quantization, discussed earlier as a form of nonlinearity, is sometimes regarded as a form of noise.

Other spurious signals may be by-products of the operation of sensors in the system: Some sensors are designed on the principle of modulation of a high-frequency carrier by the low-frequency information relevant to the control system. The operations of modulation and subsequent demodulation are never ideal. The resulting errors constitute noise at the modulation frequency and its harmonics. In a well-designed system, you choose the carrier frequency to be much higher than the operating bandwidth of the control system, and the effects of modulation and demodulation are negligible. In some applications, however, you may want to "push" the performance of the system to its limits, for example, to achieve the highest bandwidth possible with a given sensor. In such applications, the spectrum of the noise due to modulation may not fall far outside the operating bandwidth and it may be necessary to examine its effect and to deal with it, if necessary.

For simulating the effect of noise, a model is needed. The most accurate model would be based on the underlying physical principles of the process that produces the noise. To simulate the effects of modulation and demodulation accurately, for example, you could use a mathematical representation of the actual operations of modulation and demodulation. In most situations, however, it would be impractical to develop a detailed physical model for each source of noise. The physical principles may be too complicated or not well enough understood to be represented mathematically. And even when it is possible to develop an accurate model of the noise process, doing so simply may not be worth the time and effort. Fortunately it is rarely necessary to have very high-fidelity noise models; approximate models that capture the essential characteristics of the noise are usually adequate.

In many situations, a noise source can be adequately represented as the response of a low-order linear system to white noise, as illustrated in Figure 7.3. The transfer function of the linear system, which can be called a *noise shaping filter* is adjusted to produce a spectrum that approximates the observed spectrum of the noise. For narrow-band noise caused by the power supply or the by-product of modulation and demodulation, a narrow-band filter

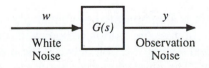

Figure 7.3 Representation of observed noise as output of linear system excited by *white* noise.

$$G(s) = \frac{1}{s^2 + 2\zeta\Omega s + \Omega^2} \tag{7.1}$$

would be appropriate, with Ω set to the frequency of the noise and the damping factor ζ set to some small value, perhaps in the range of .1 to .01.

It is appropriate to represent each physical source of noise by its own white noise source and shaping filter.

If simulation studies with these simplified noise models reveal possible deleterious effects, the problems can often be handled by incorporating the noise model into the design model of the plant. If the plant in the absence of noise is linear (or linearized for purposes of design), the design model that includes the noise will also be linear and the compensator design can be addressed by linear, quadratic, Gaussian (LQG) theory such as are presented in *Control System Design* [1]. The application of LQG theory will result in a design that will tend to attenuate the effect of the noise. In particular, if the narrow-band noise, represented by (7.1), is included in the model, the compensator will have a transfer function with a "notch" at the frequency of the noise to avoid amplifying the noise in the closed-loop system. The depth of the notch will depend on the assumed intensity of the noise. Including a model of the noise in the design model has an advantage over the use of an ad-hoc notch filter. The latter may affect the stability of the closed-loop system, whereas a compensator design that explicitly includes the noise model guarantees (local) stability of the closed-loop system. (See Problem 7.2.)

7.3 FRICTION

Friction is essential for the operation of common mechanical systems (wheels, clutches, etc.). But in most control systems it is a nuisance: It is difficult to model and troublesome deal with in control system design. Manufacturers of components for precision control systems take pains to minimize friction and charge dearly for their exertions. Notwithstanding efforts at reducing friction, however, its problems remain and it is necessary to contend with them in precision control systems. The possible undesirable effects include "hangoff" and limit cycling. The former phenomenon prevents the steady-state error from becoming zero with a step command input. The latter is behavior in which the steady-state error oscillates or hunts about zero. (Hangoff can be interpreted as a d-c limit cycle.)

It is generally accepted that friction is a complicated nonlinear phenomenon in which a force is produced that tends to oppose the motion of bodies in contact (rubbing against each other) in a mechanical system. Motion between the bodies causes the dissipation of energy in the system. The physical mechanism for the dissipation depends on the materials of the rubbing surfaces, their finish, the lubrication applied, and other factors, many of which are not yet fully understood. (Much of the research on friction in the field of tribology is concerned with the reduction of friction through improved lubrication and other methods.)

As a control engineer your concern is not reducing friction, but dealing with friction that cannot be reduced. At the very least, you should appraise the effect of friction in a proposed control system design by simulation. You may find that the effect (i.e., hangoff, limit cycling) is tolerable and does not require taking any special measures. But if the simulation predicts that the effect of friction is unacceptable, you must do something about it. Remedies can include simply modifying the design parameters (gains) or more complex measures such as using integral control action or estimating the friction and cancelling its effect. Several methods are described below.

7.3.1   Modeling Friction

The classical model of friction is a force (or torque) that depends on relative velocity and acts instantaneously in the direction that opposes motion. At its simplest, it is represented by the nonlinear function

$$f(v) = a \, \mathrm{sgn}(v) \tag{7.2}$$

where v is the (linear or angular) relative velocity and $f(v)$ is the corresponding force or torque. In this model, known as classical *Coulomb* friction, and depicted in Figure 7.4(a), the magnitude of the friction force or torque is constant. Physical experiments, however, have shown that in many cases the force required to initiate relative motion is larger than the force that opposes the motion once it starts. This effect is sometimes called "stiction." Modeling stiction effects is accomplished by use of a nonlinearity of the form shown in Figure 7.4(b). If viscous effects are included, the force of friction will increase with velocity after initially dropping because of stiction, with a resulting nonlinear model having the form shown in Figure 7.4(c). An empirical formula sometimes used for expressing the dependence of the friction force upon velocity is

$$f(v) = (a - b e^{-c|v|} + d|v|) \, \mathrm{sgn}(v) \tag{7.3}$$

in which the parameters a, b, c and d are chosen to impart the desired shape to the friction function. The dip and subsequent rise in the friction force with increasing velocity is sometimes called the "Stribeck" effect [2].

The model of (7.3) is satisfactory for representing the behavior of a process that operates in the vicinity of a nonzero reference velocity. But the discontinuity at the origin is problematic when the velocity of the process crosses zero frequently, as it does if the process exhibits a tendency to oscillate. The problems are both theoretical and practical. The discontinuity at the origin is distasteful for physical reasons, since it suggests an infinite rate-of-change of force or torque, hence an infinite acceleration. From the practical standpoint there are several reasons why the discontinuous model is less than satisfactory: In simulation, the discontinuous force creates a system with infinite stiffness which is not readily amenable to many popular variable-step-size numerical integration algorithms with error control, since the step size may be reduced to a tiny number in an effort to pass the error test. Moreover, even if the numerical integration is not a problem, the simulation may lack verisimilitude: It may predict

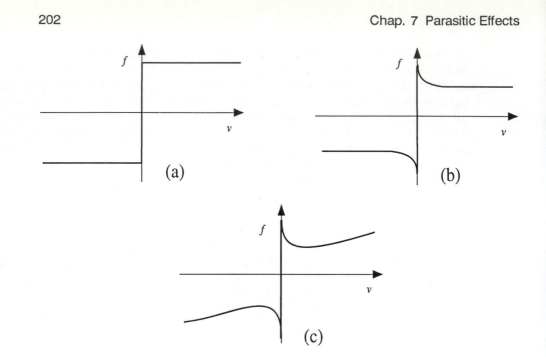

Figure 7.4   Representations of static friction.   (a) Classical Coulomb friction.   (b) Representation of "stiction."   (c) Including viscous effects.

dynamic behavior that is qualitatively different from that of the actual process being simulated—it may predict a limit cycle, for example, that is absent in the actual system [3] or fail to predict one that actually does occur.

To overcome the problem of the discontinuity at the origin, the model of (7.2) is usually modified to have a small linear region in the neighborhood of the origin, i.e., the friction force is represented by

$$f^*(v) = \begin{cases} kv, & |v| < 1/k \\ f(v), & |v| > 1/k \end{cases} \tag{7.4}$$

where the width $1/k$ of the linear region is made large enough to avoid difficulties with numerical integration but small enough not to change the results significantly. While this procedure may overcome numerical integration problems it may not overcome the lack of verisimilitude. Several ways to avoid the lack of verisimilitude are described in the technical literature. Karnopp [4] proposed a technique that entails reducing the order of the system when the rubbing members of a system do not exhibit any relative motion. This technique agrees well with our understanding of the physics and is efficient in the use of computer time, but it is difficult to program.

Considering the possibility that the effect of friction at very low relative velocity may qualitatively appear as an elastic effect, Dahl [5] proposed a dynamic model such as shown in Figure 7.5 to represent friction in ball bearings. This model has been used successfully in various studies since its introduction. Similar considerations motivated Haessig and Friedland [3] to introduce the "reset integrator" model, shown in Figure 7.6, which gives results similar to the Dahl model. The Dahl and the reset integrator models entail *increasing* the order of the system and hence are less efficient in the use of computer time than the Karnopp model.

The presence of the integrator in the forward loop of the Dahl and the reset integrator models makes these models dynamic. If the integrator in the reset integrator model is regarded as a static element with a very high gain, its effect is to invert the nonlinearity in the feedback path. (See Problem 7.1.) Hence, the static input-output characteristic of the reset integrator model agrees with the classical model as the slope k of the nonlinear element in the feedback path tends to infinity. On the other hand, for very small oscillatory velocity, the friction force in the reset integrator model depends locally on the integral of velocity, which may agree better with an intuitive conception of how friction should behave. The Dahl model exhibits similar behavior.

To summarize: There are now several different dynamic models all of which are apparently superior to the classical static models. Which of the models, if any, is the best representation of the true physics of friction remains an open question. Interest in this question [6], [15] is increasing as the problem of friction becomes acute with higher performance goals for various applications. Optical techniques now make it possible to measure distances with submicron accuracy, and there is a recognized need for precise position control at this level of accuracy. (One applications is advanced semiconductor manufacturing.) Modeling and simulation of friction can be expected to contribute significantly to improving the performance of motion control systems.

7.3.2   Friction compensation

An early method of avoiding the hangoff effect of friction has been to inject noise ("dither") as an input to the system at the location at which the friction force or torque appears. The dither keeps the velocity oscillating around zero and tends to make the average torque due to friction small. While this method can be effective, it tends to create oscillation in exchange for hangoff. In most applications neither is desirable, and other methods of dealing with friction are preferable.

If the force due to friction is known, and smaller in magnitude than the control force, one attractive possibility is to use a portion of the control force to cancel the friction. The problem with this approach, however, is that the force due to friction is rarely known. The parameters of friction models (even the improved models) depend on factors that are all but impossible to predict *a priori*. Hence cancelling the friction will always require estimating the friction.

One method of simultaneously estimating and cancelling friction is based on the premise that the friction force is constant in magnitude, and switches sign at the in-

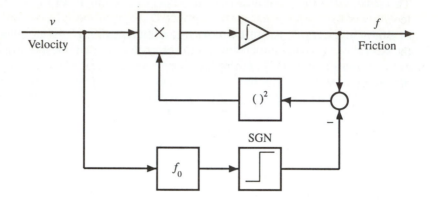

Figure 7.5   Dahl model of friction.

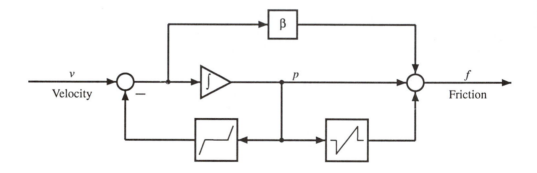

Figure 7.6   Reset integrator model of friction.

stants that the velocity crosses zero. These instants of course depend on the friction
and the dynamics of the entire system in which the friction is present. But, if we suppose that the zero crossings are a random process independent of the dynamic process,
the effect of friction resembles that of a random square-wave disturbance, which, in
turn, can be approximated as a Wiener process—the output of an integrator excited
by white noise. This chain of assumptions and approximations leads to the conclusion
that the effect of friction is represented by a disturbance force that is the output of an
integrator excited by white noise [8]. The level of the white noise is proportional to the
amount of friction. When this disturbance model is accounted for in the compensator
design, the result is that the compensator will have integral control action, with the
gain in the integral control path being proportional to the supposed level of friction.

This result is hardly surprising to veterans in the field: Integral control is a venerable technique for countering the effects of friction.

Integral control has its limitations, however, because it usually results in reducing the system bandwidth and stability margins. Linear observers based on other considerations were reported by Brandenburg [9], [10]. Brandenburg showed the benefits of these observers by use of describing function analysis.

A more direct method of estimating and compensating for friction is based on the use of a reduced-order observer to estimate the level of friction as expressed by a parameter in a postulated friction model, and then to cancel the estimated friction by a term in the control [11]. Suppose, for example, that the friction is represented by the simple classical model of (7.2) in which a is an unknown parameter. An observer for undermined coefficient a in the friction model is postulated in the form of a nonlinear reduced-order observer, namely:

$$\hat{a} = z - k|v| \tag{7.5}$$

where the variables z is given by

$$\dot{z} = k\mathrm{sgn}(v)(w - f(v, \hat{a})) \tag{7.6}$$

The gain k is determined experimentally.

A block-diagram representation of the observer is shown in Figure 7.7. A more general form of this observer for estimation of parameters of dynamic systems is considered in Chapter 10, where it is shown that the error in estimation of the parameter a can be made to converges to zero.

Having estimated the friction coefficent, you have an estimate of the friction force (or torque). If the total force (or torque) that can be supplied by the actuator is large enough, you can add a term that cancels the estimated friction. In other words, the control force (or torque) would be computed using the formula:

$$u = \tilde{u} - u_f \tag{7.7}$$

where $\tilde{u}$ is the control in the absence of friction and u_f is the control needed to cancel the friction estimated by the observer.

The above method for estimating the friction coefficient depends on the knowledge of the velocity. It was shown subsequently, however, that the velocity can be estimated concurrently [12] using another reduced-order observer if only the position can be measured. Moreover, it was shown that the method can be applied to the estimation of the coefficients of friction between different masses in a multi-mass system [13].

This approach to cancellation of friction is still experimental, although early results have been favorable. Simulation results have shown significant improvement in performance under the assumption that the "truth model" is expressed by (7.2) and favorable results have been obtained in the laboratory in a simple application [14]. But, as argued above, the classical zero-memory model of (7.2) may have inadequacies. How well this approach works with more faithful representations of friction, and of course with real-world friction remains to be more fully investigated.

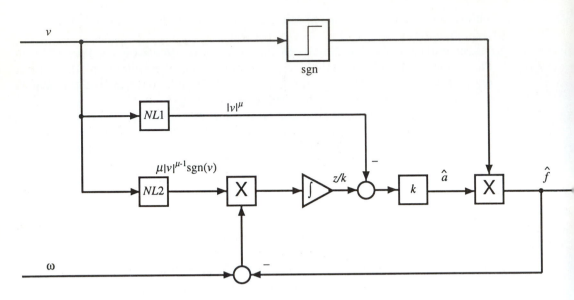

Figure 7.7   Friction observer.

Other methods of estimating friction have been proposed and demonstrated. Canu-das de Wit, *et al.* [15], in particular, displayed impressive results using a maximum likelihood method of friction estimation.

The availability of a variety of methods for estimating and compensating for friction is encouraging: It suggests that friction in control systems is a problem that you can really do something about.

7.4 BACKLASH

Gears and similar drive systems generally exhibit an effect called "backlash." This effect is illustrated in Figure 7.8, a geared drive system. Suppose that the actuator is a d-c motor connected to the drive gear and that the driven gear is connected to an inertial load. The equations of motion of the system are

$$J_1 \ddot{\theta}_1 + \tau = K i_m \tag{7.8}$$
$$J_2 \ddot{\theta}_2 - \tau = 0 \tag{7.9}$$

where τ is the torque transmitted through the gears and i_m is the motor current.

When the gears are not engaged, the torque τ is zero; (7.8) and (7.9) are uncoupled. When the gears are engaged, however, the motion is constrained such that $\theta_2 = \theta_1$ and the torque τ is whatever value it must be to maintain the constraint.

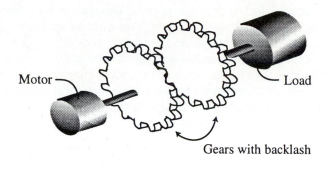

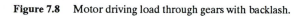

Figure 7.8   Motor driving load through gears with backlash.

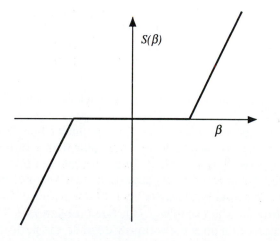

Figure 7.9   Nonlinear function used in representation of back-lash.

Once the teeth are in contact, they will remain in contact until the relative motion of the gears changes in direction.

It is difficult to translate this physical description into differential equations, because the order of the system depends on whether or not the gears are engaged: If they are not engaged the system has two degrees of freedom and is fourth-order; it they are engaged, the system has only one degree of freedom and the system is second-order. This paradox can be resolved by representing the effect of the gears by a highly nonlinear spring, in which the torque is near zero for small angular displacements and becomes very large when the relative displacement exceeds the backlash level b, as shown in Figure 7.9. Mathematically

$$\tau = f(\beta) = \begin{cases} 0, & |\beta| < b \\ S(\beta - b), & \beta > b \\ S(\beta + b), & \beta < -b \end{cases} \tag{7.10}$$

where $\beta = \theta_1 - \theta_2$ is the "backlash angle."

Equations (7.8) and (7.9) can be rewritten in terms of β and the nonlinear function $f(\beta)$ as

$$\ddot{\theta}_1 = -\frac{1}{J_1}f(\beta) + \frac{K}{J_1}i_m \tag{7.11}$$

$$\ddot{\theta}_2 = \frac{1}{J_2}f(\beta) \tag{7.12}$$

The differential equation for β is obtained from (7.11) and (7.12):

$$\ddot{\beta} = -\frac{1}{J}f(\beta) + \frac{K}{J_1}i_m \tag{7.13}$$

A block-diagram representation of backlash, using (7.8) and (7.10), is shown in Figure 7.10.

As the slope S of the spring in the nonlinear function $f(\beta)$ tends to infinity, the backlash angle is constrained to be smaller than b in magnitude. For purposes of simulation, however, a very large but finite value of S may be used. Note that (7.13) defines the dynamics of a nonlinear oscillator. The presence of the oscillation can be confirmed in actual systems, and its frequency ω_B can be measured. (In view of the nonlinearity, ω_b is not simply $\sqrt{S/J}$. See Problem 7.4.)

Backlash often does not seriously degrade the performance of a system in which it is present, particularly if the bandwidth of the system is much less than the gear resonance frequency ω_b. If the application requires that the closed-loop bandwidth be relatively high, on the other hand, the presence of the backlash can be troublesome. To mitigate (but not to eliminate) the problem of backlash, spring-loaded split gears are sometimes used. The two gears are arranged so that one or the other is always in contact with the input gear and the two gears are connected to each other with a spring. The effect of this arrangement is to change the slope of the backlash function

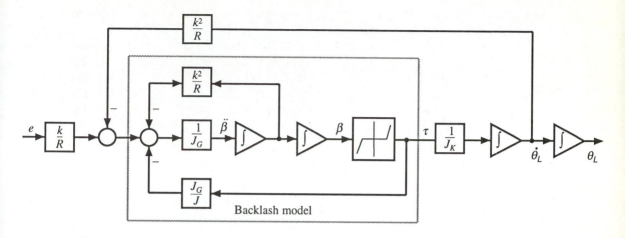

Figure 7.10   Schematic representation of backlash.

$f(\beta)$ for $|\beta| < b$ from zero to some nonzero value. Since spring-loaded "anti-backlash" gears tend to be relatively costly, it may be worthwhile to simulate the performance enhancement such gears would achieve.

The phenomenon of backlash illustrates how complicated the accurate representation of a nonlinear physical phenomenon can be. Moreover, the representation of Figure 7.10 is itself only an approximation; a more accurate representation would also include friction between the gears when they are not in contact and between the gears and the supporting structure.

7.5   HYSTERESIS

Hysteresis is a nonlinear phenomenon occurring in many physical systems in which the output depends not only on the input, but also on whether the input is increasing or decreasing, as shown in Figure 7.11. Perhaps the most familiar example of hysteresis is ferromagnetism. In this phenomenon, the relationship between flux density and current exhibits the behavior depicted by Figure 7.11.

Whether a function is increasing or decreasing is determined by its time derivative. Hence hysteresis is a nonlinear function of not only the input, but also its derivative. This dependence is typically expressed by

$$y = g(x, \dot{x}) = \begin{cases} f_1(x), & \dot{x} > 0 \\ f_2(x), & \dot{x} < 0 \end{cases} \tag{7.14}$$

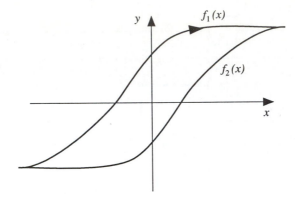

Figure 7.11   Conventional representation of hysteresis.

Usually the hysteresis function is symmetric, i.e.,

$$f_1(x) = f(x - a)$$
$$f_2(x) = f(x + a)$$

Note that the representation of hysteresis by (7.14) makes the output a discontinuous function of velocity $\dot{x}$ and introduces the same types of problems that arise in the case of friction, namely the physical unacceptability of any quantity in nature changing discontinuously, and the practical difficulty that numerical integration methods may have in dealing with discontinuous variables. To avoid both difficulties, as in the case of friction, you might postulate that hysteresis is governed by an underlying differential equation. One method of representing symmetrical hysteresis is by a system of the form depicted in Figure 7.12. It can be established by simulation that the input-output behavior of the system approaches the hysteresis curve of Figure 7.11 as the loop gain K becomes infinite. The reset integrator model of friction has the form of Figure 7.12 suggesting that friction may exhibit a hysteresis effect. Similarly, the dynamic backlash model of (7.13) exhibits the hysteresis phenomenon.

The deleterious effects of backlash are often manifested as a limit cycle of the closed-loop system. One possible remedy is to reduce the loop gain and thereby the bandwidth of the closed-loop system. To maintain system bandwidth in the presence of hysteresis (or backlash) you might consider including a model of the backlash in an observer of the process used in the implementation of the control algorithm.

7.6 UNMODELED RESONANCES

Accurate modeling of the dynamic behavior of mechanical systems will result in a dynamic system of higher order than you probably would want to use for the design model. A shaft that connects a drive motor to a load, for example, has finite compli-

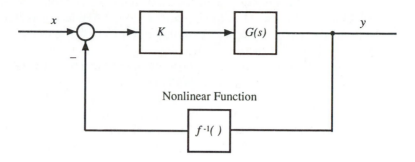

Figure 7.12   Dynamic representation of hysteresis.

ance. If you are realistic and don't assume that the stiffness is infinite, your model will include a resonance at a frequency $\Omega = \sqrt{k/J}$ where k is the spring constant and J is the effective inertia. If you are more realistic, and represent the shaft by the correct partial differential equation, you will get an infinite number of resonances.

A gear train, such as described in Section 7.4, is another example of a resonance phenomenon—in this case a nonlinear resonance.

In most situations, the frequencies of these resonances will be orders of magnitude above the operating bandwidth of the control system and there will be enough natural damping present in the system to prevent any trouble. Difficulties can arise, however, in applications that require the system to have a bandwidth that approaches the lowest resonance frequency. A control system based on a design model that does not account for the resonance may not provide enough loop attenuation to prevent oscillation and possible instability at or near the frequency of resonance.

In the complex plane, the resonance is revealed by the appearance of a pole (or a pole-zero pair) near the imaginary axis. A branch of the root locus emanates from this pole and terminates either at a zero or goes off to infinity. If the control algorithm does not account for the presence of this pole, its root-locus branch might cross over into the right half plane and invite instability. The control system design must avoid the possibility of this "spillover" effect.

If the precise nature (i.e., pole location) of the resonances are known, they can be modeled and included in the design model of the plant. The compensator design would thereby include the effect of the resonances. This approach, however, is impractical in many applications in which the frequencies of the poles (and the neighboring zeros) of the resonances are not known with precision or may shift during the operation of the system because of environmental factors. A small error in a resonance frequency, damping, or distance between the pole and the zero might result in a compensator design that is even worse than a compensator that ignores the resonance phenomenon.

As alternatives to including the resonances in the design model, several approaches are available. One approach is to increase the loop attenuation at frequencies above the desired operating frequency by adding one or more stages of low-pass filtering to the compensator. If the cut-off frequencies of the low-pass filters are well above the loop gain crossover frequency, the additional phase lag introduced by these filters should not seriously compromise overall stability. Since the phase lag due to these filters starts to be come effective well before their crossover frequency, however, it is important that the phase margin of the loop without the filters be large enough to handle the additional phase shift caused by the filters. New computer-aided design techniques (H_∞ techniques, for example) can be very effective in automating the design process.

Another approach is to "cover" the resonance with narrow-band noise, instead of attempting to include an accurate model of the actual resonance in the design model. This is accomplished by assuming the presence of a disturbance input d defined by

$$\ddot{d} + 2\zeta\Omega\dot{d} + d = w \tag{7.15}$$

where w is white noise. The noise bandwidth (which is adjusted by the damping factor ζ) is chosen to be broad enough to encompass the entire range of possible resonance frequencies. The spectral density of the white noise w is selected produce as much of an effect as that of the resonance. The differential equation (7.15) of the narrow-band noise is used in the control system design model. The presence of the noise, prevents the compensator from relying upon the precise resonance frequency but rather places a broad notch in the vicinity of the resonance.

Whatever method you elect to use, it is important to evaluate the stability of the resulting design in the presence of resonances not included in the design model.

Example 7.1   Resonance in belt-driven servo.

Motion control systems are frequently designed with a belt drive connecting the servo motor with the load. A typical configuration is illustrated schematically in Figure 7.13. For simplicity, assume that the ratio of pulley diameters is 1. (There is almost no difference in the analysis for other ratios.) If the belt resilience is modeled by an ideal spring, the equations of motion of the system are:

$$J_1\ddot{\theta}_1 = k(\theta_2 - \theta_1)$$
$$J_2\ddot{\theta}_2 = k(\theta_1 - \theta_2) - B\dot{\theta}_2 + Lu$$

where θ_1, θ_2 are the angles of the motor and load shafts, J_1, J_2 are the corresponding inertias, k is the spring rate of the belt, B is the damping torque constant, and L is the control torque coefficient.

Simplifying further (to facilitate exposition), assume that both inertias are equal. The transfer functions from the input u to the shaft angles are given by

$$H_1(s) := \frac{\Theta_1}{U(s)} = \frac{\Omega^2 b}{s^4 + ds^3 + 2\Omega^2 s^2 + d\Omega^2 s}$$

$$H_2(s) := \frac{\Theta_2}{U(s)} = \frac{(s^2 + \Omega^2)b}{s^4 + ds^3 + 2\Omega^2 s^2 + d\Omega^2 s}$$

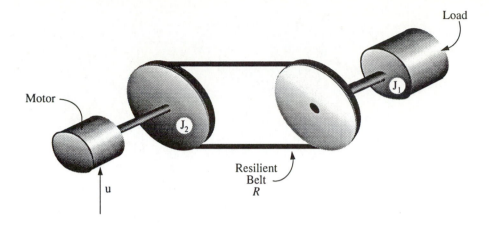

Figure 7.13 Motor driving inertial load with belt may have un-modeled resonance.

where

$$\Omega^2 = k/J_1 = k/J_2, \quad d = B/J_2, \quad b = L/J_2$$

As the resonance frequency Ω becomes infinite, the two transfer functions $H_1(s)$ and $H_2(s)$ coalesce to the single transfer function

$$H(s) = \frac{b/2}{s(s + d/2)}$$

which is the transfer function for a noncompliant coupling between the motor and the load. (The factor of 2 in $H(s)$ arises because the motor is driving two inertias.) For finite values of inertia, the transfer functions $H_1(s)$, $H_2(s)$ have a pole at the origin, a pole on the negative real axis at $s \in (d/2, d)$ and resonance poles near $s = \pm j\Omega$. Figure 7.14 shows the locus of poles as Ω is varied.

The transfer function from the input to the motor shaft has a zero on the imaginary axis at $s = \pm j\Omega$, which is absent from the transfer function from the input to the load shaft. The zero is important.

The compensator design is based on the model that does not include the resonance effect. Assume the shaft position and angular velocity can be measured. (Since the coupling is rigid, it does not matter *which* shaft.)

With a control law

$$u = -G_1(\theta - \theta_r) - G_2\dot{\theta}$$

we can place the closed-loop poles corresponding to the design model at any desired location in the s plane.

As a numerical illustration, take $b = 2$, $d = 0.2$ and suppose that we want the closed-loop poles to lie at

$$s = -2 \pm j1 \tag{7.16}$$

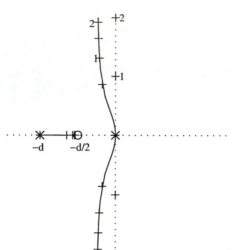

Figure 7.14 Dependency of plant poles on resonance frequency.

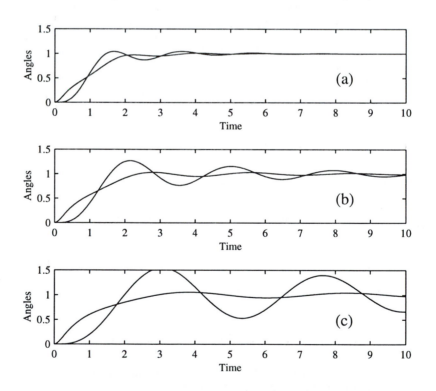

Figure 7.15 Step responses of belt-driven instrument servo with control designed without regard to belt compliance Ω. (a) $\Omega = \infty$. (b)$\Omega = 5$. (c) $\Omega = 2$.

This condition is achieved for $G_1 = 5$, $G_2 = 3.9$. The step response corresponding to these gains is shown in Figure 7.15(a).

With a compliant belt drive, the shaft at which the measurements are made is important. If the measurements are made at the motor shaft, the sensor and actuator are said to be *collocated*. In this case the effect of the resonance is relatively benign. For a resonance frequency Ω higher than about 10 rad/sec the response is essentially the same as that of the system with a rigid coupling. At $\Omega = 5$ rad/sec, the difference between the motion of the two shafts from each other and from the ideal case is just perceptible (Figure 7.15(b)). At $\Omega = 1$ rad/sec the difference is much more pronounced: Although the oscillation is damped, it is quite evident (Figure 7.15(c)). Moreover, the motion of the load shaft (not the one used for the measurements) is more oscillatory than the motor shaft. This is not unexpected: There is no direct feedback from the load shaft to the motor, so it is relatively free to flop about.

Alas, when the feedback is taken from the load shaft, the performance is unsatisfactory. Even when the resonance frequency is 100 rad/sec the system is unstable. The difference between collocated sensors and non-collocated sensors can be explained by the presence of the zero on the imaginary axis in the former case. The zero draws the root locus emanating from the resonant pole further into the left half plane as shown in Figure 7.16(a). Without the zero, the root locus plunges almost immediately into the right half plane, as shown in Figure 7.16(b)

Reality is not quite so grim as this analysis would make it seem. (Otherwise one would never be able to stabilize systems with drive belts or flexible shafts.) The effects of structural damping in the resilient belt and the friction elsewhere in the system are likely to be stabilizing. Moreover, the bandwidth of the amplifier that furnishes the voltage to the motor is likely to be lower than the resonance frequency and will provide additional attenuation at this frequency, thus further stabilizing the system.

If you don't want to rely on the uncertain characteristics of the hardware to stabilize the system, you must attend to the stabilization in your compensator design. A simple low-pass filter might do the job. Suppose, for example, that $\Omega = 5$, and consider a low pass filter with the transfer function

$$L(s) = \frac{\alpha}{s + \alpha}$$

between the control u and the input to the motor.

Since the operating bandwidth, determined by the low-frequency pole locations specified in (7.16), is of the order of 1 rad/sec (gain crossover of the Bode plot), a reasonable value of α would be seem to be about 5. Simulation results, however, show that a more suitable value of α is around 1. The step responses for $\Omega = 10$ and $\alpha = 1.0$ and 1.2 are shown in Figure 7.17. The system is stable, although the ringing at the resonance remains. More effective elimination of the ringing would require a more sophisticated compensator design. (See Problem 7.6.)

To illustrate the possible deleterious effects of unmodeled resonances, the resonance frequency in the foregoing was intentionally chosen relatively close to the operating bandwidth. If you want the system to operate with a bandwidth near the lowest resonance frequency, either you must include the resonance in the design model or be prepared to consider other measures to avoid the possible unfavorable consequences.

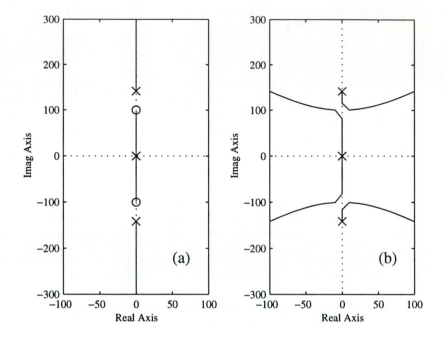

Figure 7.16   Closed-loop root loci of belt-driven instrument servo. (a) Sensors on motor shaft (collocated). (b) Sensors on load shaft (non-collocated).

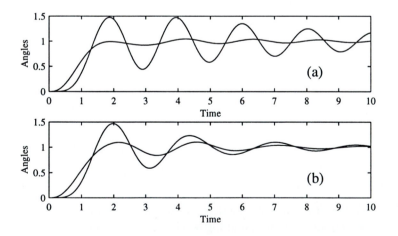

Figure 7.17   Step responses of belt-driven servo with low-pass filter compensation. (a) $\alpha = 1.2$ (b) $\alpha = 1.0$.

7.7 LIMIT STOPS

Stops (bumpers) are often used to prevent some parts of a machine from moving beyond limits of motion imposed by the physical size and construction of the machine. Rarely will these parts run into the stops in a well-designed control system for the machine, but it is always possible. Hence it is worth considering the effect of the stops in a simulation of the system.

You might be tempted to represent the presence of limit stops by adding a saturation nonlinearity at the output of the limited variable as shown in Figure 7.18. While the saturation will indeed limit the output to lie within the limits, it is *not* a correct representation of the effect of the stops. The state variable itself must be limited, not a signal derived from the state variable. The correct way to model limit stops can be understood by considering how they work. If a moving part runs into a stop, the latter pushes back on the part to decellerate the it and thereby prevent further motion: The stop behaves like a very stiff spring. Theoretically, the spring constant of an ideal stop is infinite. But a real stop will yield somewhat, depending upon its physical size and the material of which it is constructed. Thus the correct way to model a limit stop is to use a nonlinear spring with a spring constant of zero until the limit is reached and a very large spring constant thereafter. This representation is similar to the spring used in modeling the mechanism of backlash.

Mathematically, the motion of a component with a limit stop would be represented by the differential equation

$$\ddot{x} = a - f(x) \tag{7.17}$$

where a represents the acceleration due to all sources other than the limit stops and

$$f(x) = \begin{cases} K(x - x_L), & x > x_L \\ 0, & x < x_L \end{cases} \tag{7.18}$$

where K is a large number and x_L is the limit at which the motion is to be stopped. Since K is finite, x will exceed x_L by a small amount as it would in a real physical system.

If the motion is constrained bilaterally, two stops each of the form of (7.18) are needed. They can be combined into a single function

$$f(x) = \begin{cases} K(x - x_{L1}), & x > x_{L1} \\ 0, & -x_{L2} < x < x_{L1} \\ K(x + x_{L2}), & x < -x_{L2} \end{cases} \tag{7.19}$$

where $-x_{L2}$ is the lower limit.

The block-diagram representation of (7.17) is given in Figure 7.19.

The effect of limit stops can be seen in the following example.

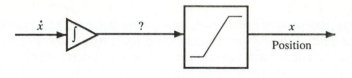

Figure 7.18   Incorrect representation of limit stops.

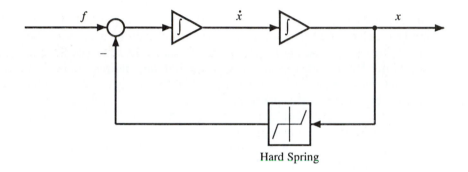

Figure 7.19   Correct representation of limit stops.

Example 7.2   Effect of Limit Stops in Second Order System

In the absence of limit stops the motion of an object is governed by

$$\ddot{x} = a = -G_1\dot{x} - G_2(x - x_R)$$

where x_R is the desired position of the object. Limit stops are assumed to be located at $x_{L1} = -x_{L2} = 1$.

With the following gains:

$$G_1 = 5, \quad G_2 = 4$$

the step response exhibits no overshoot and will thus not run into the limit stops for a step reference smaller in magnitude than 1. If the reference is greater in magnitude than 1, however, the limit stops will prevent the object from reaching the reference value. The larger the value of x_R the larger the value of a at the time the object runs into the stop and the greater the motion beyond the stop.

The simulated response for $x_R = 5$ and $K = 200000$ is shown in Figure 7.20. The response time-history is shown in (a) and the phase-plane portrait is shown in (b). It is observed that object repeatedly runs into the stop and bounces off it the way a ball bounces off a hard surface. The phase plane is even more instructive, showing a reversal of sign in the velocity each time the object runs into the stop and bounces off it.

The foregoing example shows that the limit stop is a *reflecting barrier* Another ex-

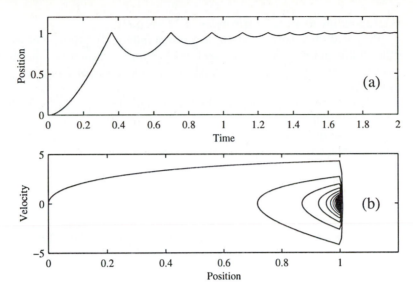

Figure 7.20   Behavior of system with elastic limit stops. (a)
Time history. (b) Phase plane.

ample is an *absorbing* barrier, in which the object sticks to the stop instead of bouncing
off it. Such a barrier can be modeled by a large damping force that goes into action
when the object motion exceeds the limit. (See Problem 7.5.)

Each collision of the moving object with the limit stops stresses the system and
invites potential failure. To avoid physical contact with the limit stops, you can include
the effect of limit stops in the control algorithm by adding a term like (7.19) to the
control law, perhaps setting x_L to a smaller value than the actual value of the physical
stops.

In some applications there is a need to limit velocity rather than (or in addition to)
position. This limit is achieved by including a damping term (i.e., velocity feedback),
which becomes very large when the velocity exceeds the limiting value.

7.8   TIME DELAY

There are many dynamic processes in which it takes a finite amount of time for a signal
to propagate from one location in the process to another. If, for example, you want to
control a process on the moon by transmitting observation data from the process to
the earth and control signals back to the moon, you must take into account the time
delay of about one sec in each direction. The transfer function you must deal with in
the control system design is $G(s)e^{-2Ts}$, where T is the one-way propagation delay and
$G(s)$ is the transfer function of the process. The factor e^{-2Ts} is the transfer function

corresponding to a round-trip delay of $2T$ seconds.

Propagation delays also occur in more mundane situations such as in chemical and other industrial processes in which it takes a finite amount of time for material to be moved through a pipe or other conduit or via a conveyor from one part of the plant to another.

Human operators also exhibit behavior that appears to include a finite time delay, often attributed to the propagation time from the point of measurement to the brain. For representing the behavior of a human operator in a control system, a transfer function of the following form is often used:

$$G(s) = \frac{e^{-Ts}}{s + \alpha}$$

The effect of a time delay on the performance of a control system is most readily explained in the frequency domain. Consider a plant having a transfer function $G(s)$ for which a compensator with the transfer function $D(s)$ has been designed without regard to a time delay. The forward loop transmission, in the absence of the delay is

$$F(s) = D(s)G(s)$$

In the presence of a time delay of T seconds, however, the plant transfer function is $G_d(s) = e^{-Ts}G(s)$ and hence the forward loop transmission is

$$F_d(s) = D(s)e^{-Ts}G(s)$$

For $s = j\omega$

$$F_d(j\omega) = D(j\omega)e^{-j\omega T}G(j\omega) = |F_d(j\omega)|e^{j\theta_d(\omega)}$$

Thus, since $|e^{-j\omega T}| = 1$,

$$|F_d(\omega)| = |F(\omega)| \quad \text{and} \quad \theta_d(\omega) = \theta(\omega) - \omega T \qquad (7.20)$$

This result can be interpreted with the aid of the Bode plot for the forward loop transmission:

- The amplitude characteristic of the Bode plot is unaffected by the time delay.

- The time delay increases the phase shift proportional to frequency, with the proportionality constant being equal to the time delay.

This interpretation shows that time delay always *decreases* the phase margin of a system. Since the gain crossover frequency ω_c is unaffected by the time delay, the phase margin is reduced by $\omega_c T$. Since ω_c is a measure of the bandwidth of the closed-loop system, one way of increasing the phase margin in the presence of time delay is to reduce the closed-loop bandwidth, for example, by reducing the loop gain of the system. The increase in phase margin achieved by this method, however, is at the expense of diminished performance.

To avoid compromising performance of the closed-loop system it is necessary to account for the time delay explicitly. Methods for accomplishing this include the "Smith Predictor", and including a model of the time-delay in the design model of the plant, as discussed below.

7.8.1 Smith predictor

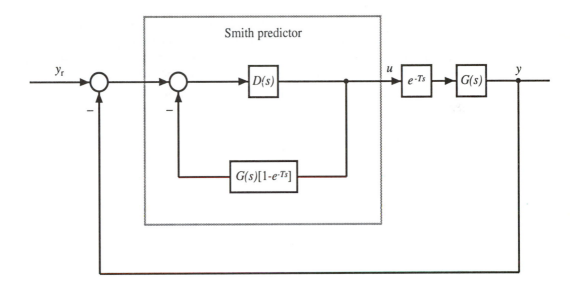

Figure 7.21   Compensating plant with time delay using Smith
Predictor.

In 1958 Professor O.J.M. Smith [16] of the University of California at Berkeley devised a clever way of dealing with the problem of designing a compensator for a plant with time delay. Suppose that you have designed a suitable compensator, with the transfer function $D(s)$, for a plant whose transfer function in the absence of time delay is $G(s)$. Then, using Smith's technique (now known as the "Smith Predictor"), the transfer function of the compensator in the presence of time delay should be

$$\bar{D}(s) = \frac{D(s)}{1 + (1 - e^{-Ts})G(s)} \tag{7.21}$$

which can be implemented as shown within the dashed box of Figure 7.21. The closed-loop transfer function of the compensated system is

$$H(s) = \frac{\bar{D}(s)e^{-Ts}G(s)}{1 + \bar{D}(s)e^{-Ts}G(s)}$$

$$= \frac{D(s)G(s)}{1 + D(s)G(s)}e^{-Ts} \tag{7.22}$$

upon use of (7.21). Except for the factor e^{-Ts} this transfer function is seen to be the same as the transfer function of the closed-loop system for the plant without the time delay and with the compensator $D(s)$. The time response of the closed-loop system with a compensator that uses a Smith predictor will thus have the same shape as the response of the closed-loop system without the time delay compensated by $D(s)$; the only difference is that the output will be delayed by T seconds. See Figure 7.22.

To implement the Smith predictor $\bar{D}(s)$, you need a method of realizing the pure time delay that appears in the feedback loop. This is easy to do in a digital computer, but difficult using only analog components. Implementation, moreover, is not the only problem with the Smith predictor. The other major problem is that you must know the plant transfer function and the time delay with reasonable accuracy. Errors in these quantities will result in a closed-loop transfer function whose denominator includes e^{-Ts} terms, the effect of which will have to be considered. (See Problem 7.10.) These issues notwithstanding, the Smith predictor finds application in process control systems.

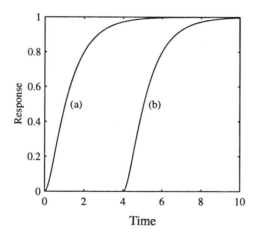

Figure 7.22 Step responses of system (a) without time delay (b) with time delay and Smith-predictor based compensator.

7.8.2 Approximating the time delay in continuous-time system by a rational function

Difficulty of implementation and problems of robustness may make you eschew the Smith predictor approach and seek an alternative approach. The standard approach based upon the separation principle is to design a full-state feedback control law and use an observer to generate an estimate of the state to be used in place of the true state. The state of a system with a time delay, however, is essentially infinite order, because the transfer function corresponding to the time delay is not a rational function of s. Hence a system containing a time delay is not directly amenable to state-space design techniques. Frequency domain techniques (e.g., loop shaping) are not restricted to rational transfer functions and, in principle, you can use them for systems that include time delays. In practice, however, you may find it difficult to carry out the required design calculations.

To apply state-space design techniques to systems with time delays, you can *approximate* their transfer functions by rational functions of s:

$$e^{-Ts} \approx T_k(s) = \frac{b_0 s^k + b_1 s^{k-1} + \cdots + b_k}{s^k + a_1 s^{k-1} + \cdots + a_k}$$

A variety of approximations are possible. One commonly used approximation is the so-called *Padé approximant* derived as follows:

$$e^{-Ts} = \frac{e^{-Ts/2}}{e^{Ts/2}} = \frac{1 - Ts/2 + T^2 s^2/8 + \cdots}{1 + Ts/2 + T^2 s^2/8 + \cdots} \tag{7.23}$$

Truncation of the infinite series in the numerator and denominator of (7.23) yields a Padé approximant to e^{-Ts}. Although the numerator and denominator of a general Padé approximant do not need to be the same order, it is customary to make them equal, thus giving

$$T_{1k}(s) = \frac{1 - Ts/2 + T^2 s^2/8 + \cdots + (-Ts/2)^k/k!}{1 + Ts/2 + T^2 s^2/8 + \cdots + (Ts/2)^k/k!} \tag{7.24}$$

Another approximation is obtained by using the definition of the exponential function:

$$e^x = \lim_{k \to \infty} \left(1 - \frac{x}{k}\right)^k \approx \left(1 - \frac{x}{k}\right)^k$$

with k being a finite number. Applying this approximation to $e^{-Ts/2}/e^{Ts/2}$ yields

$$T_{2k}(s) = \frac{(1 - Ts/2k)^k}{(1 + Ts/2k)^k} \tag{7.25}$$

For $k = 1$ it is seen that T_{1k} and T_{2k} are identical:

$$T_{11} = T_{21} = \frac{1 - Ts/2}{1 + Ts/2}$$

Although they are not identical for $k > 1$, they share a number of interesting properties:

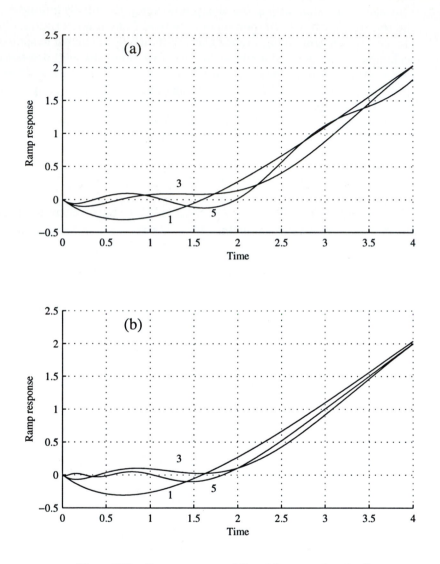

Figure 7.23   Ramp responses of time-delay approximants of different orders. (a) Padé approximants. (b) Exponential approximants.

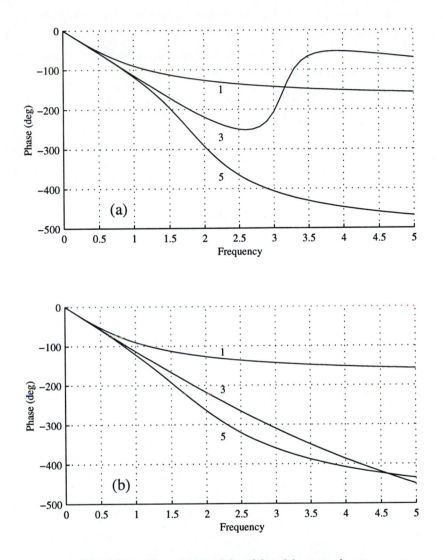

Figure 7.24   Phase characteristics of time-delay approximants of different orders. (a) Padé approximants. (b) Exponential approximants.

- The transfer functions are *all pass*, i.e., the magnitude of the transfer function is 1 for all frequencies: $|T(j\omega)| = 1$.

- The transfer functions are non-minimum phase, i.e., they have zeros in the right-half plane.

- As the order of the approximation is increased, they approximate the low frequency phase characteristic with increasing accuracy.

For purpose of comparison, the ramp response and phase characteristic of T_{1k} and T_{2k} are shown in Figure 7.23 and Figure 7.24, respectively, for $k = 1, 3$, and 5. (The ramp response is more revealing than the step response, because the discontinuity of the step give rise to high frequencies which are poorly handled by the approximations.)

7.9 PROBLEMS

Problem 7.1 High gain inverts nonlinearity

Consider a zero-memory system consisting of a high gain K in the forward loop and a nonlinear element with the characteristic $z = \phi(y)$ in the feedback loop as shown in Figure 7.25. Show that the characteristic of the closed-loop system approaches

$$y = \phi^{-1}(x)$$

as $K \to \infty$. How large must K be to ensure that the error achieving the inversion is less than 10 percent?

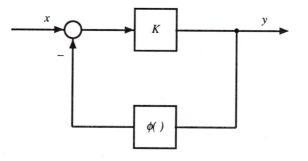

Figure 7.25 Use of high gain to invert nonlinearity.

Problem 7.2 Reduced-order filter for sensor noise

Consider a plant modeled by linear dynamics

$$\dot{x} = Ax + Bu$$

in which all the state variables are measured, but narrow band noise is present on some or all of the observations:

$$y = x + Cz$$

where

$$\dot{z} = A_n z + B_n w$$

with w being white noise with spectral density W. Find the compensator transfer function for the quadratic performance criterion $x'Qx + u'Ru$ and compare it with that for the system without the sensor noise.

Problem 7.3   Compensator for instrument servo with narrow-band noise

Consider an instrument servo comprising a servomotor and an inertial load.

Let the transfer function of the motor and load be given by

$$G(s) = \frac{1}{s(s + 25)}$$

(a) Design a compensator for the system, without considering the noise, to achieve an operating bandwidth of 50 Hz without overshoot to a step reference input.

(b) The motor position is measured by a sensor operating by modulating and demodulating a carrier at a frequency of 60 Hz. Assume noise can be represented by a full-wave sinusoid of half period of 1/120 sec. Simulate the performance of the servo for various levels of sensor noise.

(c) Design a notch filter to ameliorate the effects of the sensor noise by modeling the noise as a narrow-band random process with a center frequency of 120 Hz.

Problem 7.4   Backlash in instrument servo

For the instrument servo of Problem 7.3, assume that the load inertia and motor armature inertia are coupled by a one-to-one ratio gear train with a backlash of 0.2 degrees and a measured resonant frequency of 100 Hz.

(a) Simulate the performance of the system using the compensator of Problem 7.3(a), assuming that the sensor is noise-free.

(b) If the backlash effects are objectionable, modify the design of part (a) by including a model of the backlash in the observer and assess the benefits of the modification by simulation.

Problem 7.5   Absorbing barrier

Consider the system of Example 7.2 , which exhibits the behavior of a reflecting barrier. Add damping that goes into action when the object reaches the barrier so that the velocity is reduced enough to prevent the object from bouncing.

Problem 7.6   Resonance in belt-driven instrument servo

Consider the belt-driven servo discussed in Example 7.1 . Investigate the possibility of improving the dynamic response and stability margins for a resonance frequency Ω

of 10 over that obtained using a simple first-order low-pass filter, by use of the following techniques:

(a) A second-order low-pass filter with poles on the negative real axis.

(b) A reduced-order Kalman filter designed by representing the resonance as broadband noise centered near the resonance frequency.

c) A full-order Kalman filter designed by representing the resonance as broadband noise centered near the resonance frequency.

Problem 7.7   Belt-driven robot arm

Assume that the robot arm described in Problem 7.1 is separated from the drive-motor by a resilient belt and that the resonant frequency of the belt drive with its coupled inertias is 100 rad/sec. Simulate the performance of the control laws of Problem 7.5 and of Problem 7.6 when

(a) the drive shaft position is the measured variable.

(b) the arm angle is the measured variable.

Problem 7.8   Resonance in spring-coupled cars

The motion of a pair of spring-coupled cars is controlled by a single d-c motor. The dynamics are governed by the differential equations

$$\ddot{x}_1 = \Omega^2(x_2 - x_1)$$
$$\ddot{x}_2 = -bx_2 + \Omega^2(x_1 - x_2) + bu$$

where u is the motor control voltage, $\Omega^2 = k/J$ is natural frequency of the oscillation induced by the coupling spring, and b is a motor gain parameter.

The natural frequency Ω is known to be large. Hence it is proposed to design the controller by ignoring the resonance caused by the finite spring rate, i.e., by assuming $\Omega^2 = \infty$ which is equivalent to assuming $x_1 = x_2$. A full-state feedback control law under this assumption is

$$u = -G_1 x_1 - G_2 x_2$$

Assume $b = 10$.

(a) What values of G_1 and G_2 will result in closed-loop poles that lie at

$$s = -20 \pm j20$$

(b) Using these gains, assess the stability of the design for $\Omega \in [40, 400]$.

(c) Adjust the design of part (a) to account for the resonance, assuming that the resonant frequency Ω is known only with an accuracy of 25 percent.

Note. The dynamic model in this problem also represents the motion of an inertial load by a motor through an elastic shaft.

Problem 7.9   Spring-coupled cars with limit stops

Consider the two-car train of Problem 7.8.

(a) Indicate how the linear dynamics given above are to be modified to include the following constraints:

1. The carts run into each other.

2. Neither cart can run past a stop at the end of the track on which they run, as shown in Figure 7.26.

(b) Assume full-state feedback. Design a linear control law

$$u = -G_1 x_1 - G_2 \dot{x}_1 - G_3 x_2 - G_4 \dot{x}_2$$

which places the closed-loop poles at

$$s = -20 \pm j20$$

Simulate the behavior of the system starting at rest at one stop and with a command for it to move to the other stop.

(c) Repeat part (b) except select the gains G_i so that the closed-loop poles are at

$$s = -5 \pm j5$$

Explain any differences in performance in parts (b) and (c).

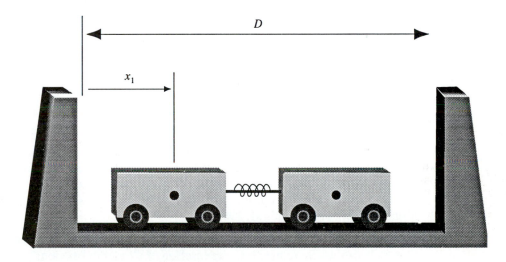

Figure 7.26   Spring-coupled carts with limit stops.

Problem 7.10   Robustness of Smith predictor

Suppose a compensator for a system with time delay is designed to incorporate a Smith predictor, as described in Section 7.8.1, but that the delay time T is not known accurately, so the Smith predictor is designed on the basis of an estimate $\bar{T}$ of the delay time.

(a) Calculate the closed-loop transfer function of the system using the Smith predictor so designed.

(b) Is it preferable to underestimate or to overestimate the time delay T?

7.10   REFERENCES

[1] B. Friedland, *Control System Design: An Introduction to State Space Methods*, McGraw-Hill, New York, 1986.

[2] B. Armstrong-Hélouvry, *Control of Machines with Friction*, Kluwer Academic Publishers, 1991.

[3] D.A. Haessig, Jr. and B. Friedland, "On the Modelling and Simulation of Friction," J. Dynamic Systems, Measurements, and Control, (Trans. ASME) Vol. 113, pp. 354–362, September 1991.

[4] D. Karnopp, "Computer Simulation of Slip-Stick Friction in Mechanical Dynamic Systems," J. Dynamic Systems, Measurements, and Control, (Trans. ASME) Vol. 107, pp. 100-103, 1985.

[5] P.R. Dahl "A Solid Friction Model," AFO 4695-67-C-0158, The Aerospace Corporation, El Segundo, CA, 1968.

[6] A. Harnoy, B. Friedland, and H. Rachoor, "Modeling and Simulation of Elastic Forces in Lubricated Bearings for Precise Motion Control," Wear, Vol. 172, pp. 155–165, 1994.

[7] C. Canudas de Wit, H. Olsson, K.J. Åström, and P. Lischinsky, "A New Model for Control of Systems with Friction," IEEE Trans. on Automatic Control, Vol. 40, No. 3, pp. 419–425, March 1995.

[8] B. Friedland, M.F. Hutton, C. Williams, and B. Ljung, "Design of Servo for Gyro Test Table Using Linear Optimum Control Thoery," IEEE Trans. on Automatic Control, Vol. AC-21, No. 2, pp. 293–296, April 1976.

[9] G. Brandenburg, H. Hertle, and K. Zeiselmair, "Dynamic Influence and Partial Compensation of Coulomb Friction in a Position and Speed Controlled Elastic Two-Mass System," Preprints, 10th IFAC World Congress, Vol. 3., pp. 91-99, Munich, Germany, July 27–31, 1987.

[10] U. Schäfer and G. Brandenburg, "Model Reference Position Control of an Elastic Two-Mass System with Compensation of Coulomb Friction," Proc. 1993 American Control Conference, Vol. 2., pp. 1937–1941, San Francisco, CA, June 1993.

[11] B. Friedland and Y.-J. Park, "On Adaptive Friction Compensation," IEEE Trans. on Automatic Control, Vol. 37, No. 10., pp. 1609– 1612, 1992.

[12] B. Friedland and S. Mentzelopoulou, "On Adaptive Friction Compensation Without Velocity Estimation," Proc. 1992 Conference on Control Applications, Vol. 2, pp. 1076–1081, Dayton, OH, September 1992.

[13] B. Friedland, S. Mentzelopoulou, and Y.J. Park, "Friction Estimation in Multi-Mass Systems," Proc. 1993 American Control Conference, Vol. 2, pp. 1919–1924, San Francisco, CA, June 1993.

[14] S. Mentzelopoulou and B. Friedland, "Experimental Evaluation of Friction Estimation and Compensation Techniques," Proc. 1994 American Control Conference, Baltimore, MD, pp. 3132–3136, June 1994.

[15] C. Canudas de Wit, K.J. Åström, and K. Braun, "Adaptive Friction Compensation," IEEE Jour. of Robotics and Automation, Vol. RA-3, No. 6., pp. 681–685, December 1987.

[16] O.J.M. Smith, *Feedback Control Systems*, McGraw- Hill, New York, 1958.

Chapter 8

DISCRETE-TIME SYSTEMS

Modern feedback control systems rely increasingly on the digital computer not only as a tool for design of the control algorithm, but also to implement the control algorithm in real time. The configuration of a typical computer-controlled system is shown in Figure 8.1. The controller comprises the digital computer, an analog-to-digital converter (ADC) which converts the continuous-time signal to the digital format required by the computer, and a digital-to-analog converter (DAC) which converts the numerical value of the control into the physical signal that drives the plant.

The physical control input u to the plant is a continuous-time variable, but its values change only at the discrete-time instants that the computer produces its outputs, as shown in Figure 8.2. While the digital computer is calculating a new control input, the digital-to-analog converter (DAC) "holds" or "latches" the old value. The behavior at the input to the digital computer is similar: Upon a signal from the computer, the analog-to-digital converter (ADC) initiates a sampling operation; when the sampling operation is completed the result is transmitted to the digital computer. Hence the input y_s to the digital computer is a sampled version of the continuous-time signal $y(t)$ at the input to the ADC. The digital computer generally reads the value of y_s only once in each sampling cycle, at the instant governed by the program of the computer, and the value of $y(t)$ at other instants does not influence the behavior of the closed-loop process.

Use of a digital computer for real-time feedback control has a number of advantages, including the following:

- **Flexibility**   Because a digital computer can perform complicated logical and nonlinear operations just as easily as linear operations, advanced control algorithms are not especially difficult to implement.

- **Accuracy**   The accuracy of the calculation performed by a digital computer is not dependent on the stability of the components used in the computer. If the computer is working, it will always give the same numerical result for the same inputs, unlike an analog implementation in which the accuracy of the output depends on the stability of resistors, capacitors, diodes, transistors, and similar components used in the implementation. (Note, however, that the ADC and the DAC are inherently analog components and hence subject to the same drift problems as other analog components.

- **Stable, drift-free integration**   It is all but impossible to construct a drift-free analog integrator. Prior to the advent of digital computers, achieving stable long-term integration for processes with long time constants (measured in hours) required special measures (e.g.,"chopper stabilization"), which were expensive and somewhat unreliable. In contrast, perfect digital integration is a routine operation.

- **Remote operation**   Operation of a process in which the sensors and actuators are at a location remote from the computer is much more reliable with digital communication, using error-correcting coding, between the computer and the other hardware in the system.

Nevertheless, digital-computer control has some disadvantages. Cost is one. In some applications, the cost of the digital control hardware, including the ADC and the DAC, is simply not justified by the performance requirements. The decreasing cost of such hardware, however, increases the justification for its use in more and more applications. Another disadvantage of digital control is the time delay between the measurements of the plant variables and the output of a new control signal. This delay includes the time required to convert from analog to digital format and vice versa, and the time it takes for the computer to perform the required computation. The sum of these times is an upper limit on the overall sampling frequency, and hence upon the closed-loop bandwidth of the system. As the speed of digital processors and associated hardware continues to increase, fewer and fewer processes are subject to this limitation.

Although systems controlled by digital computers now constitute the largest class of discrete-time systems, discrete-time systems arise in other contexts, some of which antedate the widespread use of digital computers. One of the earliest examples of a discrete-time system is one in which the sensor is a mechanically scanned radar such as might be used in an instrument landing system for aircraft. The radar antenna scans its field of view, usually at a uniform rate, and the reflection from the aircraft, from which its coordinates (e.g., range, azimuth, and elevation) are determined, occurs only once per scan. (It is worth noting that the times at which the reflection is received are not uniformly spaced as the aircraft moves in the field of view of the radar, and this

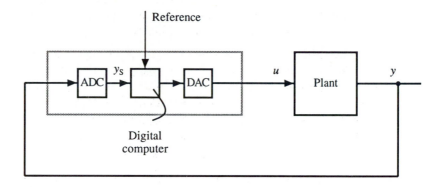

Figure 8.1   Typical computer-controlled feedback system.

means that the sampling interval is not constant. If the velocity of the aircraft is high it may be necessary to account for this lack of uniformity.)

Another situation in which a discrete-time system can arise is in a chemical process in which the data on some variable may depend on off-line laboratory analysis. It may be necessary to draw physical samples of some variable and analyze them using some external apparatus. The results of this analysis arc thcn uscd to control changes to the process. Although this method of sampling may seem archaic, it may be unavoidable. Consider a brewery or a winery, for example, in which the analysis is performed in the mouth of a brewmaster or vintner.

From the mathematical viewpoint, discrete-time systems are those in which the independent variable *time* takes on only discrete values, for example

$$t = \{t_1, t_2, t_3, \cdots\}$$

Most frequently these discrete values of time are multiples of a fixed time interval T, called the *sampling interval*:

$$t_n = nT$$

The analysis and design of systems in which sampling occurs at a uniform time interval is much simpler than if the sampling intervals $t_{n+1} - t_n$ are not uniform. Some, but not all, of the techniques developed for uniform sampling can be used with nonuniform sampling. This book, however, is concerned primarily with uniform sampling.

8.1   DEALING WITH DISCRETE-TIME

There are two basic ways to deal with a discrete-time systems: If the sampling frequency is high relative to the desired closed-loop bandwidth, you can deal with it as a

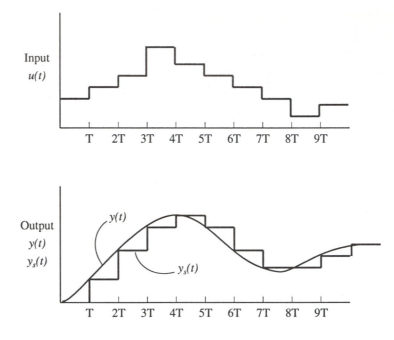

Figure 8.2   Input and output of computer-controlled feedback
system.

continuous-time system, simply digitizing a continuous-time design. The alternative
is to use methods specifically designed for discrete-time systems.

A rule of thumb for deciding whether continuous-time methods are adequate is
that the sampling frequency should exceed the closed-loop bandwidth by at least an
order of magnitude. But don't be dogmatic using this rule: Sometimes you may be able
to deal with higher bandwidth systems; sometimes you may have difficulties with lower
bandwidth systems. The best way to find out is to simulate the system in question.

The most important factor to consider in using continuous-time methods for sampled-data systems is the effect of sampling delay. This effect is seen in the system
illustrated in Figure 8.1. Suppose, for the sake of discussion, that the only function
of the computer is to "sample and hold" the input at an interval of T seconds. This
operation would produce a piecewise-constant signal at the output, as shown in Figure
8.3. If the rough edges of the piecewise-constant output $y_s(t)$ of the sample-and-hold
device were smoothed out, the resulting signal would look something like the curve
drawn with the broken line. You can see that it resembles the unsampled input $y(t)$

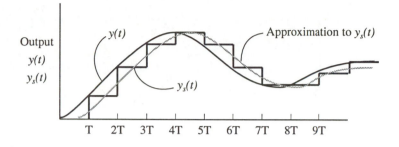

Figure 8.3 Sampling effect is approximated by delay of $T/2$.

in general shape, but delayed by approximately $T/2$ seconds. The major effect of sampling is thus a time delay of about half a sampling period.

The time delay due to sampling has a destabilizing effect on the system. In particular, since the transfer function corresponding to a time delay of $T/2$ seconds is $D(s) = e^{-Ts/2}$, which corresponds to pure phase shift (i.e., no amplitude attenuation) in the frequency domain, the result is that the phase shift in the loop increases with frequency. By considering either the Nyquist diagram or the Bode plots for the return difference, you can see that the stability margins are reduced. Since the phase shift keeps increasing with frequency, it must ultimately exceed 180 degrees. Consequently there is no way that a system with sampling can have an infinite gain margin: If the gain is made large enough, the system must become unstable. (This fact will be demonstrated more rigorously in Section 8.5, by means of the root-locus in the z-plane.)

Example 8.1 Servomotor Control with Sampling

A servo motor with an inertial load has the well-known transfer function

$$G(s) = \frac{\alpha}{s(s + \alpha)}$$

If the loop is closed by a sample-and-hold device, the approximate loop transmission is

$$F(s) = e^{-Ts/2}G(s)$$

The Nyquist and Bode plots for $F(s)$ are shown in Figure 8.4 for $\alpha = 4., T = 12$. The gain and phase margins are found to be 9.1 dB and 50 degrees, respectively. (See Problem 8.1.) These should be contrasted with an infinite gain margin and a phase margin of 75 degrees for the unsampled system.

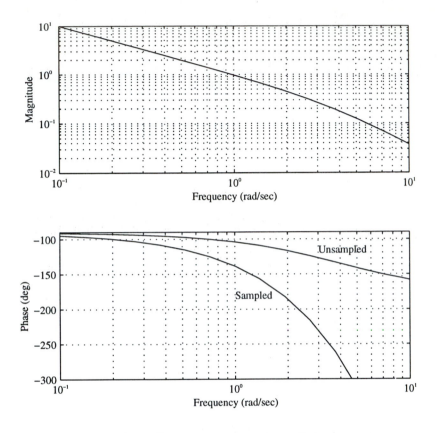

Figure 8.4   Bode plot of computer-controlled servomotor
shows how sampling reduces stability.

8.2 DISCRETE-TIME MODELS

If the sampling frequency $1/T$ is high relative to the bandwidth of the system, the approximate analysis discussed in the last section might suffice. But if the sampling frequency is relatively low—or for a more precise analysis—the tools for the analysis (and design) of discrete-time systems are needed. To use these tools, you will need a discrete-time model of the plant. For a continuous-time process, with piecewise-constant (i.e., sampled and held) inputs, the model can be obtained from the continuous-time system

$$\dot{x} = Ax + Bu \tag{8.1}$$

The general solution to (8.1) is

$$x(t + \tau) = e^{A\tau} x(t) + \int_t^{t+\tau} e^{A(t+\tau-\xi)} Bu(\xi)d\xi \tag{8.2}$$

Now suppose the input $u(t)$ to (8.1) is piecewise-constant:

$$u(t) = u(nT) := u_n = \text{const.}, \quad \text{for } nT < t < (n+1)T \tag{8.3}$$

Evaluating (8.2) with $t = nT$ and $\tau = T$, and using (8.2) gives

$$x(nT + T) = e^{AT} x(nT) + \left(\int_{nT}^{nT+T} e^{A(nT+T-\xi)} Bd\xi \right) u_n \tag{8.4}$$

(The fact that $u(t)$ is piecewise-constant is exploited to bring u_n outside the integral.)

Equation (8.4) can be written:

$$x_{n+1} = \Phi x_n + \Gamma u_n \tag{8.5}$$

where

$$x_n := x(nT) \tag{8.6}$$

$$\Phi := \Phi(T) = e^{AT} \tag{8.7}$$

$$\Gamma := \int_{nT}^{nT+T} e^{A(nT+T-\xi)} Bd\xi = \int_0^T e^{A\lambda} Bd\lambda \tag{8.8}$$

The *difference equation* or *recursion equation* (8.5) is the fundamental representation of a discrete-time system, and plays the same role as (8.1) does in a continuous-time system. In words, it can be read as "the state at the $(n + 1)$st sampling instant is a linear combination of the state at the nth instant and the (piecewise-constant) control during the nth instant."

The discrete-time model (8.5) has in this case been developed by analyzing the response of a continuous-time system to a piecewise-constant input. But models in the form of (8.5)—economic systems, for example— can arise without an underlying continuous-time system.

Note that when (8.5) results from a continuous-time system with piecewise-constant inputs, the matrix Φ is a state-transition matrix of a continuous-time system and hence is always nonsingular. Given a transition matrix $\Phi = \Phi(T)$ and the sampling time T, in fact, we can solve for the corresponding continuous-time dynamics matrix:

$$A = \frac{1}{T} \log \Phi \tag{8.9}$$

where the matrix log function can be computed by various numerical methods.

If (8.5) represents the dynamics of an arbitrary discrete-time system, however, Φ may be singular, or the continuous-time dynamics matrix A obtained from (8.9) may not be real. In these cases it is not possible to find a continuous-time process that, after sampling, will produce the given state-transition matrix Φ.

Observe that the discrete-time control distribution matrix Γ, being defined as the integral of a continuous function, always exists: It doesn't matter whether A is singular or not. But if A is nonsingular (i.e., the continuous-time system does not have a pole at the origin), Γ can be expressed by a formula that does not involve integrals:

$$\Gamma = (e^{AT} - I)A^{-1}B = (\Phi - I)A^{-1}B \tag{8.10}$$

or by

$$\Gamma = A^{-1}(e^{AT} - I)B = A^{-1}(\Phi - I)B \tag{8.11}$$

If A is singular, Γ must of course be evaluated using the integral (8.8).

Example 8.2   D-C Servomotor

The state-space representation of the servomotor considered in Example 8.1 is

$$\dot{x}_1 = x_2$$
$$\dot{x}_2 = -\alpha x_2 + \alpha u$$

where

$$x_1 := \theta$$
$$x_2 := \dot{\theta}$$

The matrices corresponding to these differential equations are

$$A = \begin{bmatrix} 0 & 1 \\ 0 & -\alpha \end{bmatrix} \qquad B = \begin{bmatrix} 0 \\ \alpha \end{bmatrix}$$

Using (8.7) and (8.8) we find that

$$\Phi(T) = \begin{bmatrix} 1 & \dfrac{1 - e^{-\alpha T}}{\alpha} \\ 0 & e^{-\alpha T} \end{bmatrix}$$

$$\Gamma(T) = \int_0^T \begin{bmatrix} 1 - e^{-\alpha t} \\ \alpha e^{-\alpha T} \end{bmatrix} dt = \begin{bmatrix} \dfrac{\alpha T - 1 + e^{-\alpha T}}{\alpha} \\ 1 - e^{-\alpha t} \end{bmatrix}$$

The defining equation (8.5) of a discrete-time system can be used recursively to express the state x_ν at some later time νT in terms of a starting state x_n at time nT:

$$x_{n+2} = \Phi x_{n+1} + \Gamma u_{n+1}$$
$$= \Phi^2 x_n + \Phi \Gamma u_n + \Gamma u_{n+1}$$
$$\cdots$$
$$x_\nu = \Phi^{\nu-n} x_n + \sum_{n}^{\nu-1} \Phi^{\nu-1-l} \Gamma u_l \tag{8.12}$$

This expression has two terms: The first is the result of the starting state x_n, and the second is a "discrete-time convolution" of the sequence of inputs $\{u_n, u_{n+1}, \cdots u_{\nu-1}\}$ with the matrix function $\Phi^n \Gamma$, which might be termed the "unit response" of the system. It is the discrete-time counterpart of the continuous-time relation (8.2). Unlike the corresponding continuous-time relationship (8.2), which is valid for $t > \tau$ or for $t < \tau$, however, (8.12) is not meaningful for $\nu < n$ when Φ is singular.

8.3 STABILITY

The general relation (8.12) between the initial state, the input sequence, and a later state can be used to study the stability of a linear, discrete-time system.

First consider the effect of the starting state in the absence of input, as given by

$$x_\nu = \Phi^{\nu-n} x_n \tag{8.13}$$

In a first-order system Φ is a scalar, and it is clear that $x_\nu \to 0$ as $\nu \to \infty$ if and only if $|\Phi| < 1$. The extension of this to the multidimensional case is expressed by the theorem:

Theorem 8.1 (Stability) *The discrete-time system (8.13) is asymptotically stable if and only if $||\Phi|| < 1$, where $||\Phi|| := |\lambda_k|$, the eigenvalue of Φ having the largest magnitude.*

The discrete-time system is stable if and only if $||\Phi|| \le 1$ and all eigenvalues of Φ having magnitude of unity are simple.

($||\Phi||$ is the *norm* of Φ.)

The theorem is most easily proved when the eigenvalues of Φ are distinct. In this case there is always a similarity transformation T which transforms Φ to diagonal form:

$$\Phi = T\Lambda T^{-1}$$

where

$$\Lambda = \mathrm{diag}[\lambda_1, \lambda_2, \cdots \lambda_k], \quad |\lambda_1| \le |\lambda_2| \le \cdots \le |\lambda_k|$$

Let

$$x_l = T\xi_l$$

then (8.13) becomes

$$\xi_\nu = \Lambda^{\nu-n} \xi_n$$

Clearly, every component of ξ_ν converges to zero with increasing ν if the magnitude of every eigenvalue is less than unity:

$$|\lambda_i| < 0, \ \text{for all} \ i = 1, 2, \cdots k$$

which is surely the case if $||\Phi|| = |\lambda_k|$ is less than unity. On the other hand, if $||\Phi|| = 1$, then one component of ξ, namely ξ_k, will not converge to zero but will oscillate with an amplitude of unity. Finally, if $||\Phi|| > 1$, then ξ_k will grow without bound.

Since $x = T\xi$, x and ξ converge or diverge together. This establishes the theorem for the case in which the eigenvalues of Φ are distinct.

The proof for the case of repeated eigenvalues is analogous to the proof used for repeated eigenvalues in a continuous-time system. (See Problem 8.3.)

The eigenvalues of Φ also determine whether or not the discrete-time system (8.5) is stable in the presence of an input, i.e., whether or not every bounded input sequence $\{u_1, u_2, \ldots\}$ produces a bounded-state sequence $\{x_1, x_2, \ldots\}$. Using a method like that used in a continuous-time system, we can establish:

Theorem 8.2 (Bounded-Input, Bounded-Output Stability) *If the unforced system*

$$x_{n+1} = \Phi x_n$$

is asymptotically stable, then every bounded input sequence $\{u_1, u_2, \ldots\}$ *produces a bounded state sequence* $\{x_1, x_2, \ldots\}$.

For reasons related to controllability, the converse of this theorem is not always true. Moreover, the unforced system must be *asymptotically* stable, not merely stable. The first-order system with $\Phi = 1$ (the discrete-time equivalent of an integrator), for example, is stable, but not every bounded input sequence, e.g. $\{1, 1, 1, \ldots\}$, produces a bounded state sequence.

The eigenvalues of Φ being smaller than unity in magnitude means that they all lie within the unit circle of the complex plane. Hence the unit circle (Figure 8.5) in discrete-time systems plays the same role as the imaginary axis of continuous-time systems: It is the boundary between asymptotic stability and instability.

Example 8.3   Computer Control of D-C Servomotor

Computer control of a d-c servomotor provides a simple example of stability analysis. A block diagram of the control system is shown in Figure 8.6. In this application the reference position θ_R is an input to the digital computer, the function of which is to subtract the sampled-and-held value of the actual angular position from the reference value, multiply the difference by a gain g and convert the result to an analog signal.

The control law for this process is simply

$$u_n = g(\theta_R - \theta_n) = g\theta_R - Gx_n$$

where

$$G = \begin{bmatrix} g & 0 \end{bmatrix}$$

The closed-loop discrete-time dynamics are expressed by

$$x_{n+1} = \Phi x_n + \Gamma(g\theta_R - Gx_n)$$
$$= (\Phi - \Gamma G)x_n + \Gamma g\theta_R$$

For stability analysis it is necessary only to consider the characteristic equation of

$$\Phi_c = \Phi - \Gamma G = \begin{bmatrix} 1 - g(T - 1 + e^{-\alpha T}) & 1 - e^{-\alpha T} \\ -g(1 - e^{-\alpha T}) & e^{-\alpha T} \end{bmatrix}$$

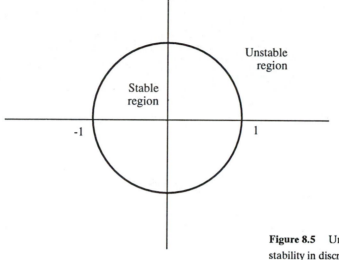

Figure 8.5 Unit circle is the boundary of stability in discrete- time systems.

The characteristic equation of the closed-loop system is

$$|\lambda I - \Phi_c| = (\lambda - 1)(\lambda - e^{-\alpha T}) + K[(\alpha T - 1 + e^{-\alpha T})(\lambda - 1 - e^{-\alpha T}) + (1 - e^{-\alpha T})^2/\alpha]$$

The variation of the stability with the gain g can be analyzed by plotting a root locus. For this purpose, the open-loop "poles" lie at

$$\lambda = 1, \quad e^{-\alpha T}$$

and the open-loop "zero" lies at

$$\lambda = 1 - (1 - e^{-\alpha T})^2/(\alpha T - 1 + e^{-\alpha T})$$

The corresponding root-locus is shown in Figure 8.7.

The maximum allowable gain g_{max} is the gain at which the root-locus crosses the unit circle. Following through with the necessary algebra, we find that the maximum allowable gain is

$$g_{max} = \frac{\alpha(1 - e^{-\alpha T})}{1 - e^{-\alpha T} - \alpha T e^{-\alpha T}}$$

It would be of interest to compare the exact expression for g_{max} with the value obtained by approximating the sample-and-hold operation by a simple delay of $T/2$ seconds. (See Problem 8.2.)

8.4   THE Z-TRANSFORM

In the 1950s, before the flowering of state-space methods, a number of investigators sought a transform method that would serve the role for discrete-time systems

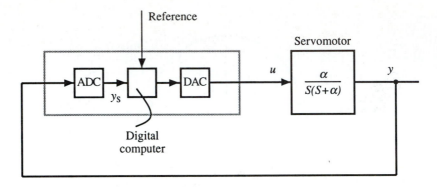

Figure 8.6   Digitally controlled servomotor.

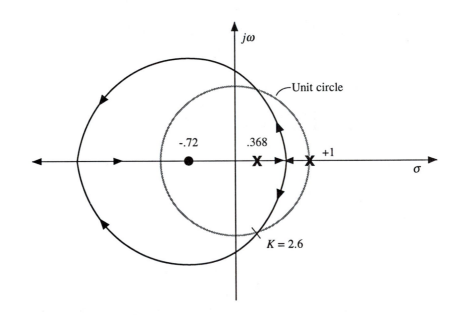

Figure 8.7   Root locus plot for digitally controlled servomotor.

that the Laplace transform served for continuous-time systems. Such a transform would make it possible to represent each discrete-time subsystem by a transfer function which could be combined with the transfer functions of the other subsystems by algebraic methods to obtain the transfer function of the overall system. The latter transfer function could then be used to calculate the response of the system to an arbitrary input, and the familiar methods developed for continuous-time systems could be used to study the properties (e.g., stability, margins) of closed-loop discrete-time systems. After a period of uncertain terminology and notation, the Z-transform ultimately emerged. Although state-space methods have reduced the importance of the Z-transform in control-systems analysis and design, it nevertheless remains a useful tool. Moreover, the Z-transform has become increasingly useful in the field of digital signal processing.

Historically (see Note 8.1), the variable z of the Z-transform has been associated with the continuous-time shift operator:

$$z = e^{sT}$$

This association may aid those who have developed an intuitive grasp of continuous-time systems to transfer some of that intuition to discrete-time systems. It is not necessary, however, to identify z with e^{sT} to understand the theory of the Z-transform; for some this association may even prove to be an impediment.

The Z-transform of a discrete-time sequence $\{x_0, x_1, x_2, \ldots\}$ is defined by

$$X(z) := x_0 + x_1 z^{-1} + x_2 z^{-2} + \cdots = \sum_{n=0}^{\infty} x_n z^{-n} := \mathcal{Z}[x_n] \qquad (8.14)$$

The Z-transform of the sequence is sometimes called its *generating function* since the expansion of $X(z)$ in a power series in the variable z^{-1} "generates" the sequence. In the literature of numerical analysis the generating function is usually defined as a power series of positive powers of some variable, say q. This creates an unfortunate discrepancy between formulas of numerical analysis and of discrete-time systems engineering, but the notations are too firmly established in their realms to be changed.

Extensive tables of Z-transforms and their properties appear in many textbooks. (See, for example [1] and [2].)

The property of the Z-transform that is most useful for analysis of discrete-time systems is the Z-transform of x_{n+1}. Using the definition (8.14):

$$\mathcal{Z}[x_{n+1}] = \sum_{n=0}^{\infty} x_{n+1} z^{-n}$$

$$= \sum_{n=0}^{\infty} x_{n+1} z^{-(n+1)} z$$

$$= z \sum_{k=1}^{\infty} x_k z^{-k}$$

$$= z(X(z) - x_0) \qquad (8.15)$$

Note the similarity of this expression to the expression for the Laplace transform of the derivative of a continuous-time function. But also note the difference: Unlike the continuous-time case, here the initial state x_0 is multiplied by z. Because of the similarity, many relations for discrete-time systems are similar in appearance to their continuous-time counterparts; but there are also numerous frustrating differences.

Equation (8.15) obviously applies to vectors as well as to scalars. Thus it can be applied to the basic discrete-time dynamic model (8.1). Take the Z-transform of both sides of (8.1) to obtain

$$zX(z) - zx_0 = \Phi X(z) + \Gamma U(z)$$

Solve for $X(z)$:

$$X(z) = (zI - \Phi)^{-1}\Gamma U(z) + (zI - \Phi)^{-1}zx_0 \qquad (8.16)$$

This expression is the Z-transform of the time-domain relation (8.12). (The presence of the extra z spoils the complete analogy with continuous-time systems.)

Suppose the discrete-time system has an output given by

$$y_n = Cx_n + Du_n \qquad (8.17)$$

The Z-transformed relation is

$$Y(z) = CX(z) + DU(z) \qquad (8.18)$$

Using (8.16) and (8.18) gives the general input-output relation

$$Y(z) = [C(zI - \Phi)^{-1}\Gamma + D]U(z) + C(zI - \Phi)^{-1}zx_0 \qquad (8.19)$$

If the initial state $x_0 = 0$, the input-output relation (8.19) becomes

$$Y(z) = H(z)X(z) \qquad (8.20)$$

where
$$H(z) = C(zI - \Phi)^{-1}\Gamma + D \qquad (8.21)$$

The matrix $H(z)$ that relates the Z-transforms of the output and the input can be termed the *discrete-time transfer function*. An earlier   term is the *pulse or pulsed transfer function*. (See Note 8.1.) If the discrete-time system has l inputs and m outputs, $H(z)$ is an $m \times l$ matrix.

In exactly the same way that transfer functions are combined in continuous-time systems, the discrete-time transfer functions of discrete-time subsystems can be combined to obtain the discrete-time transfer function of the overall system in which they appear. A system consisting of a discrete-time plant with a transfer function (matrix) $G(z)$ and a discrete-time compensator (matrix) $D(z)$, as shown in Figure 8.8, has the closed-loop transfer function

$$H(z) = [I + G(z)D(z)]^{-1}G(z)D(z) = G(z)D(z)[I + G(z)D(z)]^{-1} \qquad (8.22)$$

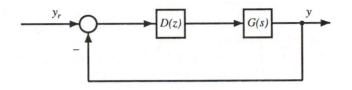

Figure 8.8   Closed-loop system including plant and digital compensator.

The system shown in Figure 8.9 could represent a computer-controlled system; $D(z)$ represents the discrete-time transfer function implemented in the computer. Implicit in $D(z)$ is the analog-to-digital conversion, the sampling of the result, and the generation of the piecewise-constant output (vector) and its digital-to-analog conversion. The transfer function $G(z)$ represents the discrete-time model of the continuous-time plant with the piecewise-constant excitation produced by the digital-to-analog converter "hidden" in $D(z)$.

It is important to recognize that the discrete-time transfer functions of subsystems can be combined only if they are truly discrete-time systems. If two continuous-time subsystems G_1 and G_2 are connected in tandem, for example, as shown in Figure 8.9, it is necessary to combine their continuous-time transfer functions into a single continuous-time transfer function

$$G(s) = G_2(s)G_1(s)$$

and get the discrete-time transfer function $G_d(z)$ from $G(s)$. It would be wrong to first get the discrete-time transfer functions G_{1d} and G_{2d} corresponding to $G_1(s)$ and $G_2(s)$ and then combine them, because, in general

$$G_d(z) \neq G_{2d}(z)G_{1d}(z)$$

There are some exceptions. For example, if either G_1 or G_2 consists of only gain elements (i.e., no dynamics), then $G_d(z) = G_{2d}(z)G_{1d}(z)$.

In combining transfer functions of discrete-time subsystems that originate from continuous-time subsystems, it is customary in some circles to use the same symbol for the discrete-time transfer function and for the continuous-time transfer function from which it comes. In other words, if $G(s)$ is the transfer function of a continuous-time system with a piecewise-constant input, its discrete-time transfer function would be designated by $G(z)$. This makes the resulting expressions simpler and saves writing, but strictly it is an abuse of notation, since the discrete-time transfer function cannot be obtained from the continuous-time transfer function by simply replacing s by z. It would be preferable to use a distinctive symbol, for example, adding a subscript d, as in $G_d(z)$, or using a distinctive type font such as $\mathsf{G}(z)$. But this is not always done.

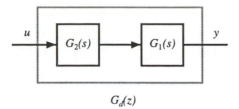

$$G_d(z)$$

Figure 8.9   Subsystems in tandem.

Example 8.4   Analysis of Servomotor Control System by $\mathcal{Z}$-transform

Using (8.21) the (open loop) discrete-time transfer function of the servomotor considered in Example 8.2 is

$$G(z) = \begin{bmatrix} 1 & 0 \end{bmatrix} \begin{bmatrix} z - 1 & -\dfrac{1-e^{-\alpha T}}{\alpha} \\ 0 & 1 - e^{-\alpha T} \end{bmatrix}^{-1} \begin{bmatrix} \dfrac{\alpha T - 1 + e^{-\alpha T}}{\alpha} \\ 1 - e^{-\alpha T} \end{bmatrix}$$

$$= \frac{(\alpha T - 1 + e^{-\alpha T})z + 1 - e^{-\alpha T}(1 + \alpha T)}{(z-1)(z - e^{-\alpha T})}$$

8.5 STABILITY ANALYSIS BY THE Z-TRANSFORM

You can use the Z-transform to analyze the stability of discrete-time systems the way you use the Laplace transform in continuous-time systems. There are tabular algorithms analogous to the Routh-Hurwitz algorithms of continuous-time systems, and graphical techniques similar to the root-locus, Nyquist, and Bode plot of continuous-time systems.

Suppose the transfer function of a system is

$$H(z) = \frac{N(z)}{\Delta(z)} = \frac{b_0 z^k + b_1 z^{k-1} + \cdots + b_k}{z^k + a_1 z^{k-1} + \cdots + a_k} \tag{8.23}$$

By partial fractions, for example, it is easy to see that the roots of $\Delta(z)$ are the poles, or eigenvalues, of the system. If all the poles are within the unit circle of the z plane, the system is asymptotically stable; if one or more poles lie outside the unit circle, the system is unstable. The unit circle of the z-plane thus corresponds to the imaginary axis of the s plane of continuous-time stability analysis. The goal of stability analysis in the z plane is thus to establish whether or not the characteristic polynomial $\Delta(z)$ has any roots outside the unit circle.

The unit circle is more troublesome for computation than the imaginary axis, hence the tabular and graphical techniques for discrete-time systems are messier than their continuous-time counterparts.

There are two ways to approach the stability analysis, namely

1. Bilinear transformation of the unit circle to the imaginary axis.

2. Direct analysis in the z-plane.

The bilinear-transformation approach is based on the observation that z^{-1} represents a delay of one sampling interval. This suggests identifying points in the s-plane with points in the z-plane by the transformation

$$z^{-1} = e^{-sT}$$

or

$$z = e^{sT} \tag{8.24}$$

It is easy to establish that the transformation (8.24) transforms the interior of the unit circle of the z-plane into the left half of the s-plane. But, alas, using e^{sT} for z in $\bar{H}(z)$ results in a transcendental expression of no benefit in tabular or graphical analysis. Instead of using 8.24, however, we can use the Padé approximant (see Section 7.8.2)

$$z = \frac{1 + sT/2}{1 - sT/2} \tag{8.25}$$

It is not too difficult to show that the bilinear transformation (8.25) also transforms the interior of the unit circle of the z-plane to the left half of the s-plane. But in this case the resulting transformed transfer function is a rational function of s, i.e.,

$$H\left(\frac{1 + sT/2}{1 - sT/2}\right) = \bar{H}(s) = \frac{c_0 s^k + c_1 s^{k-1} + \cdots + c_k}{s^k + d_1 z^{k-1} + \cdots + d_k} \tag{8.26}$$

(Hand calculation of $\bar{H}(s)$ can be formidable.)

You can use the resulting transformed transfer function $\bar{H}(s)$ for frequency domain analysis. For example, its denominator

$$\bar{\Delta}(s) = s^k + d_1 z^{k-1} + \cdots + d_k$$

can be analyzed for stability using the well-known, continuous-time Routh- Hurwitz algorithms. Or, if $G(z)$ is the forward loop transmission of a discrete-time system, you can investigate the equivalent return difference

$$\bar{T}(s) = 1 + K\bar{G}(s)$$

by the root-locus method or by using Nyquist or Bode plots. (Interpreting the resulting gain and phase margins, however, can be problematical.)

The alternative to using the bilinear transformation (8.25) is to work directly in the z-domain. A tabular algorithm, due to Jury and Blanchard [3], analogous to the Routh algorithm, is available. The algorithm is rather cumbersome, however, and rarely used, especially since the roots of the characteristic polynomial can nowadays be extracted numerically with no great difficulty.

You can also use the root-locus algorithm directly on the discrete- time return difference

$$T(z) = 1 + KG(z)$$

The critical gain $K_{\max}$ is the value of K at which the root-locus crosses the *unit circle*. This has an important implication for stability: Recall that the number of branches of the root locus plot that go to infinity is equal to the excess e of poles over zeros $G(z)$. If $G(z)$ is a proper rational function, then $e \geq 1$. Hence, if $G(z)$ is a proper rational function, at least one branch of the root locus will cross the unit circle enroute to infinity. This leads to the conclusion that for every system having a proper transfer open-loop transfer function, it is not possible to raise the gain beyond a definite value $K_{\max}$ without the system becoming unstable.

You can also plot a Nyquist diagram, using the unit circle in the z plane in place of the $j\omega$ axis. The result is interpreted as you would in a continuous-time system: If the map of the exterior of the unit circle of the z-plane by $G(z)$ does not cover the point $-1/K$, the closed-loop system is stable, and vice-versa.

8.6 CONTROLLABILITY AND OBSERVABILITY

The concepts of controllability and observability in discrete-time systems are as important as they are in continuous-time systems. Discrete-time systems have an advantage over the latter, however, in one respect: The concepts are easier to explain and the criteria are easier to derive.

8.6.1 Controllability

A discrete-time process is said to be controllable if and only if there exists a sequence of inputs $u_n, u_{n+1}, \ldots u_{n+N}$ that can take the system from any starting state x_n to any other state $x_N = x_d$ in a finite number N of steps.

We saw earlier that in a time-invariant system of order k

$$x_{n+k} = \Phi^k x_n + \Phi^{k-1}\Gamma u_n + \Phi^{k-2}\Gamma u_{n+1} + \cdots + \Gamma u_{n+k}$$

$$= \Phi^k x_n + \begin{bmatrix} \Gamma & \Phi\Gamma & \cdots & \Phi^{k-1}\Gamma \end{bmatrix} \begin{bmatrix} u_{n+k} \\ \vdots \\ u_{n+1} \\ u_n \end{bmatrix}$$

which can be written

$$\mathbf{C} \begin{bmatrix} u_{n+k} \\ \vdots \\ u_{n+1} \\ u_n \end{bmatrix} = x_{n+k} - \Phi^k x_n \tag{8.27}$$

where

$$\mathbf{C} := \begin{bmatrix} \Gamma & \Phi\Gamma & \cdots & \Phi^{n-1}\Gamma \end{bmatrix} \tag{8.28}$$

In a single-input system, $\mathbf{C}$ is a square $(k \times k)$ matrix. Hence, if it is nonsingular (i.e., its rank is k), the sequence of inputs that takes x_n to the desired state $x_{n+k} = x_d$ in exactly k steps is explicitly given by

$$\begin{bmatrix} u_{n+k} \\ \vdots \\ u_{n+1} \\ u_n \end{bmatrix} = \mathbf{C}^{-1}[x_d - \Phi^k x_n]$$

Thus we see that a single-input time-invariant system is controllable if the *controllability matrix* $\mathbf{C}$ defined by (8.28) is of rank k. If the system has more than one input, but the rank of $\mathbf{C}$ is k then (8.27) in general has at least one solution; it may be possible to reach the desired state in fewer than k steps or to minimize some performance criterion in the process of going to the desired state. (See Problem 8.4.) But in any case, we see that a sufficient condition for the controllability of a time-invariant system is that the rank of the controllability matrix is k.

This condition is also necessary. For suppose it were possible to reach the desired state in $N > k$ steps. Then we would have to be able to solve the following equation for the input sequence $u_n, \ldots u_{n+N}$:

$$\begin{bmatrix} \Gamma & \Phi\Gamma & \cdots & \Phi^{N-1}\Gamma \end{bmatrix} \begin{bmatrix} u_{n+N} \\ \vdots \\ u_{n+1} \\ u_n \end{bmatrix} = x_d - \Phi^N x_n$$

which would require that the matrix

$$\check{\mathbf{C}} = \begin{bmatrix} \Gamma & \Phi\Gamma & \cdots & \Phi^{N-1}\Gamma \end{bmatrix}$$

be of rank k, i.e., that the columns added to $\mathbf{C}$ to obtain $\check{\mathbf{C}}$ would raise the rank of the latter to the required number k. But, by the Cayley-Hamilton Theorem [4],

$$\Phi^k = -a_1\Phi^{k-1} - a_2\Phi^{k-1} - \cdots - a_k I$$

so any new columns in $\check{\mathbf{C}}$ are linearly dependent on the columns of $\mathbf{C}$ and hence the rank of the former cannot be raised by using a longer input sequence.

These results can be summarized by the following theorem:

Theorem 8.3 *The necessary and sufficient condition for the controllability of a time-invariant, discrete-time process $x_{n+1} = \Phi x_n + \Gamma u_n$ is that the controllability matrix* $\mathbf{C} := \begin{bmatrix} \Gamma & \Phi\Gamma & \cdots & \Phi^{n-1}\Gamma \end{bmatrix}$ *be of rank k = order of the process.*

It is worth emphasizing that this theorem is *not* valid for time-varying systems; even the notion of Φ^k is ambiguous.

8.6.2   Observability

An unforced, discrete-time system is said to be observable if it is possible to determine the initial state of the system by making a finite sequence of observations of the output of the system.

Consider the time-invariant system

$$x_{n+1} = \Phi x_n \tag{8.29}$$

$$y_n = C x_n \tag{8.30}$$

The sequence of outputs for the initial state x_0 are given by

$$y_0 = C x_0$$

$$y_1 = C x_1 = C \Phi x_0$$

$$\cdots$$

$$y_{k-1} = C x_{k-1} = C \Phi^{k-1} x_0$$

which can be arranged in matrix form

$$\begin{bmatrix} y_0 \\ y_1 \\ \vdots \\ y_{k-1} \end{bmatrix} = \mathbf{O}' x_0 \tag{8.31}$$

where

$$\mathbf{O} := \begin{bmatrix} C' & \Phi'C' & \cdots & (\Phi')^{k-1}C' \end{bmatrix} \tag{8.32}$$

is defined as the discrete-time *observability matrix*.

If the rank of $\mathbf{O}$ is k, then (8.31) can be solved for x_0; for a scalar observation $\mathbf{O}$ is a square matrix and we can solve explicitly for the initial state:

$$x_0 = (\mathbf{O}')^{-1} \begin{bmatrix} y_0 \\ y_1 \\ \vdots \\ y_{k-1} \end{bmatrix}$$

On the other hand, if the rank of $\mathbf{O}$ is less than k, taking more observations will not avail; as a consequence of the Cayley-Hamilton Theorem, these observations will be linearly dependent on the first k observations.

Thus we have the following theorem:

Theorem 8.4 *The necessary and sufficient condition for the observability of a time-invariant, discrete-time process $x_{n+1} = \Phi x_n$ with observations $y_n = C x_n$ is that the observability matrix*

$$\mathbf{O} := \begin{bmatrix} C' & \Phi'C' & \cdots & (\Phi')^{k-1}C' \end{bmatrix}$$

be of rank k = order of the process.

This theorem does not make sense for a time-varying system.

8.7 FINITE TIME DELAY IN DISCRETE-TIME SYSTEMS

In some applications there is a finite time delay (in addition to the time delay introduced by the implementation of the digital control) between the instant the output of the plant is measured and the instant a control signal is applied to the actuator in the plant. If the plant is a process on a spacecraft, for example, and the loop is closed by a computer on the earth, the finite time that it takes for signals to be transmitted between the earth and the spacecraft often must be taken into consideration. Finite time delays frequently occur in chemical and industrial processes (where they are sometimes called "transport lags") and are the result of the amount of time it takes for material to be moved from one part of the physical plant to another. Finite time delays are also present in processes that include human operators: A time delay (called "reaction time") of the order of tens of milliseconds occurs between the instant of perception of a physical quantity and the earliest reaction to it.

A time delay that is an exact multiple of the sampling period T is easily accommodated within the framework of discrete-time system theory.

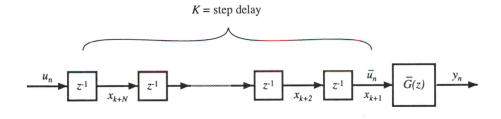

Figure 8.10 Process with K step input delay.

Figure 8.10 shows a process in which there is a delay of K steps between the application of an input command u_n and the physical control signal $\bar{u}$. Suppose that the discrete-time transfer function from the physical control to the output is

$$\bar{G}(z) = \frac{Y(z)}{\bar{U}(z)}$$

Since the discrete-time transfer function for a one-step delay is z^{-1}, and a K-step delay is equivalent to K one-step delays in cascade, the discrete-time transfer function of the K-step delay is

$$\frac{\bar{U}(z)}{U(z)} = z^{-K}$$

Thus the transfer function from the command input to the output is

$$G(z) = \frac{Y(z)}{U(z)} = z^{-K}\tilde{G}(z) = \frac{1}{z^K}\tilde{G}(z)$$

Although this transfer function has a pole of order K at the origin, it is perfectly acceptable for purposes of analysis and design. Because of the pole at the origin, however, this process is an example of one in which the state transition matrix Φ is singular and hence cannot be derived from a finite-order continuous-time state-space model.

A state-space model for the process can be obtained directly, without use of the Z-transform, by assigning state variables to $\bar{u}_n, \ldots \bar{u}_{n-K-1}$, and adjoining them to the state-space description of the physical process. Suppose that the latter is kth order and governed by

$$x_{n+1} = \Phi x_n + \Gamma \bar{u}_n$$

We would define

$$\bar{u}_n := x_{k+1}$$

Then we would adjoin the following equations

$$(x_{k+1})_{n+1} = (x_{k+2})_n$$
$$(x_{k+2})_{n+1} = (x_{k+3})_n$$
$$\vdots$$
$$(x_{k+N-1})_{n+1} = (x_{k+N})_n$$
$$(x_{k+N})_{n+1} = u_n$$

In vector-matrix form the above system would be expressed as

$$\mathbf{x}_{n+1} = \underline{\Phi}\mathbf{x}_n + \underline{\Gamma}u_n \tag{8.33}$$

where

$$\mathbf{x} = \begin{bmatrix} x_1 \\ \vdots \\ x_k \\ \cdots \\ x_{k+1} \\ \vdots \\ x_K \end{bmatrix} \qquad \underline{\Phi} = \begin{bmatrix} \Phi & \vdots & \Gamma \\ \cdots & \cdots & \cdots \\ 0 & \vdots & \Delta_K \end{bmatrix} \qquad \underline{\Gamma} = \begin{bmatrix} 0 \\ \cdots \\ \Lambda_K \end{bmatrix}$$

with

$$\Delta_K = \begin{bmatrix} 0 & 1 & \cdots & 0 & 0 \\ 0 & 0 & \cdots & 0 & 0 \\ & & \cdots & & \\ 0 & 0 & \cdots & 0 & 1 \\ 0 & 0 & \cdots & 0 & 0 \end{bmatrix} \qquad \Lambda_K = \begin{bmatrix} 0 \\ 0 \\ \vdots \\ 0 \\ 1 \end{bmatrix}$$

i.e., Δ_K is a $K \times K$ matrix that contains a $(K - 1) \times (K - 1)$ identity matrix in the superdiagonal position and zero everywhere else; and Λ_K is a K-vector that has a 1 in the Kth position and zero elsewhere.

The system (8.33) is controllable. To show this, consider first the-one-step-delay case, for which

$$\underline{\Phi} = \begin{bmatrix} \Phi & \vdots & \Gamma \\ \cdots & \cdots & \cdots \\ 0 & \vdots & 1 \end{bmatrix} \qquad \underline{\Gamma} = \begin{bmatrix} 0 \\ 0 \\ \vdots \\ 1 \end{bmatrix}$$

and the controllability matrix is given by

$$\underline{C} = \begin{bmatrix} 0 & \vdots & \Gamma & \Phi\Gamma & \cdots & \Phi^{k-1}\Gamma \\ \cdots & & \cdots\cdots\cdots\cdots & & & \\ 1 & \vdots & 0 & 0 & \cdots & 0 \end{bmatrix} = \begin{bmatrix} 0 & \vdots & C \\ \cdots & & \cdots \\ 1 & \vdots & 0 \end{bmatrix}$$

where C is the observability of the process with undelayed input. Clearly the first column of $\underline{C}$ is independent of its other columns; hence if C is of rank k, then $\underline{C}$ is of rank $(k + 1)$.

It follows by induction that the system is controllable for a K-step delay.

A similar analysis can be performed for observability. If the output of the system with undelayed input is

$$y_n = Cx_n$$

then for the state of the augmented system

$$y_n = \underline{C}\mathbf{x}_n = \begin{bmatrix} C & \vdots & 0 \end{bmatrix} \begin{bmatrix} x_1 \\ \vdots \\ x_k \\ \cdots \\ x_{k+1} \\ \vdots \\ x_K \end{bmatrix}$$

To demonstrate observability, form the observability matrix for $K = 1$ and show that its rank is $k + 1$, then extend the result to arbitrary K by induction. (See Problem 8.5.)

8.8 NONLINEAR DISCRETE-TIME SYSTEMS

Discrete-time systems can of course be nonlinear. Such systems typically arise when continuous-time processes are controlled by piecewise-constant inputs such as would be produced by a digital computer.

The evolution of a nonlinear discrete-time system can be characterized by a transition equation of the form

$$x_{n+1} = \phi(x_n, u_n) \tag{8.34}$$

The function $\phi(\ ,\ )$ may be termed the *transition function* and is a generalization of (8.5) to a nonlinear, discrete-time system. But a nonlinear discrete-time process differs from a linear discrete-time process in that a formula like (8.5) for the transition function can rarely be obtained from the nonlinear differential equations that represent the underlying continuous-time process. Rather, the transition function is defined implicitly as the result obtained by integrating the differential equation

$$\dot{x} = f(x, u)$$

of the process over the interval $t \in [nT, (n+1)T]$ starting in the state

$$x(nT) = x_n$$

with constant input

$$u(t) = u_n \quad \text{for } nT \le t \le (n+1)T$$

An example of an application in which it is possible to characterize a nonlinear, discrete-time process without the need for numerical integration of the continuous-time differential equations is the following.

Example 8.5   Orbital Elements of a Satellite

The equations of motion of a satellite (regarded as a rigid body in a central inverse-square-law force field) can be expressed in many coordinate systems and its equations of motion can be readily integrated numerically to obtain the transition function. Although an analytical expression for the solution of the differential equations is not available, but the qualitative features of the orbit are known:

- The orbit lies in a plane containing the point of attraction.

- The orbit is elliptical, with the center of attraction being at one of the foci of the ellipse.

- *Kepler's Law*—The motion of the satellite on its orbit is such as to sweep out equal areas in equal times.

In the orbital plane the satellite motion can thus be described by the shape of the elliptical orbit and its position on the orbit. The quantities that define the position of the satellite are known as *orbital elements*. In a plane, four orbital elements, corresponding to the two components of position and velocity, are required. Various sets of orbital elements are used for different purposes; one set in common use is the following:

$$a = \text{semi-major axis}$$
$$e = \text{eccentricity}$$
$$\omega = \text{argument of perigee}$$
$$\theta = \text{eccentric anomaly}$$

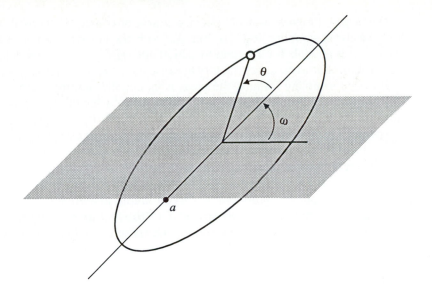

Figure 8.11 Orbital elements of a satellite.

See Figure 8.11.

In the absence of forces other than the central gravitational force, the first three of the orbital elements are constant. The first two characterize the shape of the orbit and the third specifies its orientation. Only the fourth, the eccentric anomaly, varies with time. Thus we can characterize motion of the satellite by the following difference equations:

$$a_{n+1} = a_n$$
$$e_{n+1} = e_n$$
$$\omega_{n+1} = \omega_n$$
$$\theta_{n+1} = \phi(\theta_n, e_n, a_n, T)$$

A closed-form expression for the nonlinear function $\phi(\ )$ is not available; it is defined implicitly as the solution of the transcendental equation known as *Kepler's Equation*:

$$\theta_{n+1} - \theta_n - e_n(\sin \theta_{n+1} - \sin \theta_n) - \sqrt{\mu a_n^3}\, T = 0$$

where T is the sampling time and μ is a gravitational constant.

8.9 PRACTICAL CONSIDERATIONS

The discrete-time representation (state space or Z-transform) developed above describes the behavior of the system only at sampling instants. Any control system design based on a discrete-time representation is likewise limited to the behavior at sampling instants.

When you design a control system you are, of course, concerned not only with the behavior at sampling instants, but also with the behavior between the sampling instants. It is possible for a system to exhibit apparently satisfactory behavior at sampling instants but to be unsatisfactory because of excessive *"intersample ripple."*

Between sampling instants the plant is essentially running open-loop, since the control signal does not depend on the process state during the current sampling interval. Thus the intersample behavior is governed entirely by the open-loop dynamics of the system. The plant state is a combination of the unforced response (i.e., the impulse response) and the step response (assuming that the input is piecewise-constant). If the plant is linear, both of these depend on its open-loop eigenvalues. Moreover, unless the plant is high-pass (i.e., the D matrix is nonzero), the response will be continuous through the sampling instants and hence continuous at all times.

If the open-loop eigenvalues are close to the real axis, giving the plant a well-damped natural response, the intersample behavior is not likely to be problematical. If, on the other hand, the "dominant" eigenvalues of the open-loop plant are near the imaginary axis, and hence likely damped, the intersample behavior can be oscillatory or have excessive overshoot.

Since the notion of dominant poles and their damping is somewhat vague, simulation is still the best way to determine the actual response of a discrete-time system. Nevertheless, these concepts can be helpful in thinking about practical issues such as selecting a sampling interval and use of *"anti-aliasing" filters.*

8.9.1   Choice of sampling interval

One of the important practical considerations in the overall design of a discrete-time control system is the choice of the sampling rate. One question is whether there is an "optimum" sampling rate. If performance is the only consideration, the answer is that the optimum sampling rate is infinity. This is intuitively apparent from consideration of the effect of sampling delay on the phase shift of the system.

But a more rigorous argument can be advanced. Suppose there is an optimum sampling rate, in the sense that it minimizes some performance criterion that does not depend explicitly on the sampling rate, such a minimum mean-square error, or overshoot. If so, a higher sampling rate would *increase* the value of the performance criterion regardless of the control algorithm. Now consider doubling the sampling rate, but using a control algorithm that ignores the additional data and keeps the control signal constant for two of the new sampling intervals. This algorithm will obviously produce identical behavior to that of the presumed optimum sampling interval, contradicting the original hypothesis. Hence it is clear that no finite sampling frequency can be optimum.

Although an infinite sampling rate may theoretically be optimum, it has practical limits. Cost is one. The cost of analog-to-digital and digital-to-analog converters tends to increase in proportion to the sampling frequency, hence the use of a higher than "necessary" sampling interval could add unwarranted cost to the system. Similarly, use of too high a sampling frequency could impose an unnecessary computational burden on the control computer.

Another possible consideration is the number of bits of precision of the sampler. (This is also a cost consideration.) If the sampling frequency is high relative to the bandwidth of the open-loop system, the system may be doing almost nothing for a number of sampling intervals, and the major component of the input to the control computer would be due to the quantization noise. Thus trouble-free use of a high sampling rate may require high sampling precision as well.

How high a sampling frequency is necessary? One oft-cited guideline is the so-called *sampling theorem*, which asserts [5] that if a signal has a spectrum that vanishes above W radians per second, it can be reconstructed from its values at a denumerable set of sampling intervals of $T = \pi/W$ sec apart.

The frequency in Hz of the band-limited signal is $F = W/2\pi$, hence the sampling theorem asserts that the band-limited signal can be reconstructed from samples taken at a frequency

$$F_n = 2F$$

This frequency is often called the *Nyquist frequency*.

Using the Nyquist frequency as a design guideline requires interpreting the definition of a *band-limited* signal, i.e., a signal with a spectrum that vanishes outside a given frequency interval, in terms of a real, physical system. (It can be shown that a band-limited signal is non-causal because its impulse response must occur prior to the exciting impulse.) One way of interpreting F is as the 3 db bandwidth (i.e., the "corner" frequency) of the system. This interpretation might be reasonable for moderately damped systems (i.e., for systems with dominant poles near the 45 degree line) but may be too high for well-damped systems and too low for oscillatory systems. Suppose, for example that the dominant poles are near the imaginary axis. The impulse response would be a lightly damped sinusoid. Sampling at the Nyquist frequency would produce only about two samples for each cycle of oscillation, which would probably be too low for practical reasons. If cost is not an important factor, it would be prudent to sample the oscillatory signal no less than 16 times per cycle, i.e., to use a frequency of 8 times the Nyquist frequency when the open loop system is oscillatory. If the cost of hardware precludes using a sampling frequency this high, it might be possible to get by with a lower sampling frequency, but you should carefully evaluate the design by including a suitable representation of the sampling operation in the simulation of the process.

The real issue in selection of a sampling frequency is not the bandwidth of the plant but its uncertainty. In theory, if intersample behavior is not considered, any sampling frequency that is not commensurable with the imaginary part of the open-loop poles could be used as a sampling frequency. Suppose, however, that you were to select a sampling frequency for a lightly-damped system only slightly higher than the Nyquist frequency, as indicated in Figure 8.12(a). A slight change in the oscillation frequency of the open-loop system could completely change the phase relation between the input and the output, erroneously predicting, for example, a positive output for a given positive input when in fact the output is really negative. The situation would be exacerbated by a sampling frequency lower than the Nyquist frequency. On the other hand, if you use a sampling frequency that is much higher than the resonant frequency, it would be reasonable to suppose that slight changes in the resonant frequency would have minimal effect. (See Problem 8.7.)

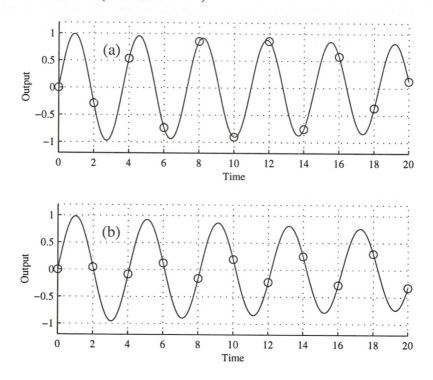

Figure 8.12   Effect of sampling too slowly. (a) Nominal plant
(b) Plant frequency slightly changed.

8.9.2 Aliasing

Since a discrete-time model does not reveal the behavior of a system between sampling instants, it is possible for a periodic signal having the same frequency as the sampling frequency to go undetected at the input to the sampling device. This possibility may produce the phenomenon called *"aliasing."* Such a signal could arise in a feedback system, for example, when an unmodeled resonance (recall Chapter 7.) is present near a harmonic of the sampling frequency. The combined phenomena of resonance and aliasing could produce a signal of sufficient amplitude to significantly degrade the performance of the closed-loop system, or even drive it unstable.

To avoid the possible deleterious effects due to aliasing, an *anti-aliasing filter* is often used. This filter, which must be implemented by analog means (why?), is a low-pass filter with a cut-off frequency of about twice the sampling frequency. (Twice the sampling frequency is justified on the theory that no information above the Nyquist frequency would pass the sampling operation in any event). An anti-aliasing filter is rarely necessary if the sampling frequency is much higher than the bandwidth of the plant, since the filtering action of the plant itself provides the loop attenuation that the filter would give. If the sampling frequency is relatively low, however, and the consequent need for an anti- aliasing filter is suspected, it is important to include the dynamics of the filter in the model of the plant. Although the attenuation and phase shift at frequencies below the cut-off frequency of the filter may be small, they may not be entirely negligible, and may affect the behavior of the system.

8.10 NOTES

Note 8.1   Terminology

There was considerable debate in the early days of discrete-time control (before 1960) about terminology. Some argued that the field should be called *sampled*-data systems, while others preferred to use *sample*-data. The former term seems to have won out.

Likewise, the title of $G(z)$, the ratio of the Z-transform of the output to the Z-transform of the input, generated animated discussion. There were those who favored *pulsed* transfer function and others who favored *pulse*. Strictly, both terms are misnomers, since pulses are almost never physically present in computer-controlled discrete-time systems. The term "discrete-time transfer function" avoids the contention.

The definition of the Z-transform itself was somewhat controversial. Originally—almost in the prehistory of control theory—someone used the symbol z to denote e^{sT}. This use was propagated in early writings, although the corresponding pulse- (or pulsed-) transfer functions turned out to be functions of z^{-1}. To avoid the negative powers of z it was suggested that z should have been defined as e^{-sT}. Since this would have caused considerable confusion, different symbols (e.g., x) were used for a time. The current definition of the Z-transform is now all but universal.

8.11 PROBLEMS

Problem 8.1   Stability margin for servomotor

Verify the gain and phase margins for the servomotor of Example 8.1 .

Problem 8.2   Stability analysis of computer-controlled servomotor

Suppose the sample-and-hold operation in the control of the servomotor of Example 8.3 is approximated by a delay of $T/2$. Find the maximum loop gain for which the closed-loop system is stable. Compare the result with the exact gain computed in Example 8.3 and discuss the result.

Problem 8.3   Stability with repeated eigenvalues

Show that a discrete-time system with repeated eigenvalues on the unit circle is unstable and that repeated eigenvalues anywhere else on the $\mathcal{Z}$ plane do not alter the stability.

Problem 8.4   Controllability in multiple input systems

Show that the condition for controllability of a multiple-input discrete-time system is the same as for a continuous-time system.

Problem 8.5   Observability in systems with delayed input

Show that observability of a discrete-time system is unaffected by a K step delay.

Problem 8.6   Motor-driven robot arm

(a) Obtain the state-space representation of the motor-driven robot arm of Problem 4.2, assuming $\sin(\theta) \approx \theta$, and using a sampling period $T = 0.01$ sec.

(b) Obtain the discrete-time transfer function corresponding to (a).

Problem 8.7   Sensitivity of discrete-time plant model to sampling frequency

Consider the plant of Problem 8.6 for each of the following sampling intervals:

(a) $T = 0.1$ sec.

(b) $T = 0.01$ sec.

For each sampling interval determine the sensitivity of the discrete-time eigenvalues to changes $\delta\Omega$ in the resonant frequency of the plant.

Problem 8.8   Temperature control

Consider the temperature control application considered in Example 1.6 *et. seq.*

(a) Obtain the matrices for the discrete-time representation of the system, using a sampling interval $T = .1$ sec.

(b) Obtain the discrete-time transfer function corresponding to the transfer function $G(s)$.

Problem 8.9   Remote control of satellite

The attitude of a satellite can be represented by

$$\ddot{\theta} = u$$

It is desired to control the satellite using an Earth-based controller. The propagation time between the Earth and the satellite cannot be assumed negligible. Develop the discrete-time state-space model of the process, assuming that the propagation time between the Earth and the satellite is exactly

(a) one sampling instant;

(b) two sampling instants.

8.12 REFERENCES

[1] J.R. Ragazzini and G.F. Franklin, *Sampled-Data Control Systems*, McGraw-Hill, New York, 1958.

[2] E.I. Jury, *Sampled-Data Control Systems*, John Wiley and Sons, Inc., New York, 1958.

[3] E.I. Jury and D. Blanshard, "A Stability Test for Linear Discrete Systems in Table Form," Proc. IRE, Vol. 49, pp. 1947–1948, 1961.

[4] F.R. Gantmacher, *The Theory of Matrices*, Chelsea Publishing Co., New York, 1959.

[5] R.J. Schwarz and B. Friedland, *Linear Systems*, McGraw-Hill, New York, 1965.

Chapter 9

CONTROLLING DISCRETE-TIME SYSTEMS

The simplest method of designing a digital controller for a continuous-time process is to "digitize" an acceptable analog design: Use a digital computer to implement the functions of the analog design, with an appropriate numerical integration algorithm in place of the analog integrators. Naturally, the digital computer and interface hardware must be fast enough to appear continuous to the process under control. The speed of computers and associated interfaces (analog-to-digital and digital-to-analog converters, etc.) is increasing to the point that they can be orders of magnitude faster than the fastest time constant of the process under control, so this design approach is realistic in many applications.

An alternative to simply digitizing an analog design is to obtain a discrete-time representation of the analog compensator and then to digitally implement the equations of this representation.

There are applications, however, in which digitizing an acceptable analog control algorithm may not be judicious, or even possible. In some situations a continuous-time model may be unavailable, because the process is characterized directly by a discrete-time model, the parameters of which are determined by parameter estimation methods (see Chapter 10).

Even if a continuous-time model of the plant is available, it may not be feasible to use a computer that is fast enough to implement the analog control algorithm with

negligible error. If the closed-loop system is required to have a bandwidth comparable
to the capability of the available computer, the phase margin afforded by an analog
design may not be sufficient to allow its use for digital control. (Review Chapter 8 for a
discussion of the role of phase margin in digital control.) Although it may be possible
to modify the analog design to provide a greater phase margin, it may be preferable
to design the digital control law directly.

Theory for the design of digital control algorithms parallels the theory developed
for continuous-time systems. In particular, you can start with a control law of fixed
structure but variable parameters, and adjust ("tune") the parameters to achieve the
required performance (assuming, of course, that the required performance can be
attained with a control law of this structure.) Or the selection of the control algorithm
can be included in the design. When you are free to select the control algorithm,
you will probably find that the "separation principle," which is highly effective for the
design of continuous-time systems, can be used to advantage for discrete-time systems.
(See Note 9.1.)

When you use the discrete-time representation of a system to design the control
algorithm, you are only considering the behavior between sampling instants; the behavior between sampling instants is concealed. You must verify that the intersample
behavior of the design is satisfactory. For this purpose simulation is a valuable tool.

9.1   DIGITIZING AN ANALOG CONTROL LAW

Digitizing an analog control law expressed in state-space form is easy. Suppose, for
example, the analog compensator D has the state-space representation

$$- u(t) = C_c x_c(t) + D_c y(t) \tag{9.1}$$
$$\dot{x}_c(t) = A_c x_c(t) + B_c y(t) \tag{9.2}$$

This compensator could be implemented by simply using the discrete time equivalent of (9.1):

$$- u_n = C_c (x_c)_n + D_c y_n \tag{9.3}$$

in which the state x_c of the compensator is obtained by use of an appropriate numerical
integration algorithm. If the sampling time T is short enough to allow the use of the
simple Euler integration algorithm in which dx/dt is approximated by $(x_{n+1} - x_n)/T$,
then (9.2) is approximated by

$$(x_c)_{n+1} = (x_c)_n + T[A_c(x_c)_n + B_c y_n] \tag{9.4}$$

For greater accuracy, a more advanced numerical integration algorithm, such as a
Runge-Kutta algorithm, can be used.

Alternatively, (9.2) can be converted directly to a discrete-time system

$$(x_c)_{n+1} = \Phi_c(x_c)_n + \Gamma_c y_n \tag{9.5}$$

where

$$\Phi_c = e^{A_c T}$$

$$\Gamma_c = \int_0^T e^{A_c t} B_c \, dt$$

Example 9.1   Temperature Control

Consider designing a temperature control system for the process

$$\dot{x}_1 = -2x_1 + x_2$$
$$\dot{x}_2 = x_1 - 2x_2 + u$$

with

$$y = x_1$$

The transfer function of the plant is readily found to be

$$G(s) = \frac{1}{(s+1)(s+3)}$$

The plant is "type zero"; hence there will be a steady-state error for a step input unless the compensator provides integral action. Thus, for this plant, proportional plus integral control would be appropriate:

$$D(s) = d_0 + d_1/s = (d_0 s + d_1)/s \tag{9.6}$$

This process is not difficult to control; a wide range of values for d_0 and d_1 would yield acceptable performance. In particular, we can pick $d_0 = d_1 = 1$. This would place a zero of the compensator at the pole at $s = -1$ and result in a loop transmission

$$F(s) = 1/s(s+3)$$

(Since the pole cancelled by the zero is in the *left* half plane, there is no stability problem associated with the resulting uncontrollable system.) The step response of the continuous-time system with $D(s) = (s+1)/s$ is shown in Figure 9.1.

The state-space implementation of (9.6) is as follows:

$$u_n = y_n + (x_c)_n \tag{9.7}$$
$$(x_c)_{n+1} = (x_c)_n + Ty_n \tag{9.8}$$

where

$$y_n = y_r - (x_1)_n$$

is the error between the desired reference y_r and the measured temperature x_1.

The discrete-time transfer function corresponding to (9.7) and (9.8) is

$$D(z) = \frac{z - 1 + T}{z - 1} \tag{9.9}$$

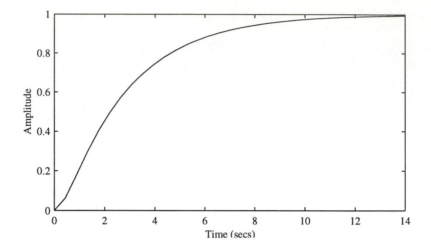

Figure 9.1   Step response of temperature control system with
PI control.

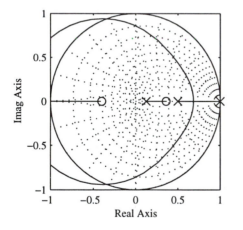

Figure 9.2   Root locus for discrete-time
temperature control.

For a sampling time of, say, $T = \log 2 = .6391$ sec, the plant and compensator transfer
functions are, respectively:

$$G(z) = 0.10417 \frac{z + 0.4}{z^2 - 0.625z + 0.0625} \tag{9.10}$$

$$D(z) = \frac{z - 0.3609}{z - 1} \tag{9.11}$$

Note that the pole-zero cancellation that was accomplished in the continuous-time design
is not achieved by the discrete time design: The zero of $D(z)$ at 0.3609 does not occur
at either of the poles of $G(z)$, which are at $z = .5$ and at $z = 0.0625$. Nevertheless, the

performance of the system is excellent. The root locus, shown in Figure 9.2, indicates a gain margin of about 17.5.

The step responses, for two values of loop gain $K = 1$ and 2.5, with the corresponding control signals, are shown in Figure 9.3. For $K = 1$ the response exhibits virtually no overshoot; for $K = 2.5$ the response, as expected, is somewhat faster but exhibits an overshoot of about 20%. For most applications the performance of the system with a gain in the range [1, 2.5] would be judged acceptable. Thus, in this application, nothing more than direct digitization of the analog PI control is necessary.

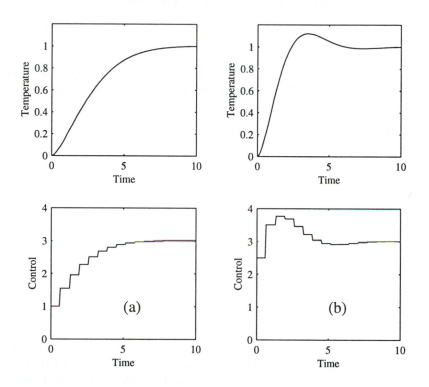

Figure 9.3 Step responses of temperature control systems with digitized PI control. (a) $K = 1$. (b) $K = 2.5$.

9.2 DIGITAL CONTROL LAWS OF FIXED STRUCTURE

Next in simplicity after digitization of an analog control law is the design method based on selection of the parameters of a digital compensator of fixed structure. Since a discrete-time "integrator" is really a "summer," with the output given by

$$y_n = \sum_{k=0}^{n} x_k$$

it has the discrete-time transfer function

$$D(z) = 1/(1 - z^{-1})$$

Hence the generic digital PI control has the transfer function

$$D(z) = K_1 + K_2 \frac{1}{1 - z^{-1}}$$

The compensator of Example 9.1 has the transfer function of a generic digital PI control, except that its parameters were chosen to correspond to the continuous-time PI control.

The generic single-input, single-output compensator has a discrete-time transfer function:

$$D(z) = \frac{n_0 z^N + n_1 z^{N-1} + \cdots + n_N}{z^N + d_1 z^{N-1} + \cdots + d_N}$$
$$= K \frac{(z - z_1) \cdots (z - z_N)}{(z - p_1) \cdots (z - p_N)}$$

The coefficients of the numerator and denominator of $D(z)$ or, equivalently, its gain and poles and zeros can be chosen by any method that you use for designing the transfer function of a continuous-time system. These would include shaping the root locus, lead-lag compensation, as well as the state-space methods discussed below.

9.3   DESIGN BY THE SEPARATION PRINCIPLE

The separation principle, by which the control system is designed in two phases, is not restricted to continuous-time systems; it can be used for discrete-time systems just as readily. (In fact, the separation principle was first stated for discrete-time systems. See Note 9.1.) The separation principle can even be used for design of control systems with both sampled and continuous-time measurements, as will be explained subsequently.

The two phases of a design by the separation principle are:

1. Design the full-state feedback (FSFB) control, assuming that all state variables are available for measurement.

2. Design an observer to estimate the state variables using the measurements that are available.

After completing the above two phases of the design, you simply combine the full-state feedback control with the observer to achieve the desired compensator.

9.3.1 Full-State Feedback

The first step in applying the separation principle to a discrete-time system in the standard form:

$$x_{n+1} = \Phi x_n + \Gamma u_n \tag{9.12}$$

is to find the gain matrix G in the linear control law

$$u_n = -Gx_n \tag{9.13}$$

Substituting (9.13) into (9.12), we see that the closed-loop system is given by

$$x_{n+1} = (\Phi - \Gamma G)x_n$$

Hence the state transition matrix is

$$\Phi_c = \Phi - \Gamma G \tag{9.14}$$

This gain matrix can be determined in various ways, such as

- Pole placement of the closed-loop system.

- Optimization of a quadratic performance measure.

If the system is controllable (using the discrete-time controllability criterion of Chapter 8) the closed-loop poles can be placed wherever in the complex plane you want them. For a single-input, single-output system a number of algorithms are available for computing the gain matrix given the desired pole locations. For example, the Bass-Gura algorithm [1] can be used:

$$G = \mathbf{C}\mathbf{W}'^{-1}(\mathbf{a_c} - \mathbf{a}) \tag{9.15}$$

where $\mathbf{a} = [a_1, \cdots, a_k]$ is the vector of coefficients of the characteristic polynomial (i.e., the denominator of the plant transfer function):

$$\Delta(z) = |zI - \Phi| = z^k + a_1 z^{k-1} + \cdots + a_k$$

$\mathbf{a_c} = [a_{1c}, \cdots, a_{kc}]$ is the vector of the desired closed-loop characteristic polynomial,

$$\Delta_c(z) = |zI - \Phi_c| = z^k + a_{1c}z^{k-1} + \cdots + a_{kc}$$

$$\mathbf{W} = \begin{bmatrix} 1 & a_1 & \cdots & a_{k-1} \\ 0 & 1 & \cdots & a_{k-2} \\ & \cdots & \cdots & \\ 0 & 0 & \cdots & 1 \end{bmatrix}$$

and $\mathbf{C}$ is the controllability matrix as defined in Chapter 8.

If you elect to design the control law gain matrix by pole placement, where should you place the closed-loop poles? This question arises in discrete-time systems as well as in continuous-time systems. Since the closed-loop system must be stable, its poles, of course, must be within the unit circle. More precise specification of the pole locations depends on the performance requirements and the available control authority. The more you move the poles, the larger the control gains and hence the control signal will be. On the other hand, the closer the poles are to the origin, the faster the response will be. Hence the choice of proper pole locations usually entails a tradeoff between performance and control authority. Another issue is robustness of the design, as measured by the stability margins of the closed-loop system. Robustness characteristics are related to the closed-loop pole locations, but they also depend upon the open-loop poles and zeros. The relationship between closed-loop poles and robustness may thus not be obvious. To meet all operating requirements, you may find it necessary to experiment with a range of designs.

If you have reason to know suitable locations of the closed-loop poles for the continuous-time system, a reasonable start for locating the poles of the corresponding discrete-time system might be at

$$z_i = e^{s_i T}$$

where s_i are the known favorable locations for the poles of the continuous-time system.

Pole placement for a multi-input plant is possible, but often not convenient. Optimization of a quadratic performance index is usually a preferable option. It is often preferable even for single-input plants. The theory for computing the gain matrix by quadratic optimization is given in Section 9.6.

9.3.2 Observer Design

The next step, after having selected the full-state feedback gain matrix G is the observer design. As in a continuous-time system, the observer is designed to provide estimates either of *all* the state variables—those that can be measured directly and those that cannot be—or only of the state variables that cannot be measured directly. The "full-order" observer, which provides estimates of all the state variables, is less messy to derive; but the "reduced-order" observer, which estimates only those state variables that cannot be measured, requires fewer storage locations for implementation and, in most cases, provides superior performance.

A full-order observer is a dynamic system of the same order as the process whose state is to be estimated. It is excited by the inputs and outputs of that process, and has the property that the estimation error, i.e., the difference between the state x_n of the process and the state $\hat{x}_n$ of the observer, converges to zero as $n \to \infty$ independent of the state of the process or its inputs and outputs.

Let the observer be defined by the general linear difference equation

$$\hat{x}_{n+1} = F\hat{x}_n + Ky_n + Hu_n \tag{9.16}$$

The goal is to find conditions on the matrices F, K, and H such that the requirements stated above are met. To find these conditions subtract (9.16) from (9.12)

$$x_{n+1} - \hat{x}_{n+1} = \Phi x_n + \Gamma u_n - F\hat{x}_n - K y_n - H u_n \qquad (9.17)$$

Letting

$$e_n = x_n - \hat{x}_n$$

and noting that the observation y_n is given by

$$y_n = C x_n$$

we obtain from (9.17)

$$e_{n+1} = F e_n + (\Phi - KC - F)x_n + (\Gamma - H)u_n \qquad (9.18)$$

Thus, in order to meet the requirements stated above, the transition matrix F of the observer must be stable, i.e., the eigenvalues of F must lie within the unit circle, and, moreover,

$$F := \hat{\Phi} = \Phi - KC \qquad (9.19)$$
$$H = \Gamma \qquad (9.20)$$

By virtue of these relations, the observer can be expressed as

$$\hat{x}_{n+1} = \Phi\hat{x}_n + \Gamma u_n + K(y_n - C\hat{x}_n) \qquad (9.21)$$

It is seen from (9.19) that the observer has the same dynamics as the underlying process, except that it has an additional input

$$K(y_n - C\hat{x}_n)$$

the role of which can be interpreted as that of driving the error e_n to zero.

The observer design thus reduces to the selection of the gain matrix K which makes the eigenvalues of $\hat{\Phi} = \Phi - KC$ lie at suitable locations within the unit circle.

If the discrete-time system is observable, the eigenvalues of $\Phi_c = \Phi - KC$ can be put anywhere. For a single-output plant, the Bass-Gura formula or other well-known algorithm can be used. For both single- and multiple-output processes the observer gain matrix can be selected to make the observer a Kalman filter, i.e., a minimum variance estimator, as discussed in Section 9.7.

9.3.3 Completion of the Compensator Design

The compensator design is completed by combining the full-state feedback control with the observer. The resulting compensator has the block-diagram representation shown in Figure 9.4. What assurance do we have that the poles of the closed-loop system comprising the plant and the compensator will be at desirable locations?

To determine the pole locations of the closed-loop system with the compensator present, suppose that the gain matrix K satisfies (9.19) and that (9.20) holds. Then the estimation error converges to zero:

$$e_{n+1} = \hat{\Phi} e_n \tag{9.22}$$

Also

$$\begin{aligned}
x_{n+1} &= \Phi x_n - \Gamma G \hat{x}_n \\
&= \Phi x_n - \Gamma G(x_n - e_n) \\
&= (\Phi - \Gamma G)x_n + \Gamma G e_n
\end{aligned} \tag{9.23}$$

Combine (9.22) and (9.23) into the $2k \times 2k$ system:

$$\begin{bmatrix} x_{n+1} \\ e_{n+1} \end{bmatrix} = \begin{bmatrix} \Phi - \Gamma G & \Gamma G \\ 0 & \Phi - KC \end{bmatrix} \begin{bmatrix} x_n \\ e_n \end{bmatrix} \tag{9.24}$$

The triangular nature of the $2k \times 2k$ matrix in (9.24) means that its eigenvalues are the eigenvalues of $\Phi - \Gamma G$ and of $\Phi - KC$. Thus the poles of the closed-loop system are those of the full-state feedback design and those of the observer. If these sets of poles have been placed at favorable locations, we are assured that the poles of the closed-loop system with the observer-based compensator in place will be at the *same* favorable locations. This is the separation principle for discrete-time systems.

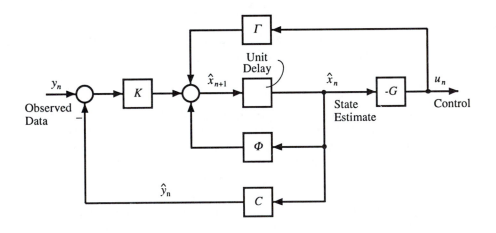

Figure 9.4   Schematic of discrete-time compensator design based on the separation principle.

It is not necessary to implement the compensator in the observer form shown in Figure 9.4. Any implementation that has the same transfer function will give the same result. To determine the compensator transfer function, combine the full-state feedback control with the observer:

$$D(z) = G(zI - \Phi + \Gamma G + KC)^{-1}K \tag{9.25}$$

9.4 FINITE-SETTLING TIME ("DEADBEAT") CONTROL

The closer the closed-loop poles of a system are to the origin of the unit circle, the faster the time response. What if the poles are all placed *at the origin*? The origin of the z-plane in discrete-time systems corresponds to the point at infinity for continuous-time systems, which can be approached but not achieved in the latter. But in discrete-time systems, if the poles can be placed at any location in the z-plane, they can be placed at the origin.

To examine the consequence of all the closed-loop poles being at the origin, note that the characteristic equation of such a system will be

$$z^k = 0$$

By the Cayley-Hamilton Theorem, the closed-loop state transition matrix $\Phi_c = \Phi - \Gamma G$ satisfies its own characteristic equation. Thus

$$\Phi_c^k = 0$$

But, the solution of the closed-loop difference equation

$$x_{n+1} = \Phi_c x_n$$

at step k is

$$x_k = \Phi_c^k x_0$$

where x_0 is the initial state. Thus, we conclude that if the poles of the closed-loop system (with full-state feedback) lie at the origin, the state will go to zero in at most k steps. In the absence of disturbance, the state will remain at the origin thereafter! This type of behavior, which cannot be achieved with a linear control in continuous-time systems, has been called *finite-settling-time control* or *deadbeat* control. (The latter term is not intended to be pejorative.)

If not all the state variables are measurable, an observer is necessary to provide an estimate of the state. In order for the closed-loop system with the observer present to retain finite-settling time performance, the observer must also be deadbeat: The eigenvalues of $\hat{\Phi} = \Phi - KC$ must also be placed at the origin.

You might think that a deadbeat performance is optimal: After all, is it not better for the error to go to zero in a finite time than in an infinite time? Not necessarily. In evaluating the advantages of deadbeat control over some other control law, consider the following:

- The performance is deadbeat only when the control gains are tuned precisely to the parameters of the plant. If the plant deviates slightly (and unavoidably) from the design model, the resulting closed-loop system will no longer be deadbeat. Deadbeat performance, moreover, is not robust, and the performance may degrade significantly when the plant changes.

- In order to make a sluggish process move fast, you have to push hard. Hence if the product kT of the process order and sampling time is short relative to the time constant of the plant, the control effort required may be larger than desirable or available. Of course, you can slow down the response time by increasing the sampling interval, but it might be preferable to simply settle for less than deadbeat performance.

Example 9.2   Control of Satellite

Consider the discrete-time control of a satellite, which can be idealized as a free mass with piecewise-constant acceleration. The differential equations of the plant are

$$\dot{x}_1 = x_2$$
$$\dot{x}_2 = u$$

where u is the acceleration. The observer output is the position x_1.

The plant is the so-called "double integrator" and is also appropriate for the study of the angular motion of a free rotational inertia.

For this plant it is readily determined that

$$\Phi = \begin{bmatrix} 1 & T \\ 0 & 1 \end{bmatrix} \quad \Gamma = \begin{bmatrix} T^2/2 \\ T \end{bmatrix} \quad C = \begin{bmatrix} 1 & 0 \end{bmatrix}$$

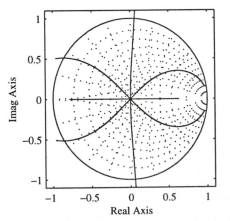

Figure 9.5   Root locus for deadbeat control of satellite.

The full-state feedback gain matrix to place the poles at the origin is given by

$$G = \begin{bmatrix} 1/T^2 & 3/2T \end{bmatrix}$$

and the observer gain matrix to place the poles at the origin is

$$K = \begin{bmatrix} 2 \\ 1/T \end{bmatrix}$$

With these gain matrices in (9.25) we find the compensator transfer function:

$$D(z) = \frac{1}{T^2} \left(\frac{3.5z - 2.5}{z^2 + 2z + 1.25} \right)$$

which has a zero at $z = 0.7142$ and poles at $z = -1 \pm j0.5$. Thus the compensator is unstable! (Although an unstable compensator may be undesirable in practice, there is nothing in the theory to preclude it.) Because the compensator is unstable, the closed-loop system is only conditionally stable with the stability range given by

$$0.35 < K < 1.32$$

The root locus for this example is shown in Figure 9.5 and tells an unpleasant story: Although the root locus passes through the origin when the loop gain is at its nominal value of unity, a small change in the gain (around 10%) moves the closed-loop poles very far from the origin, demonstrating that the design is not robust.

The lack of robustness is confirmed by the transient response of the system. Shown in Figure 9.6 are responses to an initial position offset of -1.0, using a sampling period T of 1 sec. In (a) the loop gain is unity and the system exhibits the expected deadbeat response, although the overshoot is about 200%. In (b) the response is shown with the gain raised by 15% above its nominal value. The response is now quite oscillatory.

This example confirms that the cost of deadbeat performance can be a transient response with excessive overshoot and poor robustness characteristics. These deficiencies may be too high a price to pay for deadbeat performance.

9.5 PULSE-WIDTH MODULATED CONTROL

The size and cost of an actuator increase with the power it must be capable of supplying to the process being controlled. High-power linear actuators can account for most of the hardware cost of a large control system, because it is expensive to produce a device in which the output must be proportional to the input. It is much less expensive to control the process by simply turning the power on and off at appropriate time instants. This can be accomplished in a discrete-time system by a control law that turns the power on at the beginning of the sampling interval and turns it off before the end of the interval. The control variable is the fraction of the sampling interval that the power is on. (This is sometimes called the *"duty cycle."*) If positive and negative control action is needed, two sources of power can be used with the control algorithm determining both the sign and the duration of the control action. This method of control is known as "pulse-width modulation."

The control signal in a discrete-time control system of the type considered up to now may be regarded as a train of rectangular pulses each of constant width T and of varying height u_n as shown in Figure 9.7(a). Suppose the maximum level of the control

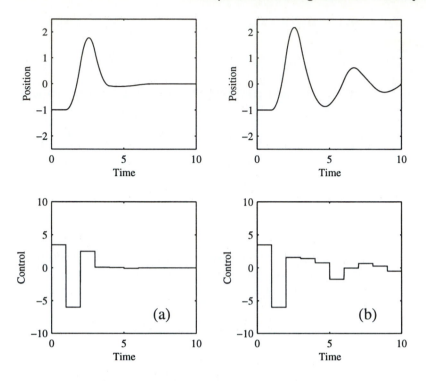

Figure 9.6   Step response of satellite with control designed to be deadbeat. (a) Loop gain = 1.0 (b) Loop gain = 1.15.

(the "control authority") is limited in magnitude to a value of U. Then the maximum area of each pulse is UT. Another way of achieving a train of pulses of varying area with a maximum area of UT is by keeping the amplitude of each pulse fixed at $\pm U$ and controlling the pulse width δ_n, as shown in Figure 9.7(b). If you can design a control law for determining the pulse amplitudes u_n, you should be able to design a control law to determine the sign and the duration of the pulse-width modulated control.

In order to design the control law, you need a model of the plant in which the control inputs are the sign and duration of the control in each sampling interval. This relation is obtained from the basic input-output relation for a linear system:

$$x(nT + T) = e^{AT}x(nT) + \int_{nT}^{nT+T} e^{A(nT+T-\xi)}Bu(\xi)d\xi \qquad (9.26)$$

Since the control is turned off after δ_n time units into the nth control cycle and

$$u(t) = U_n := U\,\mathrm{sgn}(u(t))$$

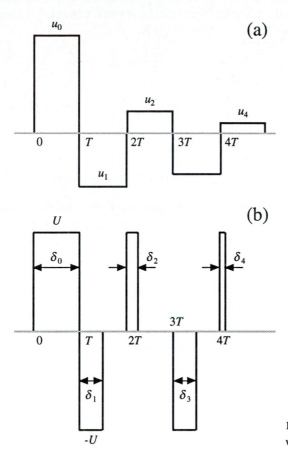

Figure 9.7 Pulse amplitude and pulse width modulation. (a) Pulse amplitude modulation. (b) Pulse width modulation.

(9.26) can be written

$$x(nT + T) = e^{AT} x(nT) + \left(\int_{nT}^{nT+\delta_n} e^{A(nT+T-\xi)} B d\xi \right) U_n \qquad (9.27)$$

which can be written

$$x_{n+1} = \Phi x_n + \gamma(u_n) \qquad (9.28)$$

where $\gamma(\ )$ is a vector of nonlinear functions defined by

$$\gamma(u_n) = U_n \int_0^{\delta_n} e^{A(T-\xi)} B d\xi \qquad (9.29)$$

Although the components of the vector $\gamma(\ )$ are often quite nonlinear, it might be possible to design the control law by linearizing $\gamma(u_n)$ at the origin on the assumption that as the error is reduced, the duty cycle δ_n approaches zero. Based on this assumption the design model derived from (9.29) is

$$x_{n+1} = \Phi x_n + \Gamma_p u_n \qquad (9.30)$$

where

$$\Gamma_p = \partial\gamma(u_n)/\partial u_n$$

$$= U_n \frac{\partial}{\partial\delta_n} \int_0^{\delta_n} e^{A(T-\xi)} B d\xi$$

$$= U_n e^{A(T-\delta_n)}|_{\delta_n=0} B = U_n e^{AT} B = U_n \Phi B \qquad (9.31)$$

Since there is no theoretical basis for the assumption that the error will become small for the control law based on the linearized control, it is necessary to verify the performance of the design through simulation or by another method of analysis.

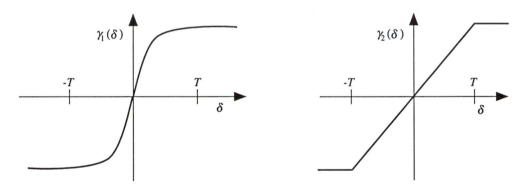

Figure 9.8   Nonlinear functions in pulse-width modulated control of spacecraft attitude dynamics.

Example 9.3   Pulse-width Modulated Attitude Control of a Satellite

The motion of a rigid spacecraft about a single axis is represented by the dynamics of a free mass. One of the methods for controlling the attitude of the spacecraft is by the use of reaction jets, which can produce torques about the vehicle's axes. Control by pulse width modulation is the preferred method of operation in this case, for several reasons:

- Valves for controlling the expulsion of gas are much more reliable if operated in an on-off mode.

- Continuous operation of the reaction jet will ultimately deplete the fuel supply. To conserve fuel, it is necessary to limit the intervals in which control is used and to allow the spacecraft to drift uncontrolled most of the time. (Fortunately, the disturbances that cause the error in space are very small, and performance with this mode of control is usually acceptable.)

The vector $\gamma(\ )$ in the pulse-width modulated dynamics of the satellite, as considered in Example 9.2, is

$$\gamma = \left[\begin{array}{c} \gamma_1 \\ \gamma_2 \end{array} \right] = U_n \int_0^{\delta_n} \left[\begin{array}{c} T - \xi \\ 1 \end{array} \right] d\xi = U_n \left[\begin{array}{c} \delta_n T - \delta_n^2/2 \\ \delta_n \end{array} \right]$$

which has the appearance shown in Figure 9.8.

The linearizing matrix Γ_p is given by

$$\Gamma_p = \left[\begin{array}{c} T \\ 1 \end{array} \right]$$

If both state variables, position and velocity, can be measured, a full-state feedback control law is

$$u_n = -g_1(x_1)_n - g_2(x_2)_n$$

The gains can be chosen, for example, to achieve deadbeat performance. This is accomplished by placing the eigenvalues of the closed-loop system

$$|zI - \Phi + \Gamma_p G| = \left| \begin{array}{cc} z - 1 & -T \\ g_2 & z - 1 + g_2 \end{array} \right|$$

at the origin. A simple calculation gives

$$g_1 = 1/T \quad g_2 = 1$$

Figure 9.9 shows the step response for two reference levels. In (a) the reference level is small enough to not require the whole duty cycle; the response is very nearly deadbeat. The response for a larger reference level is shown in (b). In this case the duty cycle is saturated to an appreciable interval; but ultimately the error is reduced to the level requiring only a fractional duty cycle, and the error converges to zero rapidly thereafter.

Example 9.4   Pulse-width Modulated Control of a Servomotor

Pulse-width modulation can be effectively used for control of a servomotor with an inertial load.

The vector $\gamma(\ )$ for the model developed in Example 8.2 is

$$\gamma = \left[\begin{array}{c} \gamma_1 \\ \gamma_2 \end{array} \right] = U_n \alpha \left[\begin{array}{c} \int_0^{\delta_n} (1 - e^{-\alpha(T-\xi)}) d\xi \\ \int_0^{\delta_n} e^{-\alpha(T-\xi)} d\xi \end{array} \right] = U_n \left[\begin{array}{c} \alpha \delta_n - e^{-\alpha T}(e^{\alpha \delta_n} - 1) \\ e^{-\alpha T}(e^{\alpha \delta_n} - 1) \end{array} \right]$$

The components of γ are shown in Figure 9.10.

The control distribution matrix for design purposes is given by

$$\Gamma_p = U\alpha \left[\begin{array}{c} 1 - e^{-\alpha T} \\ e^{-\alpha T} \end{array} \right]$$

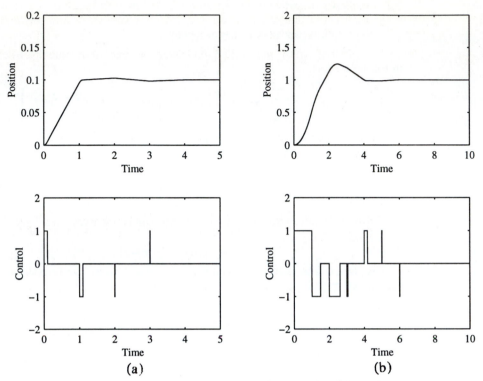

Figure 9.9   Step responses of satellite with pulse width modulated control. (a) Reference level does not require full duty cycle.
(b) Reference level causes saturation of the duty cycle.

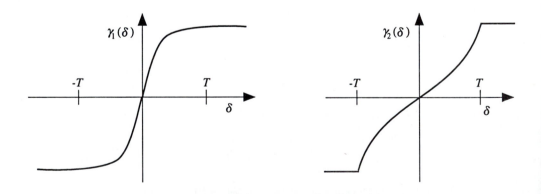

Figure 9.10   Nonlinear functions in pulse-width modulated
control of a servomotor.

For numerical values

$$\alpha : 1.$$
$$T : 0.5$$
$$U_m : 1.$$

we obtain

$$\Phi = \begin{bmatrix} 1 & 0.3935 \\ 0 & 1 \end{bmatrix} \qquad \Gamma_p = \begin{bmatrix} 0.3935 \\ 0.6065 \end{bmatrix}$$

It is readily determined that the gain matrix for deadbeat performance is given by

$$G = \begin{bmatrix} 2.5413 & 1.0000 \end{bmatrix}$$

The nonlinearities in $\gamma(\ )$ prevent perfect deadbeat performance. Nevertheless the performance is excellent, as the simulation results show.

Figure 9.11(a) shows the response and the corresponding control for a reference angle of 0.1 radian. The required duty cycle is never as much as 100%, and the transient response is very nearly deadbeat.

Figure 9.11(b) shows the response and corresponding control for a reference angle of 1.0 radian. Now we observe that the duty cycle is saturated positive for the first 1.5 sec. The sign then reverses, and switches a few times as the error is driven to zero. The error is essentially zero after 5 sec.

Figure 9.11(c) shows the response for a 10 radian reference input. A constant positive control is now generated for about the first 12 sec. The control reverses for about 2 sec, and then, after a few control intervals of less than full duty cycle, the error is essentially zero.

The foregoing examples showed the performance of a system with full-state feedback. But the approach is also applicable to a system which requires an observer because not all the state variables can be measured. (See Problem 9.3 and Problem 9.4.)

9.6 QUADRATIC OPTIMIZATION

Pole placement is a convenient way of selecting the control gains, but it may not always be satisfactory, for several reasons. One reason is that a "good" set of pole locations is not always obvious. Another is that in systems with more than one input, pole placement does not yield a unique solution for the gain matrix. Moreover, the design that results when the poles are placed in arbitrary locations may not be robust with respect to variation of the loop gain.

One way of selecting the control gain matrix G that alleviates these difficulties is to minimize a quadratic performance criterion:

$$J(n) = \sum_{l=n}^{\infty} (x_l' Q_l x_l + u_l' R_l u_l) \tag{9.32}$$

(There is no added complexity in assuming that the weighting matrices depend on the time instant l, although in most applications they are constant.) The matrices Q_l are

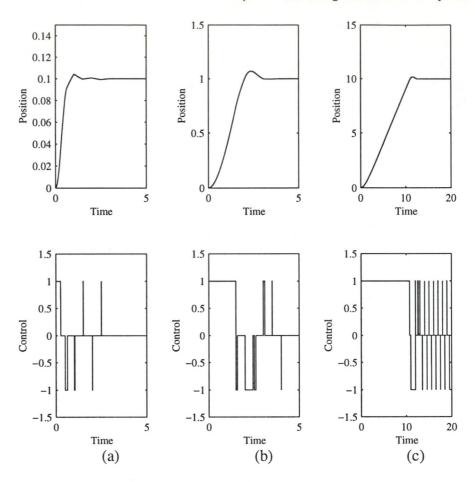

Figure 9.11   Pulse-width modulated control of a servomotor.
(a) $x_r = 0.1$ (b) $x_r = 1.0$ (c) $x_r = 10$.

selected to give appropriate weighting of the deviation of the state from the origin, and
the matrices R_l are chosen to limit the control effort. As in continuous-time control,
the performance criterion (9.32) seldom is what you really want to optimize. Even if
you could translate the actual engineering requirements into a performance criterion,
the resulting control law might be too difficult to determine and implement. The performance criterion (9.32) is a compromise between an often unattainable ideal and a
practical reality. The performance criterion of (9.32) is practical in that

- It results in a linear control law

$$u_n = -G_n x_n \tag{9.33}$$

- The gain matrices G_n can be readily calculated.

It is not difficult to show that the control law that minimizes (9.32) is linear. (See Problem 9.1.) To simplify the development of the equation that defines G_n, however, assume that a linear control law is optimum. In addition, it is convenient to let the final time be a finite value N and then let $N \to \infty$. For a finite N, the performance criterion depends on N as well as on the initial time instant n:

$$J(n, N) = \sum_{l=n}^{N} (x_l' Q_l x_l + u_l' R_l u_l) \tag{9.34}$$

Suppose that the control law of (9.33) is used to control the discrete-time (possibly time-varying) dynamic process

$$x_{n+1} = \Phi_n x_n + \Gamma_n u_n$$

Then the closed-loop process with the control present is

$$x_{n+1} = \Phi_n x_n - \Gamma_n G_n x_n = \Psi_n x_n \tag{9.35}$$

where

$$\Psi_n = \Phi_n - \Gamma_n G_n \tag{9.36}$$

Also the performance criterion (9.34) becomes

$$J(n, N) = \sum_{l=n}^{N} (x_l' Q_l x_l + x_l' G_l' R_l G_l x_l) = \sum_{l=n}^{N} x_l' S_l x_l \tag{9.37}$$

where

$$S_l = Q_l + G_l' R_l G_l \tag{9.38}$$

Consider this criterion one step later:

$$\begin{aligned} J(n+1, N) &= \sum_{l=n+1}^{N} x_l' S_l x_l \\ &= \sum_{l=n}^{N} x_l' S_l x_l - x_n' S_n x_n \\ &= J(n, N) - x_n' S_n x_n \end{aligned} \tag{9.39}$$

which is a recursive relation for $J(n+1, N)$ in terms of $J(n, N)$ and the state x_n. Since (9.35) defines a linear, homogeneous system, we can express the states x_l in (9.37) at later times as linear transformations of the current state:

$$x_l = \Psi_{l-n} \cdots \Psi_n x_n = Z_{l,n} x_n$$

where

$$Z_{l,n} = \Psi_{l-n} \cdots \Psi_n x_n$$

Hence the summation in (9.37) can be expressed as a quadratic form in the initial state:

$$J(n, N) = \sum_{l=n}^{N} x_n' Z_{l,n}' S_l Z_{l,n} x_n = x_n' M(n, N) x_n \qquad (9.40)$$

with

$$M(n, N) = \sum_{l=n}^{N} Z_{l,n}' S_l Z_{l,n}$$

Accordingly

$$J(n + 1, N) = x_{n+1}' M(n + 1, N) x_{n+1} = x_n' \Psi_n' M(n + 1, N) \Psi_n x_n$$

Thus (9.39) becomes

$$x_n' \Psi_n' M(n + 1, N) \Psi_n x_n = x_n' [M(n, N) - S_n] x_n$$

Since this must hold for any initial state x_n, the matrices of the quadratic forms on both sides must be equal:

$$\Psi_n' M(n + 1, N) \Psi_n = M(n, N) - S_n \qquad (9.41)$$

This is a *backward recursion* equation for $M(n, N)$ in terms of $M(n + 1, N)$ and S_n.

On substituting the expressions (9.36) and (9.38) into (9.41) we obtain

$$(\Phi_n' - G_n' \Gamma_n') M(n + 1, N)(\Phi_n - \Gamma_n G_n) = M(n, N) - Q_n - G_n' R_n G_n \qquad (9.42)$$

To find the optimum gain G_n we invoke Bellman's famous *Principle of Optimality* (see Note 5.1 in Chapter 5). In this application the Principle of Optimality asserts that if the sequence of gains $G_n, G_{n+1}, G_{n+2}, \ldots$ is optimum starting at step n, then the sequence of gains $G_{n+1}, G_{n+2}, \ldots$ must be optimum starting at step $n + 1$. Thus we can assume that $M(n + 1, N)$ is optimum and find the matrix G_n that minimizes

$$M(n, N) = (\Phi_n' - G_n' \Gamma_n') M(n + 1, N)(\Phi_n - \Gamma_n G_n) + Q_n + G_n' R_n G_n \qquad (9.43)$$

If $M(n, N)$ and G_n were scalar, the minimization would be achieved by differentiating the right-hand side of (9.43) with respect to G_n and setting the result to zero. This procedure cannot be used for matrices, but there is an analogous method of minimizing a matrix (actually a quadratic form) with respect to a matrix [2] that looks almost like the scalar procedure. Applying this procedure results in the optimum value of G_n:

$$G_n = [R_n + \Gamma_n' M(n + 1, N) \Gamma_n]^{-1} \Gamma_n M(n + 1, N) \Phi_n \qquad (9.44)$$

Substitution of (9.44) into (9.43) results in the backward recursion for the optimum $M(n, N)$ in terms of $M(n + 1, N)$:

$$
\begin{aligned}
M(n, N) \\
= \Phi'_n M(n + 1, N)[I - \Gamma(R_n + \Gamma'_n M(n + 1, N)\Gamma_n]^{-1} \\
\Gamma_n M(n + 1, N))\Phi_n + Q_n
\end{aligned}
\tag{9.45}
$$

This equation is analogous to the of continuous-time systems. Its solution, which can be obtained numerically iterating backward with the boundary condition

$$
M(N, N) = 0
$$

is substituted into (9.44) to obtain the optimum gain matrix G_n.

In most applications the dynamics are time-invariant, and there is no reason to make the weighting matrices depend on the sampling instant. Moreover, the final time N ("horizon") is usually infinite. In this case, if a solution to the optimization problem exists, then

$$
M(n, N) \rightarrow M(n + 1, N) \rightarrow \bar{M}
$$

which must be the solution of the steady-state version of (9.45):

$$
\bar{M} = \Phi' \bar{M} (I - \Gamma(R + \Gamma' \bar{M} \Gamma)^{-1} \Gamma \bar{M})\Phi + Q
\tag{9.46}
$$

This equation is the analog of the algebraic Riccati equation of continuous-time systems. Its numerical solution can be obtained by iterating (9.45) backwards to a steady state, but there are more efficient numerical methods patterned after their continuous-time counterparts. Control system design software packages often include algorithms [1] for solving this equation.

From the formula (9.45), it is clear that the only requirement for a solution for $M(n, N)$ to exist for all $n < N$ for the finite horizon problem is that the matrix $R + \Gamma' M(n + 1, N)\Gamma$ be nonsingular. This is assured if R is nonsingular.

The convergence of M to a steady-state solution, however, is not assured unless the plant is *stabilizable*, i.e., the subspace of the state space containing the unstable modes must be controllable. This condition is clearly satisfied if the open-loop system is either stable or controllable.

In order for the steady-state matrix to be positive-definite, the plant must be observable with respect to the performance criterion, i.e., the rank of the observability matrix for $[\Phi, C]$, where $Q = C'C$ must be equal to the order of the system.

---

[1] Some algorithms may have difficulty solving this equation when the state transition matrix is singular, as it may be, for example, in modeling a pure time delay. One way of circumventing the difficulty is to approximate the transition matrix by another which is non singular.

Table 9.1   Gains and poles for control of satellite with delay

Weighting factor r	Gain matrix	Closed-loop poles
0.01	1.45, 3.17, 2.44	0.0, -0.0557, .3820
0.10	0.97, 2.35, 1.87	0.0, $0.064 \pm j\, 0.299$
1.00	0.50, 1.50, 1.25	0.0, $0.375 \pm j\, 0.331$
10.00	0.22, 0.87, 0.76	0.0, $0.620 \pm j\, 0.263$

Example 9.5   Control of Satellite with Time Delay

Consider controlling a satellite that is separated from the controller by a time delay of one sampling instant. As explained in Chapter 8, the general form of the dynamics is

$$\mathbf{x}_{n+1} = \underline{\Phi}\mathbf{x}_n + \underline{\Gamma}u_n$$

where

$$\underline{\Phi} = \begin{bmatrix} \Phi & \Gamma \\ 0 & 0 \end{bmatrix} \qquad \underline{\Gamma} = \begin{bmatrix} 0 \\ 1 \end{bmatrix}$$

Using the state transition matrix Φ for a pure mass, as given in Example 9.2 we find

$$\underline{\Phi} = \begin{bmatrix} 0 & T & T^2/2 \\ 0 & 1 & T \\ 0 & 0 & 0 \end{bmatrix} \qquad \underline{\Gamma} = \begin{bmatrix} 0 \\ 0 \\ 1 \end{bmatrix}$$

The third row of zeros in $\underline{\Phi}$ means that it is singular; nevertheless the system is controllable for any sampling interval T. Hence the closed-loop poles can be placed at any locations desired. For deadbeat control, for example, the gain matrix is

$$g_0 = \begin{bmatrix} 1.0 & 2.5 & 2.0 \end{bmatrix}$$

For quadratic optimum control, (9.46) is solved numerically, using MATLAB. To use the algorithm in that program it is necessary for the state transition matrix to be nonsingular, and this is accomplished by replacing the zero in the (3,3) position by a very small number, say 0.000001.

The gain matrices corresponding to a sampling time of $T = 1$ and weighting matrices:

$$Q = \begin{bmatrix} 1 & 0 & 0 \\ 0 & 0 & 0 \\ 0 & 0 & 0 \end{bmatrix} \qquad R = \begin{bmatrix} r \end{bmatrix}$$

and the corresponding closed-loop pole locations are given in Table 9.1 for a range of r.

Note that the gains become larger (approaching the gains for deadbeat control) as the control weighting r is reduced. The distances of corresponding poles from the origin tend to decrease, indicating a faster response. Would you expect this behavior?

The locus of the closed-loop poles as r is varied is shown in Figure 9.12.

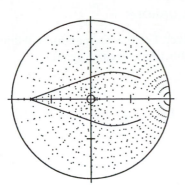

Figure 9.12 Locus of closed-loop poles for optimum control of satellite with time delay.

9.7 DISCRETE-TIME KALMAN FILTER

Choosing the gain matrix for the discrete-time observer is analogous to choosing the full-state feedback gains. Pole placement as a method of choosing the gains is subject to the same problems—definition of good observer pole locations and lack of a unique solution for systems with more than one output.

Since there is some noise present in every real-world dynamic process and all observations are subject to noise, it would make sense to optimize the observer gains for the noise that is present in the input and in the output. The theory for accomplishing this optimization was first developed by R.E. Kalman [3] in 1960, and the resulting observer is know as the *Kalman filter*. This result is widely held to be one of the crowning achievements of modern control and estimation theory. (See Note 6.1.)

Before becoming too absorbed with the theory, you should recognize that the statistical properties of the noises in the input and in the output of the system are usually not available that would be needed to calculate the true optimum gains. It is sometimes possible to make limited measurements by which the noise characteristics can be estimated, but often you must settle for a more or less educated guess. The observer will usually operate very well even when the needed statistical parameters are only crudely estimated, but you cannot honestly claim it to be optimal.

The theory of the Kalman filter can be derived by a number of methods. In his 1960 paper [3] Kalman showed that a *linear* observer is optimum for a linear process under a variety of performance criteria if the noise has a *Gaussian* probability distribution. (Otherwise, the optimum observer is generally nonlinear.) Rather than attempt to develop the general theory here, however, we shall presuppose the linear-observer structure and seek only to optimize the observer gain matrix. We will find that the result is indifferent to the nature of the noise probability distributions; only their covariance matrices are significant.

We define the optimum observer as the one that has the smallest covariance matrix of the estimation error. Our goal is to find that observer.

9.7.1   Derivation of Kalman Filter Equations

To derive the equations of the optimum observer, i.e., the Kalman filter, suppose that the plant dynamics are modeled by the (possibly time-varying) dynamics:

$$x_{n+1} = \Phi_n x_n + \Gamma_n u_n + v_n \qquad (9.47)$$

where v_n is a random noise process having the following statistical properties:

1. The noise has a mean of zero:

$$\mathcal{E}(v_n) = 0$$

2. v_n and v_k are uncorrelated for $n \neq k$, i.e.,

$$\mathcal{E}(v_n v_k') = 0, \quad \text{for} \quad n \neq k$$

3.

$$\mathcal{E}(v_n v_n') = V_n$$

A discrete-time random process, or, more simply, a random *sequence*, having these properties may be referred to as "sampled white noise" with zero-mean and covariance matrix V_n.

Similarly, suppose that the observations are given by

$$y_n = C_n x_n + w_n \qquad (9.48)$$

where w_n is sampled white noise with covariance matrix

$$W_n = \mathcal{E}(w_n w_n')$$

For convenience, we add the assumption that v_n and w_n are uncorrelated:

$$\mathcal{E}(v_n w_k') = 0 \quad \text{for all} \quad n, k \qquad (9.49)$$

If they are correlated, the cross-correlation matrix would be included in the derivation. (See Problem 9.5.)

We define the observer for this process by

$$\tilde{x}_{n+1} = \Phi_n \tilde{x}_n + \Gamma_n u_n + K_n(y_n - C_n \tilde{x}_n) \qquad (9.50)$$

where $\tilde{x}$ is the estimated state. (We have switched to using a tilde ($\tilde{\ }$) instead of a caret ($\hat{\ }$) to denote the estimated state. The reason will soon become clear.)

Upon use of (9.47) and (9.50) we determine that the estimation error, defined by

$$\tilde{e}_n := x_n - \tilde{x}_n \qquad (9.51)$$

is given by

$$\tilde{e}_{n+1} = (\Phi_n - K_n C_n)\tilde{e}_n + v_n - K_n w_n$$
$$= \tilde{\Phi}_n e_n + \nu_n \qquad (9.52)$$

where

$$\tilde{\Phi}_n = \Phi_n - K_n C_n \qquad (9.53)$$
$$\nu_n = v_n - K_n w_n \qquad (9.54)$$

Toward the goal of minimizing the covariance matrix of the estimation error, we first develop the equation that governs its propagation. Then we find the gain matrix that minimizes the covariance matrix.

In preparation for deriving the covariance propagation equation we write, using (9.52),

$$\tilde{e}_{n+1}\tilde{e}'_{n+1} = (\tilde{\Phi}_n\tilde{e}_n + \nu_n)(\tilde{e}'_n\tilde{\Phi}'_n + \nu'_n)$$
$$= \tilde{\Phi}_n\tilde{e}_n\tilde{e}'_n\tilde{\Phi}'_n + \nu_n\tilde{e}'_n\tilde{\Phi}'_n + \tilde{\Phi}_n\tilde{e}_n\nu'_n + \nu_n\nu'_n \qquad (9.55)$$

Take the expected value on both sides of (9.55), letting

$$\tilde{P}_n = \mathcal{E}(\tilde{e}_n\tilde{e}'_n) \qquad (9.56)$$

and noting that $\mathcal{E}(\tilde{e}_n\nu'_n) = 0$ to get:

$$\tilde{P}_{n+1} = \tilde{\Phi}_n\tilde{P}_n\tilde{\Phi}'_n + \tilde{V}_n \qquad (9.57)$$

where

$$\tilde{V}_n = \mathcal{E}(\nu_n\nu'_n)$$
$$= \mathcal{E}[(v_n - K_n w_n)(v'_n - w'_n K'_n)]$$
$$= V_n + K_n W_n K'_n \qquad (9.58)$$

under the assumption of (9.49) that v_n and w_n are uncorrelated.

On use of (9.53) and (9.58) we obtain from (9.57)

$$\tilde{P}_{n+1} = (\Phi_n - K_n C_n)\tilde{P}_n(\Phi'_n - C'_n K'_n) + V_n + K_n W_n K'_n \qquad (9.59)$$

Our objective now is to find the value of K_n that minimizes $\tilde{P}_{n+1}$, as we did for the optimum control law. We accomplish this by the principle of optimality, assuming that $\tilde{P}_n$ has already been optimized by the choice of $K_{n-1}, K_{n-2}, \ldots$. If we could, we would like to differentiate the right-hand side of (9.59) with respect to K_n and set the result to zero. Since we are dealing with matrices here, of course we can't do this. But, following the procedure outlined in Problem 9.2, we find the optimum gain matrix:

$$\hat{K}_n = \Phi_n\tilde{P}_n C'_n(C_n\tilde{P}_n C'_n + W_n)^{-1} \qquad (9.60)$$

After substitution of the gain matrix given by (9.60) into (9.59) and some tedious algebra, we find the equation for the propagation of the minimum covariance matrix:

$$\tilde{P}_{n+1} = \Phi_n[\tilde{P}_n - \tilde{P}_nC'_n(C_n\tilde{P}_nC'_n + W)^{-1}C_n\tilde{P}_n]\Phi'_n + V_n \qquad (9.61)$$

Although (9.60) and (9.61) suffice to establish the optimum gain and the propagation of the covariance matrix, they can be written in a way that gives a more intuitive appreciation of their significance.

Let

$$\tilde{K}_n := \Phi_n\hat{K}_n \qquad (9.62)$$

where

$$\hat{K}_n = \tilde{P}_nC'_n(C_n\tilde{P}_nC'_n + W_n)^{-1} \qquad (9.63)$$

We define the *a posteriori* covariance matrix

$$\hat{P}_n := \tilde{P}_n - \tilde{P}_nC'_n(C_n\tilde{P}_nC'_n + W_n)^{-1}C_n\tilde{P}_n = (I - \hat{K}_nC_n)\tilde{P}_n \qquad (9.64)$$

Then (9.61) becomes

$$\tilde{P}_{n+1} = \Phi_n\hat{P}_n\Phi'_n + V_n \qquad (9.65)$$

The gain matrix can also be expressed in terms of the *a posteriori* covariance matrix:

$$\tilde{K}_n = \hat{P}_nC'_nW_n^{-1} \qquad (9.66)$$

9.7.2 Interpretation

Consider the equation of the observer (9.50) using the optimized gain matrix:

$$\begin{aligned} \tilde{x}_{n+1} &= \Phi_n\tilde{x}_n + \Gamma_nu_n + \Phi_n\hat{K}(y_n - C_n\tilde{x}_n) \\ &= \Phi_n\hat{x}_n + \Gamma_nu_n \end{aligned} \qquad (9.67)$$

where

$$\hat{x}_n = \tilde{x}_n + \tilde{K}_n(y_n - C_n\tilde{x}_n) \qquad (9.68)$$

Equations (9.67) and (9.68) provide the intuitive interpretation we are seeking: The first of these is simply the equation of the dynamic process

$$x_{n+1} = \Phi_nx_n + \Gamma_nu_n$$

with $\hat{x}_n$ appearing in place of the state x_n on the right-hand side and with $\tilde{x}_{n+1}$ appearing on the left. Since the observation y_{n+1} does not appear in (9.67), we can regard this equation as a description of how the optimum state estimate $\hat{x}_n$ propagates during the time interval between the nth and the $(n+1)$st observation, an interval during which no observations have occurred.

The effect of the nth observation is reflected in the second of these expressions, (9.68). It tells how the state estimate $\tilde{x}_n$ prior to the observation is updated by the observation. First, the *residual*

$$y_n - C_n\tilde{x}_n$$

is calculated; the residual is then multiplied by the gain matrix $\bar{K}_n$; finally, the result is added to the prior state estimate. All these operations occur at the instant of the nth observation.

Thus we interpret $\tilde{x}_n$ as the optimum state estimate immediately *before* the nth observation, i.e., the *a priori* state estimate. Likewise, we interpret $\hat{x}_n$ as the optimum estimate of the state immediately *after* the nth observation, i.e., the *a posteriori* estimate. With this interpretation, (9.68) tells how the new observation is used to get the *a posteriori* estimate from the *a priori* estimate, and (9.67) tells how that estimate propagates in time to give the *a priori* estimate of the state just before the $(n + 1)$st observation.

The state update equations (9.67) and (9.68) have a counterpart in the equations for the covariance propagation: If we identify $\hat{P}_n$ with the covariance of the estimation error after taking the nth observation, and $\tilde{P}_n$ with the covariance matrix prior to taking the nth observation, then (9.64) tells how the covariance matrix is updated over a time interval in which no observations are made, and (9.65) tells how the covariance matrix is updated as a result of taking an observation.

No matter how noisy an observation might be, if it is processed optimally, it cannot increase the covariance matrix. (At worst, if the noise is infinite, the optimum action would be to ignore the observation. In this case the covariance matrix would be unchanged.) The observation update equation ought to reveal this, although it is not directly evident from (9.64) that $\hat{P}_n$ is smaller than $\tilde{P}_n$. But upon taking the inverse of both sides of (9.64) and using the Schur matrix inversion lemma, we find

$$\hat{P}_n^{-1} = \tilde{P}_n^{-1} + C_n' W_n^{-1} C_n \tag{9.69}$$

The matrix product $C_n' W_n^{-1} C_n$ is always positive semidefinite. Hence the inverse of the *a posteriori* covariance matrix is always larger than the inverse of the *a priori* covariance matrix. We can thus conclude that the *a posteriori* covariance matrix $\hat{P}_n$ is indeed smaller than the a priori covariance matrix $\hat{P}_n$. In the limit, as the observation noise becomes infinite, its inverse W_n^{-1} approaches zero and the observation does not change the covariance matrix. Also, as the observation noise becomes infinite, (9.63) asserts that the gain matrix tends to zero which implies that the observation tends to be ignored, as you would expect.

For ease of reference the important relations of Kalman filtering are summarized in Table 9.2 and in the flow chart of Figure 9.14.

The time update equations have a still broader interpretation: The state update equation can be interpreted as the solution of the continuous-time dynamic equations with a piecewise-constant input. Hence we can assert that the time update can be obtained by integrating the dynamic process equations over the time interval in which no observations occur, using the *a posteriori* state estimate as the starting state. Likewise, the covariance update can be interpreted as first solving the continuous-time variance equation

$$\dot{P} = AP + PA'$$

over the interval between observations, and then adding the covariance of the discrete-

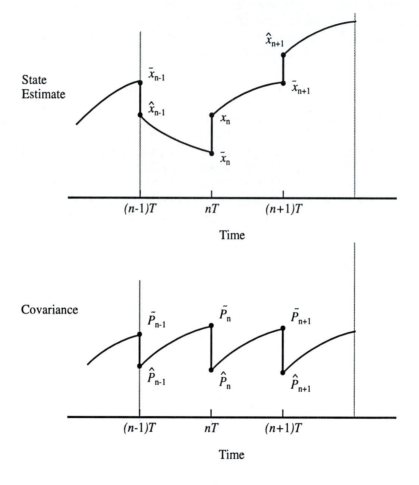

Figure 9.13   Schematic illustration of the behavior of the Kalman filter with time and observation updates.

Table 9.2 Important relations for discrete-time Kalman filter.

Update for	State	Covariance
Observation	$\hat{x}_n = \tilde{x}_n + \tilde{K}_n(y_n - C_n\tilde{x}_n)$	$\hat{P}_n = (I - \tilde{K}_n C_n)\tilde{P}_n$
Time	$\tilde{x}_{n+1} = \Phi_n\hat{x}_n + \Gamma_n u_n$	$\tilde{P}_{n+1} = \Phi_n\hat{P}_n\Phi_n' + V_n$

$$\tilde{K}_n = \tilde{P}_n C_n'(C_n\tilde{P}_n C_n' + W_n)^{-1} = \hat{P}_n C_n' W_n^{-1}$$

time process noise. If the process is continuously excited by white noise of spectral density V, then it would be appropriate to omit the discrete-time covariance matrix V_n from (9.65) but include it in the continuous-time variance equation:

$$\dot{P} = AP + PA' + V$$

This last interpretation is depicted in Figure 9.13. Notice that the covariance may go up or down between observations, depending on the nature of the state-transition matrix Φ_n and the intensity of the process noise V, but it always goes down as a result of the observation. The state transition may do anything between observations or at the instants of observation.

Note that the sequences of *a priori* and *a posteriori* covariance matrices $\tilde{P}_n$ and $\hat{P}_n$ depend only on the observation times and *not* on the observation data. Hence these matrices, and the corresponding gain matrices, in principle, can be computed before the process starts, and stored in the memory of the computer used to realize the Kalman filter. (This is not done in many practical applications, however, because the process is generally nonlinear and an "extended" Kalman filter is used in which the covariance and gain matrices do depend on the observation data, as discussed in Section 9.7.4.)

The procedure shown in the flow chart of Figure 9.14 can be used even if the schedule of observations is not known in advance. (The covariance and gain matrices of course cannot be computed in advance.) In particular, the procedure is valid for applications in which observations are made at random instants.

9.7.3 Steady-State Kalman Filter

If the process is time-invariant and the noise intensities are constant and nonzero we can expect that a steady-state condition will be reached at which the increase in the covariance matrix between intervals is exactly balanced by the decrease due to the observations. The steady-state is thus characterized by:

$$\hat{P}_n \to \hat{P} \tag{9.70}$$

$$\tilde{P}_n \to \tilde{P} \tag{9.71}$$

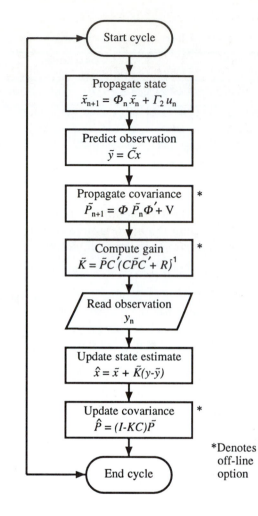

Figure 9.14   Flowchart for discrete-time Kalman filter.

with these matrices given by

$$\tilde{P} = \Phi \hat{P} \Phi' + V \tag{9.72}$$

$$\hat{P} = (I - KC)\tilde{P} \tag{9.73}$$

$$K = \tilde{P} C'(C\tilde{P}C' + W)^{-1} \tag{9.74}$$

Equations (9.72) through (9.74) are the duals of the steady-state optimum control equation (9.46). The existence of a steady-state solution is subject to observability and controllability conditions that are the "duals" of the conditions for the existence and nonsingularity of the steady-state optimum control solution, as discussed in Section 9.6.

In this case the conditions are:

1. For the steady-state covariance to exist, the process must be *detectable*, i.e., the unstable modes must be observable.

2. For the steady-state covariance matrix to be nonsingular, the process must be "controllable by the noise," i.e., the controllability matrix for the pair $[\Phi, V^{1/2}]$ must be of rank k = system order.

If condition 1 is not satisfied, the covariance matrix will become infinite as $n \rightarrow \infty$. If condition 2 is not satisfied, the steady-state covariance matrix will become singular, which implies that some states or linear combinations thereof can be estimated with vanishing error variance.

9.7.4   Extended Kalman Filter

It is often necessary to estimate the state of a nonlinear process. An *extended Kalman filter* (EKF) can be used for this purpose. The time update is computed using the non-linear transition function, and the observation update is computed using the nonlinear observation equations. The correction for the residual, however, is linear and the gain matrix is computed using a linearized covariance propagation equation.

Consider a nonlinear discrete-time process that evolves according to the transition equation

$$x_{n+1} = \phi(x_n, u_n) \tag{9.75}$$

with observations given by

$$y_n = \gamma(x_n) \tag{9.76}$$

The transition function $\phi(\ ,\ )$ may be known explicitly, or it may be defined implicitly as the solution of the nonlinear differential equations that define the process, as explained in Chapter 8. The observation function $\gamma(\ )$ is also known.

The extended Kalman filter algorithm for this process is given by

$$\hat{x}_n = \tilde{x}_n + \hat{K}_n[y_n - \gamma(\tilde{x}_n)] \tag{9.77}$$

$$\tilde{x}_{n+1} = \phi(\hat{x}_n, u_n) \tag{9.78}$$

The residual, i.e., the difference between the actual observation and the expected observation,

$$r_n = y_n - \gamma(\tilde{x}_n) \tag{9.79}$$

is computed using the nonlinear observations. The time update is computed using the nonlinear transition equations. If these are defined as the result of numerical integration of the nonlinear differential equations, (9.78) should be interpreted as the numerical integration of these equations over the interval $T_n \leq t \leq T_{n+1}$ (during which interval the control is assumed constant) starting with the initial condition

$$x(T_n) = \hat{x}_n$$

The *a priori* covariance matrix $\tilde{P}_n$, from which the gain matrix $\hat{K}_n$ is computed using (9.63), is propagated using (9.64) and (9.65). But the state transition matrix Φ_n

in (9.65) is the Jacobian matrix of the transition function $\phi(\ ,\ )$ with respect to the state, and the observation matrix C_n is the Jacobian matrix of the observation function $\gamma(\ )$, both calculated at the estimated state $\hat{x}_n$:

$$\Phi_n = \left[\ \partial\phi(x_n, u_n)/\partial x_n\ \right]_{\hat{x}_n} \tag{9.80}$$

$$C_n = \left[\ \partial\gamma(x_n)/\partial x_n\ \right]_{\hat{x}_n} \tag{9.81}$$

The required partial derivatives needed in C_n are computed directly from the transition function $\gamma(\ )$. If the transition function $\phi(\ ,\ )$ is known, the transition matrix Φ_n can be computed analytically. In most cases, however, it will be necessary to determine this matrix by numerical integration of the linearized matrix equation

$$\dot{\Phi} = A(t)\Phi \tag{9.82}$$

over the interval $T_n \leq t \leq T_{n+1}$ starting with the initial condition

$$\Phi(T_n) = I \quad \text{the } k\text{th order identity matrix}$$

and with $A(t)$ being the Jacobian matrix of the nonlinear, continuous-time dynamics:

$$A(t) = \left[\ \partial f(x, u_n)/\partial x\ \right]_{\hat{x}(t)} \tag{9.83}$$

The meaning of the subscript $\hat{x}(t)$ in (9.83) is that the partial derivatives are to be computed along the trajectory in state space defined by the numerical integration of the continuous-time process equations. This means that the process dynamics and the state transition matrix must be computed as part of the extended Kalman filtering algorithm. Thus, in contrast to the linear Kalman filtering algorithm, in which the sequence of gain matrices $\{K_n\}$ can be computed off-line if desired, this option is not available with the extended Kalman filter: the covariance matrix must be computed on-line. The flow chart of Figure 9.14 remains valid, however, except that the box labeled "**Propagate State**" should be interpreted as requiring the concurrent numerical integration of the nonlinear process equations and (9.82).

Example 9.6   Estimation of Orbital Elements of Satellite

The equations for the orbital elements of a satellite, given in Example 8.5, exemplify the case of a transition function that does not require numerical integration of the differential equations of the process. The corresponding state-transition matrix, needed for propagation of the covariance equation is given by

$$\Phi_n = \begin{bmatrix} 1 & 0 & 0 & 0 \\ 0 & 1 & 0 & 0 \\ 0 & 0 & 1 & 0 \\ \partial\theta_{n+1}/\partial a_n & \partial\theta_{n+1}/\partial e_n & 0 & \partial\theta_{n+1}/\partial\theta_n \end{bmatrix}$$

The required partial derivatives can be obtained from Kepler's equation of Example 8.5 by use of the implicit function theorem.

$$\frac{\partial \theta_{n+1}}{\partial a_n} = \frac{-\frac{3}{2} T \mu^{1/2}}{a_n^{5/2}(1 - e_n \cos \theta_{n+1})}$$

$$\frac{\partial \theta_{n+1}}{\partial e_n} = \frac{\sin \theta_{n+1} - \sin \theta_n}{1 - e_n \cos \theta_{n+1}}$$

$$\frac{\partial \theta_{n+1}}{\partial \theta_n} = \frac{1 - e_n \cos \theta_n}{1 - e_n \cos \theta_{n+1}}$$

9.7.5 Applications

The interpretation of the previous section makes it possible to deal with a multitude of problems in the practical application of Kalman filtering techniques. Included among these problems are multi-rate and delayed sampling, nonsynchronous sampling, and mixing of continuous-time and discrete-time data.

Multi-rate, delayed, and nonsynchronous sampling An application of moderate complexity might involve the use of several sensors. The sensors may not all take their observations concurrently, but it is desired to integrate the observations optimally. A typical sampling pattern involving three sensors is illustrated in Figure 9.15. The first sensor, with output y_A, takes samples at multiples of T seconds. The second sensor, with output y_B, takes samples at the same rate, but these samples are offset from the first by T_1 sec. The third sensor, with output y_C, takes samples twice as frequently as the first two and is synchronized with the first sensor.

Merging the sampling schedules of the three sensors results in dividing the sampling interval into three subintervals of lengths T_1, T_2, and T_3, respectively. (For the particular sampling pattern of Figure 9.15, $T_1 + T_2 = T_3 = T/2$.) Thus each cycle has three sampling cases separated by intervals of different duration:

- *Case 1* Sensors A and C operate simultaneously.

- *Case 2* Sensor B operates alone.

- *Case 3* Sensor C operates alone.

To merge the data from the sensors, the following relationships would thus be used:

$$\hat{x}_{n+1} = \tilde{x}_n + K_n(y_n - C_{AC} x_n)$$
$$\hat{x}_{n+2} = \tilde{x}_{n+1} + K_{n+1}(y_{n+1} - C_B x_{n+1})$$
$$\hat{x}_{n+3} = \tilde{x}_{n+2} + K_{n+2}(y_{n+2} - C_C x_{n+2})$$

Since y_n occurs for Case 1 in which two sensors operate simultaneously, the observation matrix C_{AB} has two rows; the first contains the observation sensitivity coefficients for sensor A, and the second for sensor B. The other two observation matrices

C_B and C_C correspond to Cases 2 and 3 in each of which there is only one observation and thus each has only one row.

Considering the observation cycle, the state update equation is written

$$\tilde{x}_{n+i} = \Phi_{\alpha_i}\hat{x}_{n+i-1} + \Gamma_{\alpha_i}u_{n+i-1}, \qquad i = 1, 2, 3$$

and

$$\Phi_{\alpha_i} = e^{AT_i}$$

$$\Gamma_{\alpha_i} = \int_0^{T_i} e^{A\tau}Bd\tau \quad i = 1, 2, 3$$

The covariance matrices and gains would be similarly computed. After the cycle of three updates, the cycle would revert to the beginning.

Writing the equations for the operation is more complicated than expressing the procedure in words:

- Between samples, propagate the state estimate and covariance matrix to the next sampling instant using the state transition matrix corresponding to the elapse of time between the samples.

- At sampling instants, update the state estimate using the current observation and the covariance matrix using the current observation and the corresponding noise covariance matrix.

There is no theoretical reason for the sampling periods of the sensors to be commensurable. The only requirement is that the matrices Φ and Γ correspond to the actual time intervals between the observations.

Combining continuous- and discrete-time data   Situations can arise in which it may be necessary to combine continuous-time and discrete-time data. In aircraft navigation, for example, continuous-time data may be available for some dynamic variables (e.g., airspeed, body rates), but for other variables (e.g., inertial position) data may be available at relatively infrequent discrete instants. One way of handling the continuous-time data is to time-average it over the sampling period of the discrete-time data and process it as an additional discrete-time measurement. It is not necessary to resort to this procedure, however, since it is possible to combine the two forms of data directly.

The method of combining the data is based on the fact that the optimum covariance matrix at any instant represents all the information available about the process up to that instant, and the gain matrix is based on that optimum covariance matrix. Hence during a time interval in which there are only continuous-time observations, the continuous-time Kalman filter is used:

$$\dot{\hat{x}} = A\hat{x} + Bu + K(y - C_c\hat{x}) \tag{9.84}$$

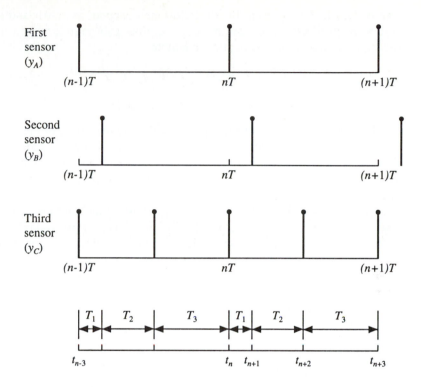

Figure 9.15 Sampling pattern that results in subdivision of the sampling period into three subperiods.

where C_c is the continuous-time observation matrix and K is the continuous-time gain matrix, computed from using the standard continuous-time Kalman filter equations

$$K = PC'W_c^{-1}$$

with

$$\dot{P} = AP + PA' - PC_c'W_c^{-1}C_cP + FV_cF' \tag{9.85}$$

in which V_c and W_c are the spectral density matrices of the continuous-time observation noise and process noise, respectively, and F is the noise distribution matrix. (As usual, both noise processes are assumed to be white.)

The continuous-time covariance matrix is propagated using (9.85) until the next instant at which a discrete-time observation is made. This matrix immediately prior to the observation is interpreted as $\tilde{P}$ and is used to compute the discrete-time gain matrix

$$\bar{K} = PC_d'(C_dPC_d' + W_d)^{-1}$$

where C_d and W_d represent the discrete-time observation matrix and the covariance matrix, respectively. The resulting discrete-time gain matrix is then used to update the state estimate and the covariance matrix:

$$\hat{x}^+ = \hat{x}^- + \tilde{K}(y - C_d\hat{x}^-)$$

$$P^+ = (I - \tilde{K}C_d)P^-$$

Here we use the superscripts $-$ and $+$ to denote the values of the same objects (i.e., the optimum state estimate and the covariance matrix) before and after the discrete-time observation and the resulting updates. The updated covariance matrix P^+ is then used as the initial condition for continued integration of the continuous-time variance equation (9.85).

The operation of the Kalman filter with continuous- and discrete-time data is illustrated schematically in Figure 9.16. The inner loop is the standard continuous-time Kalman filter. The discrete- time update is represented by an outer loop that operates only at the sampling intervals, represented by closing of the switches. The outer loop has the effect of making an instantaneous state change in the estimated state. This is denoted as an initial condition reset ("IC reset" in the figure), suggesting how the change would be achieved in an analog implementation of the filter.

One of the advantages of the implementation shown in Figure 9.16 is that it produces a continuous-time estimate of the state and hence a continuously varying (rather than a piecewise-constant) control signal input to the plant.

If the sampling interval is constant, and the process is allowed to proceed to the steady-state, we should expect that the covariance matrix P would approach a periodic function of time. The continuous-time gain matrix would also become periodic, but the discrete-time gain matrix would approach a constant. These matrices are obtained by the following procedure [4]:

1. Form the $2n \times 2n$ ("Hamiltonian") matrix

$$H = \begin{bmatrix} A & FVF' \\ C_c'W^{-1}C_c & -A' \end{bmatrix}$$

2. Compute the state-transition matrix corresponding to H for the sampling time T:

$$\Psi = e^{FT} = \begin{bmatrix} \Psi_{11} & \Psi_{12} \\ \Psi_{21} & \Psi_{22} \end{bmatrix}$$

where Ψ_{ij} are the 2×2 submatrices of Ψ.

3. Form the following matrices:

$$\Phi = \Psi_{11}$$
$$V = \Psi_{12}\Psi_{11}'$$
$$W = \Psi_{21}\Psi_{11}^{-1}$$

4. Compute the discrete-time covariance matrices and gain matrix using

$$\hat{P}^{-1} = \tilde{P}^{-1} + W + C_d W_d^{-1} C_d'$$
$$\tilde{P} = \Phi \hat{P} \Phi' + V$$
$$K_d = \hat{P} C_d' W_d^{-1}$$

5. Compute the continuous-time covariance matrix, and from it, the time-varying gain matrix over one sampling interval by numerical solution of the matrix Riccati equation (9.85) over the sampling interval $t \in [0, T]$ using $\hat{P}$ obtained in Step 4.

It can be shown [4] that the matrices V and W defined in Step 3 are symmetric, as they must be in order for $\hat{P}$ and $\tilde{P}$ to be covariance matrices. It may happen, however, that the matrix Φ is not the state-transition matrix of any continuous-time system, i.e., that there is no real matrix A for which $\Phi = e^{AT}$.

The equations for updating the covariance matrix are analogous to (9.72) through (9.74). Note, however, that the equation for updating the covariance matrix for the observation contains information about the discrete-time and the continuous-time observation noise. The former is accounted for by W_d and the latter by W.

9.8 NOTES

Note 9.1 Separation principle in discrete-time system

The separation principle, a bedrock of state-space design methods, was first articulated in the context of discrete-time control systems in 1963 by Gunckel and Franklin [5]. An earlier formulation, which has come to be known as the *Certainty Equivalence Principle*, was given by Simon [6] in the context of economics.

9.9 PROBLEMS

Problem 9.1 Optimality of linear control

Show that a linear control law

$$u_n = -G_n x_n$$

minimizes a quadratic performance criterion for a linear, discrete-time plant.

Problem 9.2 Determining optimum gains in Kalman filter

Find the optimum gain matrix of the Kalman filter by using the following procedure:

1. Assume that the optimum gain $\bar{K}$ and the corresponding covariance matrix $\bar{P}$ exist and are given by the expressions in Section 9.7.1.

2. Let $K = \bar{K} + \delta K$ and $P = \bar{P} + \delta P$. Substitute these into the variance equation and find an expression for δP.

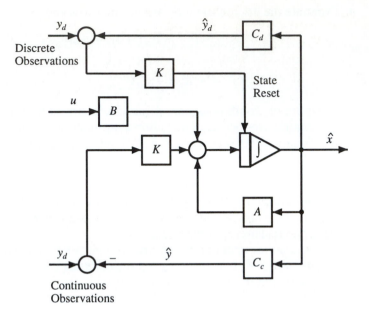

Figure 9.16   Schematic representation of Kalman filter operating with continuous- and discrete-time data.

3. Show that ΔP can be decreased unless the gain $K = \bar{K}$.

Problem 9.3   Pulse-width modulated control of satellite using observer

In Example 9.3 the pulse-width modulated control of a satellite using full-state feedback was studied. Suppose only the position and not the velocity of the satellite is measurable.

(a) Design a control system that incorporates a continuous-time observer to estimate the velocity. (Note that the observer should include a model of the PWM process, as discussed in Chapter 6.)

(b) Design a control system that incorporates a discrete-time observer, using an appropriate nonlinear function to represent the PWM process.

Problem 9.4   Pulse-width modulated control of servomotor using observer

Repeat Problem 9.3 for the servomotor of Example 9.4 .

Problem 9.5   Kalman Filter for cross-correlated noises

Derive the equations for the propagation of the state and the covariance matrices when the process noise and the observation noise in the linear estimation problem are correlated.

Problem 9.6   Motor-driven robot arm

Design a control law for the motor-driven robot arm of Problems 6.5 and 6.6 by:

(a) Digitizing a continuous-time designs.

(b) Performing direct discrete-time designs.

In each case simulate the performance and comment on the differences between the control laws.

Problem 9.7   Motor-driven robot arm with pulse-width modulated control

Design a control law for the motor-driven robot arm of Problem 9.6 using a pulse-width modulated control, with $u_{\max} = 100$.

Problem 9.8   Temperature control

Design a temperature control law for the system of Problem 9.8 using each of the following approaches, with a sampling interval of $T = .1$ sec:

(a) Digitizing the control law of Example 1.7.

(b) Using "plant inversion" in the Z-domain, (i.e., the digital version of the design approach of Example 1.7).

(c) A direct discrete-time state-space design based on use of the separation principle, and providing integral action.

Problem 9.9   Discrete-time control of autonomous vehicle

Develop a closed-loop control algorithm for the autonomous vehicle of Problem 1.2 *et seq.* using sensors that measure the total wheel rotation and the platform angular velocity continuously, but the range only intermittently, at intervals of 0.1 sec.

Problem 9.10   Navigation of an autonomous vehicle

Consider the autonomous vehicle described in Example 1.2, *et seq.* and Problem 9.9.

(a) Specify the configuration of an observer for the system in which the sensors measure the following quantities:

1. Range of the vehicle with respect to a station at a fixed location, intermittently, at intervals of T sec.

2. Angular rotation of each wheel, continuously.

(b) Simulate the performance of the observer for the parameters given in Chapter 1, and for $T = 0.25$ sec.

Problem 9.11   **Estimating state of satellite with continuous-time and discrete-time measurements**

A control system for the idealized satellite considered in Example 9.2 is to be designed. The sensor complement for the control system includes a rate gyro, which provides a continuous-time measurement of the angular velocity, and a position sensor, which can measure the attitude at fixed intervals of time.

Specify a control algorithm for the satellite that combines the data from the two sensors.

9.10   REFERENCES

[1] R.W. Bass and I. Gura, "High-Order System Design Via State-Space Considerations," Proc. Joint Automatic Control Conference, Troy, NY, June 1965, pp.311-318.

[2] M. Athans, "The Matrix Minimum Principle," Information and Control, Vol. 11, pp. 592–606, November 1967.

[3] R.E. Kalman,"A New Approach to Linear Filtering and Prediction Problems," Trans. ASME, (J. Basic Engineering) Vol. 82D, No. 1, March 1960, pp. 35–45.

[4] B. Friedland, "Steady-State Behavior of Kalman Filter with Discrete- and Continuous-Time Observations," IEEE Trans. on Automatic Control, Vol. AC-25, No. 5, pp. 988–992, October 1980.

[5] T.L. Gunckel, III and G.F. Franklin, "A General Solution for Linear Sampled Data Control," Trans. ASME (J. Basic Engineering) Vol. 85D. No.1., January 1963, pp. 197–201.

[6] H.A. Simon, "Dynamic Programming Under Uncertainty with a Quadratic Criterion Function," Econometrica, Vol. 24, 1956, 74–81.

Chapter 10

ADAPTIVE CONTROL

When you use a model of the plant as the basis of a control system design, you are tacitly assuming that this model is a reasonable representation of the plant. Although the design model almost always differs from the true plant in some details, you are confident that these details are not important enough to invalidate the design.

There are many applications, however, for which a design model cannot be developed with any reasonable degree of confidence. Such applications include processes (e.g., biological, ecological, economic) in which the underlying physical principles are not understood well enough for mathematical modeling. In addition, there are processes for which the physical principles are understood but which have parameters that cannot be measured or accurately estimated.

Moreover, most dynamic processes change with time. Parameters may vary because of normal wear, aging, breakdown, and changes in the environment in which the process operates.

The feedback mechanism provides some degree of immunity to discrepancies between the physical plant and the model that is used for the design of the control system. But sometimes this is not enough: A control system designed on the basis of a nominal design model may not behave as well as expected, because the design model does not adequately represent the process in its operating environment.

How can one deal with processes that are prone to large changes, or for an adequate design models are not available? One approach is brute force, i.e., high loop-gain: as the loop-gain becomes infinite, the output of the process tracks the input with vanishing error. Brute force rarely works, however, for well-known reasons: dynamic instability, control saturation, and susceptibility to noise and other extraneous inputs.

Modern robust control design techniques are more sophisticated and make it possible for a control system design to tolerate substantial variation in one or more parameters. But the price of achieving immunity to parameter variation may be a sacrifice in performance. Moreover, robust control design techniques are not readily applicable to processes for which no design model is available.

Adaptive control may provide another answer. The basic idea is to have the control law adapt its own behavior, as it learns about the process it is designed to control, or as the process changes with its environment. Nature provides many examples of processes that exhibit adaptive behavior, and, many believe, can teach how to include similar adaptive behavior in control system designs.

The concept of controlling a process that is not well understood, or one in which the parameters are subject to wide variations, has a history that predates the beginnings of modern control theory. (See Note 10.1.) The early theory was empirical and was developed before digital computer techniques could be used for extensive performance simulations. Prototype testing was one of the few available techniques for testing adaptive control techniques. At least one early experiment had disastrous consequences. As the more mathematically rigorous areas of control theory were developed starting in the 1960s, interest in adaptive control faded for a time, only to be reawakened in the late 1970s with the discovery of mathematically rigorous proofs of the convergence of some popular adaptive control algorithms. This interest continues unabated.

Two basic viewpoints on adaptive control have emerged over the years. The first assumes that a design model is available, but that the parameters of the model are either not known or subject to wide variation. The second assumes that no such design model is available.

The first viewpoint leads to an approach in which the control system design depends explicitly upon the parameters that are subject to variation, and then uses measurements or estimates of these parameters as inputs to the control system for the purpose of "tuning" it during the course of its operation. If the parameters can be measured directly, the control system gains can be "scheduled" for these measured parameters. (For this reason, this approach is often called *gain scheduling*.)

If it is not possible to measure the uncertain parameters directly, the relationships between these parameters and the other measurable quantities in the system might feasibly be exploited to estimate the parameters. A design based on estimates of parameters may be called a *self-tuning controller*, since it uses its own operating data to estimate the parameters and thereby tune its own behavior. A block diagram of a self-tuning control system is shown in Figure 10.1. The symbol θ is widely used to denote the set of parameters in the dynamic process that are unknown and/or subject to change and whose values are important to the favorable operation of the system. (Unknown or uncertain parameters that do not significantly affect system performance can be ignored.) The control system is designed with the parameter vector θ as one of its inputs. In the course of operation of the system, an estimate $\hat{\theta}$ of the parameter vector is computed using whatever information is available, namely the process input u, its output y and the desired output y_r, and used as an input to the adaptive control law. Note that the parameter θ can be adjoined to the process state x to give the

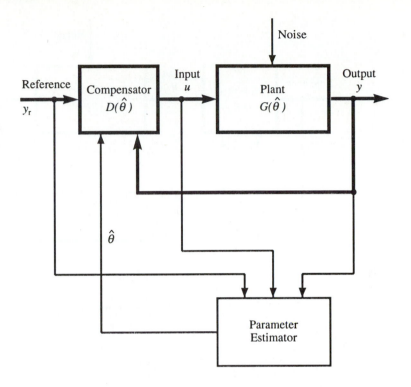

Figure 10.1   Self-tuning control based on parameter estimation.

"metastate"[1]:

$$\mathbf{x} = \left[\begin{array}{c} x \\ \cdots \\ \theta \end{array} \right]$$

Thus, if the design of the parameter estimator is merged with the state estimator in the design of the compensator, the result is a nonlinear control system in the metastate. This observation has led to the contention that self-tuning adaptive control is not more than a type of nonlinear control. This argument has validity.

A bolder approach, commonly known as *model reference adaptive control* (MRAC) is predicated upon the viewpoint that the process is not amenable to being modeled with any reasonable accuracy and hence that the control law must be designed with little or no knowledge of the process. Known only is how the closed-loop system is required to behave. This desired behavior is represented by an ideal or reference model, as shown in Figure 10.2.

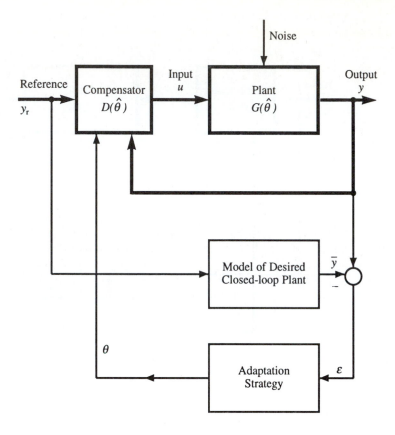

Figure 10.2   Model-reference adaptive control.

The reference input is applied to both the real closed-loop plant and the ideal model. If the performance error ϵ between the two is zero (or, in practice, sufficiently small), the closed-loop system is operating as desired and left alone. But if the error is appreciable, it becomes the input to a tuner that adjusts some vector, say q, of adjustable parameters of the controller. This parameter vector is typically a set of controller gains and has nothing to do with the physical parameters of the process being controlled. In principle, the operation of the tuner is indifferent to the dynamics of the plant. For performance evaluation, of course, a model of the plant is necessary, so that a simulation of a model-reference adaptive control may not appear very different from a self-tuning adaptive control.

Whether it is worth all the effort entailed in designing an adaptive control and the hardware required for its implementation depends very much on extrinsic considerations, sometimes not always technical. A supervisor's actual or imputed favorable or unfavorable experience with adaptive control at some time past may promote or preclude its use, notwithstanding the actual benefits; a goal to develop a product having every "state-of-the-art" feature may lead to the use of adaptive control in situations in which there is no significant benefit to be gained.

10.1 GAIN SCHEDULING

The simplest type of adaptive control is gain scheduling: If θ is a parameter vector of the process that can be measured or inferred from other measurements, you can design the controller to depend on θ. In operation, the controller uses the measured or inferred value of θ.

If θ is a constant, or varies so slowly with respect to the time scale of the closed-loop system as to be regarded as a constant, there is no problem with this approach. It may happen, however, that the rate of variation of the parameter θ is not insignificant relative to the time scale of the closed-loop system. If that is the case, the system must be considered time-varying. Its behavior may differ considerably from that predicted on the assumption that θ is constant. Conceivably the system might even become unstable. Liapunov stability theory provides a general approach to evaluation of the performance of gain-scheduled systems [2] but it may be difficult to apply and, in any case, gives very conservative results. Extensive simulation could help confirm or dispel doubt about the validity of the method in a specific application.

Example 10.1 Cargo-Moving Gantry Crane

The simplified dynamics for a cargo-moving gantry crane were given in Chapter 1, where it was shown that the length control is (essentially) independent of the truck motion, but that the truck motion depends on the length of the cable (which determines the period of the swing).

To optimize efficiency of operation of the gantry crane, we consider simultaneously controlling the length of the cable and the translation of the truck. Controlling both on the same time scale would be desirable. Doing so, however, leads to a time-varying system that makes for an interesting example of gain scheduling.

The reciprocal of the cable length is used as the scheduling parameter. Thus the linearized dynamics are given by

$$\dot{x} = Ax + Bu \qquad (10.1)$$

with the state vector defined by

$$x = \begin{bmatrix} d & \phi & d & \phi \end{bmatrix}$$

where d and ϕ are the lateral truck displacement and cable angle, respectively and where the process matrices are given by

$$\begin{bmatrix} 0 & I \\ A_{21} & 0 \end{bmatrix} \qquad B = \begin{bmatrix} 0 \\ B_2 \end{bmatrix}$$

with

$$A_{21} = \begin{bmatrix} 0 & \rho g \\ 0 & -(1+\rho)g\theta \end{bmatrix} \quad B_2 = \begin{bmatrix} 1/m_T \\ \theta/m_T \end{bmatrix}$$

with $\theta = 1/l$.

Assume full-state feedback:

$$u = -m_T(G_1 x_1 - G_2 x_2 - G_3 x_3 - G_4 x_4)$$

If the length $l = 1/\theta$ were constant, the system would be time invariant. A straightforward (if somewhat tedious) algebraic calculation gives the characteristic polynomial

$$\Delta(s) = s^4 + a_1 s^3 + a_2 s^2 + a_3 s + a_4$$

with

$$a_1 = G_3 + \theta G_4$$
$$a_2 = G_1 + \theta[G_2 + (1+\rho)g]$$
$$a_3 = \theta G_3(1+2\rho)g$$
$$a_4 = \theta G_4(1+2\rho)g$$

where

$$\rho = m_L/m_T, \quad g = \text{gravitational acceleration}$$

A reasonable design goal might be to keep the closed-loop dynamics independent of the cable length by fixing $a_1, \ldots, a_4$ and making the control gains $G_1, \ldots, G_4$ vary accordingly, i.e.,

$$G_1 = \frac{a_4}{\theta(1+2\rho)g} = \frac{a_4}{(1+2\rho)g}l$$

$$G_2 = -(1+\rho)g + \frac{a_2 - G1}{\theta} = -(1+\rho)g + (a_2 - G1)l$$

$$G_3 = \frac{a_3}{\theta(1+2\rho)g} = \frac{a_3}{(1+2\rho)g}l$$

$$G_4 = \frac{a_1 - G_3}{\theta} = (a_1 - G_3)l$$

Since we want to change the cable length while the truck is being moved, however, the system is time-varying and there is no *a priori* assurance that the above gains, based on "frozen" system dynamics, will result in the desired performance. Hence we verify performance by simulation.

The performance of the control law is illustrated in Figure 10.3 for $\rho = 0.2$, $g = 9.8\text{m/sec}^2$, and with the following values of the feedback gains:

$$G_1 = 5.0729, \quad G_2 = 10.867, \quad G_3 = 11.423, \quad G_4 = 4.8990$$

which were selected to produce a response with no overshoot. It is seen that the performance is quite satisfactory during a transient in which the length is reduced from 10m to 2m in about the same time as the truck translates a distance of 2m.

Comparison with a fixed-gain controller is the subject of Problem 10.1.

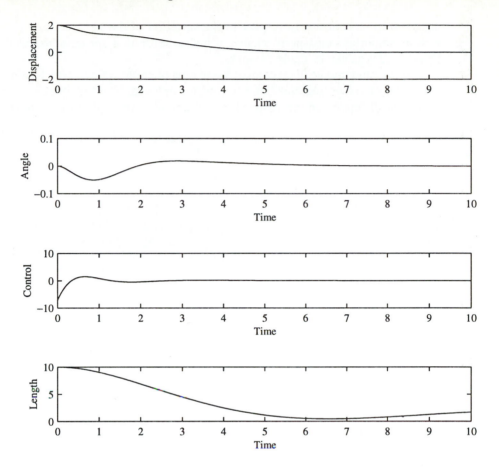

Figure 10.3 Transient response of truck control using gain scheduling on cable length.

The dynamics in the foregoing example are simple enough for you to determine the gains algebraically in terms of the desired characteristic polynomial of the closed-loop system. But this is not a limitation of the method. If it is not possible to find the gains algebraically, you can always determine them numerically for each fixed value of the parameter and tabulate them. Then implement the control law by interpolation in a lookup table. Or, alternatively, use some curve-fitting program to approximate the tabulated data by some appropriate functions and use these functions to implement the control law. At the expense of greater complexity, the method is adaptable to two or perhaps three variable parameters.

This approach is easy implement, but it may not be worth the effort, because it may be possible to achieve nearly as good performance with a time-invariant compensator designed to be robust with respect to the variation of the parameters. If simplicity is more important than performance, it makes sense to determine how well a time-invariant design can be made to work.

Aircraft and missile flight control is one area of application in which gain scheduling is almost always necessary, and routinely employed. The aerodynamic forces and moments all depend on the airspeed V of the vehicle, through the *dynamic pressure*

$$Q = \frac{1}{2}\rho V^2$$

where ρ is the density of the air surrounding the vehicle, and through the Mach number M (which is the ratio of the airspeed to the speed of sound at the operating density and temperature). During its entire envelope of operation, the aerodynamic parameters of an aircraft are subject not only to variations in magnitude of a factor of 10 or more but also to changes in sign. Attempting to cover the entire flight envelope with a fixed gain compensator, even if possible, would entail an unacceptable sacrifice in performance.

10.2 PARAMETER ESTIMATION

In many applications it may not be feasible to use gain scheduling, because it may be too expensive or impossible to measure the parameters on which the process dynamics depend. By estimating these parameters on-line (i.e., during the operation of the process), we can in principle achieve the benefits of gain scheduling without the need for an external measurement.

10.2.1 Nonlinear Observer/Extended Kalman Filter

A direct method of implementing adaptive control with parameter estimation is to model the parameter θ (possibly a vector) that is to be estimated by a differential equation:

$$\dot{\theta} = \gamma(\theta) + v \tag{10.2}$$

where $\gamma(\ )$ is an appropriate nonlinear function and v is assumed white noise. Equation (10.2) is adjoined to the original process and a nonlinear observer is used to estimate the metastate

$$\mathbf{x}' = [x' \ \theta']$$

The equations of the nonlinear observer may be designed *ad hoc* or can be implemented as an extended Kalman filter. (The role of the assumed noise v is to prevent the gain matrix of the extended Kalman filter from becoming too small.) If the parameter θ is nominally constant, it would be appropriate to make $\gamma(\theta) = 0$.

Even if the plant in which the parameter vector θ appears is linear in the state x, the "metaprocess" comprising x and θ is nonlinear, and there is no assurance that the proposed procedure will be effective. Since the parameter is presumably not influenced by the control, it is clear that the metastate is not controllable in the sense

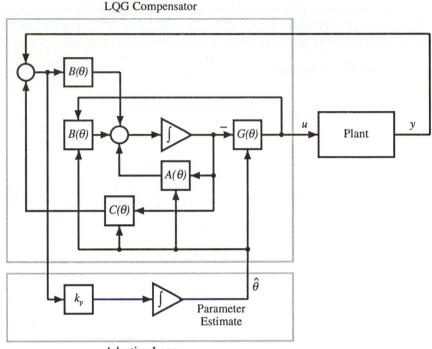

Figure 10.4   Adaptive controller using extended Kalman filter architecture for estimating state and parameters.

discussed in Chapter 4. But the parameter may affect the output and hence make the metastate observable. Even if the process is observable, the overall system may be unstable if the observer gain is chosen incorrectly.

The use of a nonlinear observer to estimate the state and parameter in a linear process would result in a closed-loop system governed by the following equations:

$$u = -G(\hat{\theta})\hat{x} \tag{10.3}$$

$$\dot{\hat{x}} = A(\hat{\theta})\hat{x} + B(\hat{\theta})u - K_x[y - C(\hat{\theta})\hat{x}] \tag{10.4}$$

$$\dot{\hat{\theta}} = K_p[y - C(\hat{\theta})\hat{x}] \tag{10.5}$$

The architecture of the control system based on this adaptive control concept is shown in Figure 10.4. The adaptive loop consists of nothing more than a bank of integrators, one for each parameter, preceded by an appropriate gain matrix and driven by the residual vector $y - C(\hat{\theta})\hat{x}$. The choice of the gain matrices K_x and K_p is, of course, crucial to the stability of the overall system. These matrices can be chosen by any method that works. One extreme would be to use constant gain matrices, the numerical values of which are chosen empirically (i.e., by cut and try); the other extreme would be to implement the full extended Kalman filter as described in Chapter 6.

Example 10.2   Robot Arm

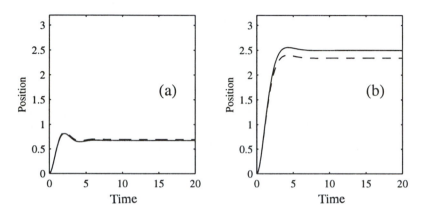

Figure 10.5   Non-adaptive control of robot arm with uncertain parameter does not achieve desired steady state. True parameter = 1. (a) Assumed parameter = 0.5. (b) Assumed parameter = 2.0

Representing a robot arm as a pendulum and linearizing about a constant desired angle $\phi_r = x_r$ results in the linear system:

$$\dot{x}_1 = x_2$$
$$\dot{x}_2 = -\theta(x_1 + x_r) + u$$

where

$$x_1 = \phi - \phi_r, \qquad x_2 = \dot{\phi}$$

and

$$\theta = \sqrt{g/l}\cos\phi_r$$

is the parameter to be estimated.

A possible nonlinear design for this process was presented in Chapter 4 under the assumption that g/l was known. In this case θ is also known once the reference input ϕ_r is specified. Suppose, however, that g/l is not known. Then the process is a candidate for adaptive control.

Suppose that the observation is the position error:

$$y = x_1$$

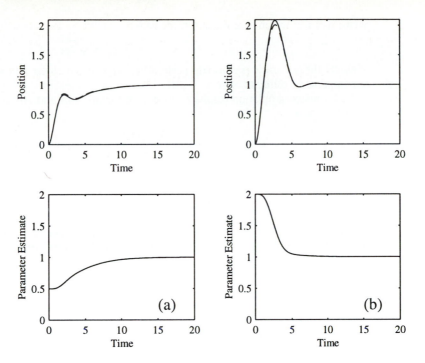

Figure 10.6   Adaptive control of robot arm shows that desired steady state is achieved and parameter is estimated. (a)
$\hat{\theta}(0) = 0.5$ (b) $\hat{\theta}(0) = 2.0$

Then, assuming a linear full-state feedback law and using an observer in the form of an
extended Kalman filter gives the following compensator equations:

$$u = -g_1(\hat{\theta})\hat{x}_1 - g_2(\hat{\theta})\hat{x}_2 - g_0(\hat{\theta})x_r \tag{10.6}$$

where

$$\dot{\hat{x}}_1 = \hat{x}_2 + k_1(y - \hat{x}_1)$$
$$\dot{\hat{x}}_2 = -\hat{\theta}(\hat{x}_1 + x_r) + u + k_2(y - \hat{x}_1)$$
$$\dot{\hat{\theta}} = k_\theta(y - \hat{x}_1)$$

The first two of these equations define observer for x, and the third is the observer for
θ.

The gains in the full-state feedback law (10.6) can be obtained by any convenient
method. It can easily be established, however, that for zero steady-state error, the "feedforward" gain must be

$$g_0 = -\theta$$

which means that this control implementation relies on knowledge of the parameter θ.
The other two gains can be designed by pole placement or by optimization of a quadratic

performance criterion. For the latter, the gains

$$g_1 = \sqrt{\theta^2 + c^2} - \theta, \qquad g_2 = \sqrt{2g_1}$$

minimize the quadratic performance criterion $c^2 x_1^2 + u^2$. Implementation of these gains also requires an estimate of θ.

Performance simulation results are shown in Figures 10.5 and 10.6 for the following numerical data:

$$\phi_r = 1.$$
$$\theta = 1.$$
$$c^2 = 4.$$
$$k_1 = 8.$$
$$k_2 = 16.$$

Figure 10.5 shows the behavior of the system in the absence of adaptation when the actual parameter is 1, but the controller is designed under the assumption that it is (a) 0.5 or (b) 2.0. Although the transient response is acceptable, the desired steady-state output is not achieved because the feedforward gain, which is based on an erroneous assumption of θ, is not correct.

Figure 10.6 illustrates the benefits of adaptation, using a constant, empirically selected gain $k_\theta = -0.5$. It is seen that the desired steady-state angle ϕ_r is achieved asymptotically, and the parameter θ is also estimated correctly. A different choice of k_θ could well improve the performance of the parameter estimation.

As impressive as this performance might seem, it is *not* the "right" engineering solution to the problem. The reason why the feedforward control is needed in this case is that the plant is "type zero" and thus cannot track a step input with zero steady-state error. The "right" engineering solution is a control law with "integral action," i.e., a compensator that has a pole at the origin of the s-plane. This can be achieved either by augmenting the plant with an integrator, or by assuming (contrary to fact) that the reference state x_r is *not* measurable, and hence must be estimated by the observer. (See Problem 10.2.)

10.2.2   Nonlinear Reduced-Order Observer

A reduced-order observer can of course be used. Based on reduced-order observer theory, the author of this book developed a novel parameter estimation method [3]. The method is simplest when it is possible to measure all the state variables and only the parameters are to be estimated. For this case, consider a dynamic process

$$\dot{x} = f(x, u, \theta) \tag{10.7}$$

where x is the state of the process, u is the control input, and θ is a vector of constant parameters to be estimated. Applying the theory developed in Chapter 5 of the nonlinear reduced-order observer to the metaprocess suggests an observer governed by the following estimation equations:

$$\hat{\theta} = \phi(x) + z \tag{10.8}$$
$$\dot{z} = -\Phi(x)f(x, u, \hat{\theta}) \tag{10.9}$$

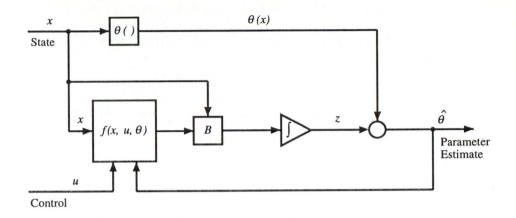

Figure 10.7 Architecture of parameter estimation based on re-duced-order observer.

where $\phi(x)$ is an appropriately chosen nonlinear function and $\Phi(x)$ is its Jacobian matrix:

$$\Phi(x) = \left[\ \partial\phi_i(x)/\partial x_j\ \right] \tag{10.10}$$

Figure 10.7 illustrates the architecture of this parameter estimation algorithm.

The function $\phi(\ )$ plays the role of the gain matrix K in a linear reduced-order observer, and it must be chosen by the user to ensure convergence of the estimation error to zero with the desired speed.

To aid in the determination of $\phi(\ )$ we analyze propagation of the estimation error e, given by

$$e := \theta - \hat{\theta} \tag{10.11}$$

Then, since θ is assumed constant,

$$\begin{aligned}
\dot{e} &= -\dot{\hat{\theta}} \\
&= -\Phi(x)\dot{x} - \dot{z} \\
&= -\Phi(x)[f(x, u, \theta) - f(x, u, \theta - e)]
\end{aligned} \tag{10.12}$$

Equation (10.12) is as far as one can proceed in general with the exact analysis of the estimation error. If the dynamics of the process (10.7) are linear in the parameter vector θ, however, then we can write, *without approximation*,

$$f(x, u, \theta) = F(x, u)\theta \tag{10.13}$$

where the matrix $F(x, u)$ is the Jacobian matrix of $f(\ )$ with respect to the parameter vector θ:

$$F(x, u) = \left[\ \partial f_i(x, u, \theta)/\partial\theta_j\ \right] \tag{10.14}$$

Note that the assumption that the dynamics are linear in θ implies that F does not contain θ.

When (10.13) applies, it can be used to reduce (10.12) to the linear, homogeneous, equation

$$\dot{e} = -\Phi(x, u)F(x, u)e \tag{10.15}$$

To ensure that the estimation error goes to zero, it is necessary to find a function $\phi(x)$ such that its Jacobian matrix $\Phi(x)$, when postmultiplied by $F(x, u)$, is a stability matrix. Since

$$A(t) := \Phi(x(t), u(t))F(x(t), u(t))$$

is time-varying, we cannot readily determine stability on the basis of the locations of the instantaneous eigenvalues of $A(t)$. To help ensure convergence of the error to zero, we can try to select $\phi(\ )$ so that

$$\Phi(x, u) = F'(x, u)Q$$

where Q is a positive-definite matrix. In this case

$$A(t) = F'(x, u)QF(x, u)$$

which is a positive-definite or -semidefinite matrix. For this matrix, a sufficient condition for asymptotic stability of (10.15) is that

$$M = \int_{t_1}^{t_2} A(t)dt > cI \tag{10.16}$$

for some $c > 0$ and t_1 and all $t_2 > t_1$ [3]. If A is a constant matrix, (10.16) is satisfied if and only if A is positive-*definite*. But when $A(t)$ is time-varying, it is possible for (10.16) to be satisfied even when $A(t)$ is singular. The nature of the time variation of $x(t)$ is important: $x(t)$ must be persistently exciting of sufficiently high order. (*Persistent excitation* is a term occurring frequently in adaptive control. The concept is discussed further in Section 10.4.2.)

In some cases it may be desirable or necessary to include the input u in the nonlinear function $\phi(\ )$. This requires modification of (10.9). In this case, let

$$\Phi_x(x, u) = [\ \partial\phi_i(x, u)/\partial x_j\ ] \tag{10.17}$$

$$\Phi_u(x, u) = [\ \partial\phi_i(x, u)/\partial u_j\ ] \tag{10.18}$$

Then instead of (10.8) and (10.9) the following estimation algorithm is used:

$$\hat{\theta} = \phi(x, u) + z \tag{10.19}$$

$$\dot{z} = -\Phi_x(x, u)f(x, u, \hat{\theta}) - \Phi_u(x, u)\dot{u} \tag{10.20}$$

An analysis similar to that carried out for the case in which $\phi(\ )$ does not contain u shows that (10.15) remains valid for (10.19) and (10.20).

The modification to the algorithm of (10.8) and (10.9) is seen to require the derivative of the control input. Although the control input is available for use in the implementation of (10.19) and (10.20), its derivative may not be readily available. In principle, it can be calculated from the known dynamics of the process. Suppose, for example, that a linear control law is used:

$$u = -Gx \qquad (10.21)$$

where G is a constant gain matrix. Then the derivative of the control is given by

$$\dot{u} = -G\dot{x}$$
$$= -Gf(x, u, \theta) \qquad (10.22)$$

which contains the unknown parameter θ. If (10.22) is implemented using the estimated parameter $\hat{\theta}$, the required cancellation of the $\dot{u}$ terms in arriving at (10.15) will not be perfect. Fortunately, however, the effect of the error is only to add a forcing term to (10.15) and not to change the dynamics. The presence of the forcing term, which is proportional to $\dot{u}$, may prevent the estimation error from converging to zero, but will not cause it to diverge. In applications in which the control input approaches a constant, the forcing term will go to zero.

If the plant is controlled by integral control, then $\dot{u}$ is explicitly generated within the controller and would be available for the implementation of (10.20).

A technique for estimating parameters of a process is often useful for purposes other than control. On-line estimation of a parameter may be a means for monitoring safe operation: Deviation of the parameter from bounds established by safety codes or standard practice might be cause for sounding an alarm rather than for tuning the control system. (Safety codes, however, usually mandate direct measurement of the parameter in question with a very reliable sensor.)

Another application of parameter estimation could be in production-line testing. The parameters of products such as sensors and actuators, which are dynamic systems, are often needed by the user for design of systems that use these products. The manufacturer can use parameter estimation techniques to measure the values of these parameters and provide the information to the customer, or in final inspection, to determine whether the product is within the tolerance limits specified.

There are few real control problems in which it is possible to measure all the state variables. Hence this technique, as it stands, may not be very broadly applicable. In some cases it may be possible to substitute estimates of the state variable, obtained by another observer, for the measured state variables. This couples the parameter estimator and the state estimator, as shown in Figure 10.8, and it is more complicated to verify stability of the entire system. Nevertheless, use of a two-observer configuration can be quite effective, as has been shown in the case of friction estimation [4], and as the following example demonstrates.

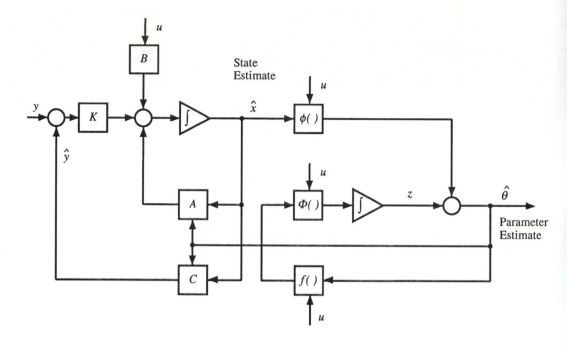

Figure 10.8   Coupled observers for state and parameter estimation.

Example 10.3   Instrument Servo

A simple instrument servo often comprises a d-c motor driving an inertial load. The transfer function of the motor and load is given by

$$G(s) = \frac{\theta_2}{s(s + \theta_1)} \qquad (10.23)$$

The parameters θ_1 and θ_2 depend upon the amplifier gain, the load inertia, and on the motor parameters. An appropriate design objective might be to achieve performance independent of these parameters. To achieve this goal you might consider adaptive control based upon parameter estimation.

In this application, estimation of the two parameters θ_1 and θ_2 is considered. A state-space representation of (10.23) is

$$\dot{x}_1 = -\theta_1 x_1 + \theta_2 x_2 \qquad (10.24)$$

$$\dot{x}_2 = u \qquad (10.25)$$

(Note that x_1 is the angular position, but x_2 is *not* the angular velocity of the motor.)

Assuming that only the position of the motor can be measured, we must use an observer to estimate x_2 in addition to the parameters. For the latter purpose, we elect to use a reduced-order observer. If the parameters were known, standard theory for reduced-order observers would give the observer as

$$\hat{x}_1 = y, \quad \text{the measured position error} \qquad (10.26)$$

$$\hat{x}_2 = Ky + z \qquad (10.27)$$

$$\dot{z} = -Kf_1(\hat{x}) + u \qquad (10.28)$$

where

$$f_1(\hat{x}) = -\theta_1\hat{x}_1 + \theta_2\hat{x}_2 \qquad (10.29)$$

Since θ_1 and θ_2 must be estimated, we propose to use the estimates in place of the actual values in (10.29).

The Jacobian matrix of the dynamics with respect to the parameters in this case is given by

$$F = \frac{\partial f}{\partial \theta} = \begin{bmatrix} -x_1 & x_2 \\ 0 & 0 \end{bmatrix} \qquad (10.30)$$

For this matrix an appropriate choice of ϕ is

$$\phi(x_1, x_2) = \begin{bmatrix} -K_1x_1^2 \\ 2K_2x_1x_2 \end{bmatrix} \qquad (10.31)$$

from which we obtain

$$\Phi = \begin{bmatrix} -2K_1x_1 & 0 \\ 2K_2x_2 & 2K_2x_1 \end{bmatrix} \qquad (10.32)$$

The parameter estimation equations based on this choice of $\phi(\ )$ are

$$\hat{\theta}_1 = \phi_1 + z_1 = -K_1x_1^2 + z_1 \qquad (10.33)$$

$$\hat{\theta}_2 = \phi_2 + z_2 = 2K_2x_1x_2 + z_2 \qquad (10.34)$$

with

$$\dot{z}_1 = 2K_1x_1\hat{f}_1 \qquad (10.35)$$

$$\dot{z}_2 = -K_2(x_2\hat{f}_1 + x_1u) \qquad (10.36)$$

where

$$\hat{f}_1 = -\hat{\theta}_1\hat{x}_1 + \hat{\theta}_2\hat{x}_2 \qquad (10.37)$$

To select the parameter estimation gains K_1 and K_2 we analyze the matrix

$$A = \Phi F = \begin{bmatrix} -2K_1x_1 & 0 \\ K_2x_2 & K_2x_1 \end{bmatrix} \begin{bmatrix} -x_1 & x_2 \\ 0 & 0 \end{bmatrix} = \begin{bmatrix} K_1x_1^2 & -K_1x_1x_2 \\ -K_2x_1x_2 & K_2x_2^2 \end{bmatrix}$$

Choosing K_1 and K_2 as positive constants will make A positive semi-definite. To satisfy (10.16) we rely on persistency of excitation, i.e., that $x_1(t)$ and $x_2(t)$ do not remain zero. This means that the reference input to the servo must be time-varying.

The control system design is completed by defining a full-state feedback control law

$$u = -G_1y - G_2\hat{x}_2 - G_0x_r$$

With this control law the closed-loop characteristic polynomial is

$$\Delta(s) = s^2 + a_1 s + a_2 \tag{10.38}$$

where

$$a_1 = \theta_1 + G_2$$
$$a_2 = \theta_1 G_2 + \theta_2 G_1$$

Thus, in order to keep the closed-loop characteristic polynomial invariant with respect to changes in the parameter, we use gains G_1 and G_2 that vary with the estimated parameters:

$$G_2 = a_1 - \hat{\theta}_1, \quad G_1 = (a_2 - \hat{\theta}_1 g_2)/\hat{\theta}_2 \tag{10.39}$$

To achieve zero steady-state error in tracking a step, the gain G_0 is given by

$$G_0 = -G_2 \theta_1/\theta_2 \tag{10.40}$$

To evaluate the performance, the system is simulated using the following parameters:

$$a_1 = 80., \quad a_2 = 2500., \quad \theta_1 = 10.0, \quad \theta_2 = 1.0, \quad K_1 = 4.0, \quad K_2 = 0.01,$$

and a square-wave reference input of unit amplitude and a period of 2 sec. Initial estimates of the parameters were taken at

$$\hat{\theta}_1(0) = 20., \quad \hat{\theta}_1(0) = 2.$$

The results of this simulation are shown in Figure 10.9. It is seen that the parameters are estimated in about 10 sec at which time the square wave response reaches its steady state.

The results of this one simulation indicate excellent performance. But in view of the highly nonlinear nature of the closed-loop process, much more simulation would be necessary before you could conclude that the performance was acceptable. Note, in particular, that two of the gains are inversely proportional to $\hat{\theta}_2$. What would happen if $\hat{\theta}_2$ were to come close to zero or become negative? What could be done to prevent this from happening? (See Problem 10.5.)

10.2.3   Self-Tuning of Discrete-Time Systems

The techniques described above for continuous-time systems are generally applicable to discrete-time systems, with appropriate changes where necessary. Although the parameter estimation method based on nonlinear reduced-order observers, described in Section 10.2.2, does not appear to be applicable to discrete-time systems, other parameter estimation methods that are suitable only for discrete-time systems are available.

In current practice it is all but certain that any adaptive control strategy will be implemented digitally, so you might ask why consider continuous-time systems in the first place. One reason is that differential equations are a more natural setting for physical processes, and hence that an intuitive appreciation of what is happening is better acquired in this setting. Another reason is that in most processes the digital hardware is fast enough to implement a very close approximation to the continuous-time design.

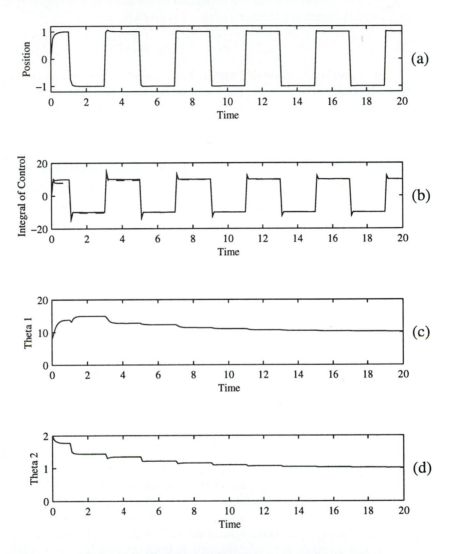

Figure 10.9 Performance of instrument servo using adaptive control based on parameter estimation. (a) Square wave response. (b) Actual and estimated values of x_2. (c) Estimate of θ_1. (d) Estimate of θ_2.

Extended Kalman Filter Architecture

A general representation for the evolution of the state in a discrete-time system with an uncertain constant parameter vector θ is given by

$$x_{n+1} = \phi(x_n, u_n, \theta) \tag{10.41}$$

The transition function $\phi(\ )$ may be known analytically or it can be defined implicitly as the result of integrating the continuous-time differential equations over the nth sampling interval with the control input assumed constant during the interval:

$$u(t) = u_n \quad \text{for} \quad nT < t < (n+1)T$$

The observation vector can be another nonlinear function of the state and the parameter vector

$$y_n = \gamma(x_n, \theta) \tag{10.42}$$

As we did with the development of the continuous-time case, attach the equation

$$\theta_{n+1} = \theta_n \tag{10.43}$$

which represents a constant parameter, to (10.41) to obtain the metastate

$$\begin{bmatrix} x_n \\ \cdots \\ \theta_n \end{bmatrix}$$

Then proceed with the resulting metasystem using the discrete-time extended Kalman filter equations. The result is

$$\hat{x}_{n+1} = \phi(\hat{x}_n, u_n, \hat{\theta}_n) + \tilde{K}_{xn}[y_n - \gamma(\hat{x}_n, \hat{\theta}_n)] \tag{10.44}$$

$$\hat{\theta}_{n+1} = \hat{\theta}_n + \tilde{K}_{pn}[y_n - \gamma(\hat{x}_n, \hat{\theta}_n)] \tag{10.45}$$

The gain matrices $\tilde{K}_{xn}, \tilde{K}_{pn}$ can, as in the continuous-time case, be obtained using the discrete-time variance equation (9.61) applied to the metastate $[x', \theta]'$. Alternatively, any convenient approximation that produces a stable closed-loop system can be used to obtain these matrices. The choice of these matrices is crucial, of course, to the satisfactory operation of the observer. Moreover, because the overall system is nonlinear even when the plant is linear in both the state and the parameter vectors, i.e., bilinear in x and θ, it is difficult to establish stability except by simulation.

Parameter Estimation in ARMA Models

Methods for estimating the parameters of a linear, discrete-time system are available that have no obvious counterparts in continuous-time systems. These methods are based on representing the system by its *autoregressive, moving average* (ARMA) model which comes from its discrete-time transfer function

$$G(z) = \frac{Y(z)}{U(z)} = \frac{b_1 + b_1 z^{-1} + \cdots + b_k z^{-k}}{1 + a_1 z^{-1} + \cdots + a_k z^{-k}}$$

Corresponding to this transfer function is the difference equation

$$y_n + a_1 y_{n-1} + \cdots + a_k y_{n-k} = b_1 u_n + b_1 u_{n-1} + \cdots + b_k u_{n-k} \qquad (10.46)$$

If the sequences of inputs $\{u_i\}$ and outputs $\{v_i\}$ are known, it is possible to form a set of linear equations from which the parameters a_i and b_i of the ARMA model can be determined. Specifically, for a kth order system it is possible in principle to determine the $2k$ parameters with $2k + 1$ samples of the input and output. From (10.46) we have

$$y_n = -a_1 y_{n-1} - a_2 y_{n-2} - \cdots - a_k y_{n-k} + b_1 u_{n-1} + b_2 u_{n-2} + \cdots + b_k u_{n-k}$$

$$y_{n+1} = -a_1 y_n - a_2 y_{n-1} - \cdots - a_k y_{n-k+1} + b_1 u_n + b_2 u_{n-1} + \cdots + b_k u_{n-k+1}$$

$$\vdots \quad \vdots$$

$$y_{n+2k} = -a_1 y_{2n} - a_2 y_{2n-1} - \cdots - a_k y_{2n-k+1} + b_1 u_{2n} + b_2 u_{2n-1} + \cdots + b_k u_{2n-k+1}$$

which can be arranged in matrix form

$$\begin{bmatrix} y_n \\ y_{n+1} \\ \vdots \\ y_{n+2k} \end{bmatrix} = M \begin{bmatrix} a_1 \\ \vdots \\ a_k \\ b_1 \\ b_2 \\ \vdots \\ b_k \end{bmatrix} \qquad (10.47)$$

where

$$M = \begin{bmatrix} -y_{n-1} & -y_{n-2} & \cdots & -y_{n-k} & u_{n-1} & u_{n-2} & \cdots & u_{n-k} \\ -y_n & -y_{n-1} & \cdots & -y_{n-k+1} & u_n & u_{n-1} & \cdots & u_{n-k+1} \\ vdots & \vdots & \vdots & \vdots & \vdots & \vdots & \vdots & \vdots \\ -y_{2n} & -y_{2n-1} & \cdots & -y_{2n-k+1} & u_{2n} & u_{2n-1} & \cdots & u2n-k+1 \end{bmatrix}$$

Thus, inverting the matrix M, the elements of which are the applied inputs and the resulting outputs, we can solve for the unknown coefficients a_i and b_i.

Instead of starting at step 0 and going to step $2k$ we can start at any step and use the data from a sequence of $2k + 1$ steps. This flexibility is one of the problems of the method. Which sequence of steps is best? If data from more than $2k + 1$ steps is available, would it not be more accurate to use all the available data? Suppose the matrix

M is ill-conditioned? Suppose the numerical data are noisy? All of these questions, and possibly a few others, suggest that this method may not be very practical.

A more practical method, which can be shown to be the recursive, least-squares version of (10.47) (see Problem 10.6), can be obtained by formulating the ARMA model estimation as a Kalman filtering problem. For this purpose suppose observation noise is present in the data and write (10.46) as

$$y_n = -a_1 y_{n-1} + \cdots - a_k y_{n-k} + b_1 u_n + b_1 u_{n-1} + \cdots + b_k u_{n-k} + w_n$$
$$= \phi_n' \theta + w_n \tag{10.48}$$

where w_n is the observation noise and

$$\phi_n' = \begin{bmatrix} -y_{n-1} & -y_{n-2} & \cdots & -y_{n-k} & u_{n-1} & u_{n-2} & \cdots & u_{n-k} \end{bmatrix} \tag{10.49}$$
$$\theta' = \begin{bmatrix} a_1 & \cdots & a_k & b_1 & \cdots & b_k \end{bmatrix} \tag{10.50}$$

We can regard (10.48) as an observation equation for the discrete- time dynamic process

$$\theta_{n+1} = \theta_n + v_n \tag{10.51}$$

where v_n is another random noise process chosen to represent the possibility that the parameters may not be true constants. (The real reason for using v_n is to have its covariance matrix V_n appear in the variance equation and thereby prevent the filter gain matrix from becoming zero. This will be discussed again below.)

Equation (10.18) and the observation equation (10.14) define a standard Kalman filtering problem with the state being the vector of $2k$ parameters to be estimated. The Kalman filter corresponding to (10.51) is given by

$$\hat{\theta}_n = \tilde{K}_n [y_n - \phi_n' \tilde{\theta}_n] \tag{10.52}$$
$$\tilde{\theta}_n = \hat{\theta}_{n-1} \tag{10.53}$$

with the gain matrix K_n given by

$$K_n = \tilde{P}_n \phi_n [\phi_n' \tilde{P}_n \phi_n + W_n]^{-1} = \frac{\tilde{P}_n \phi_n}{\phi_n' \tilde{P}_n \phi_n + W_n} \tag{10.54}$$

where W_n is the variance of the observation noise w_n. Note that W_n and $\phi_n' \tilde{P}_n \phi_n$ are scalars, which is why the inverse matrix in (10.54) can be expressed as the division operation.

The covariance matrix of the state estimation is propagated by the equations

$$\hat{\theta}_n = (I - \tilde{K}_n \phi_n') \tilde{P}_n \tag{10.55}$$
$$\tilde{P}_{n+1} = \hat{P}_n + V_n \tag{10.56}$$

If the matrix V_n in (10.56) is zero, the two equations (10.55) and (10.56) for propagation of the covariance matrix can be combined into a single equation

$$\tilde{P}_{n+1} = (I - \tilde{K}_n \phi_n') \tilde{P}_n = \left(I - \frac{\tilde{P}_n \phi_n \phi_n'}{\phi_n' \tilde{P}_n \phi_n + W_n} \right) \tilde{P}_n \tag{10.57}$$

which is the form in which the recursive parameter estimation algorithm is often presented [5]

If the noise covariance matrix V_n and the observation variance W_n are both zero, it can be shown that the covariance matrix $\hat{P}_n$ can go to zero in $2k$ steps, indicating that the state, i.e., the parameter vector has been estimated perfectly. (See Problem 10.7.) This will make the gain matrix $\bar{K}_n$ zero and prevent any further updating of the estimates. In practice we would want to continue updating the state estimates and hence would not operate the estimator with W_n set to zero.

If the covariance matrix V_n of the process noise alone is zero, the covariance matrix $\bar{P}_n$ of the parameter estimate will converge to zero as $n \to \infty$ and with it the gain matrix. This situation has undesirable consequences:

1. More recent data will be accorded less weight than older data.

2. The propagation of the covariance may tend to become ill-conditioned.

The first consequence is relatively benign; it means only that better estimates of the parameters might be obtained using more recent data. The second consequence is more serious: If the ill-conditioning is bad enough, the covariance matrix $\bar{P}_n$ may become indefinite and cause the resulting parameter estimates to diverge from their true values. Numerical problems can be circumvented to an extent by use of sophisticated covariance propagation methods such as Bierman's UDU' method [6], but use of a nonzero value of V_n is simpler and has a similar effect.

Once the plant parameter vector is estimated, it can be used to determine the control system gains. In a design that depends on only a small number of plant parameters, it is feasible to pre-calculate or tabulate the parameters of the control system as functions of the plant parameters. In this case, however, all the parameters of the plant transfer function are estimated. Even in a second-order system, the number of plant parameters is almost too large for this approach to be feasible. Instead, it is necessary for the controller design to be accomplished "on-line." This means that the compensator must incorporate the control design algorithm, using the estimated plant parameters in the model of the plant used in the design. Fortunately, with state-of-the-art microprocessors, the required computations can usually be accomplished in real time.

State-space methods in general, and the separation principle in particular, can be used for these design calculations by using the estimated parameters to formulate a state-space model in terms of the estimated parameters:

$$x_{n+1} = \Phi(\hat{\theta})x_n + \Gamma(\hat{\theta})u_n \tag{10.58}$$
$$y_n = C(\hat{\theta})x_n \tag{10.59}$$

The components of x_n appearing in this model do not necessarily have any physical significance. Only the matrices appearing in the model are used in the design; the control law is implemented using only the measured outputs and the parameter estimates.

Convenient representations are the controllable or observable canonical forms. Consider, for example, the controllable canonical form, the matrices of which are given by:

$$\Phi = \begin{bmatrix} -\hat{a}_1 & -\hat{a}_2 & \cdots & -\hat{a}_{k-1} & -\hat{a}_k \\ 1 & 0 & \cdots & 0 & 0 \\ 0 & 1 & \cdots & 0 & 0 \\ \vdots & \vdots & & \vdots & \vdots \\ 0 & 0 & \cdots & 1 & 0 \end{bmatrix} \qquad \Gamma = \begin{bmatrix} 1 \\ 0 \\ 0 \\ \vdots \\ 0 \end{bmatrix} \qquad (10.60)$$

$$\qquad \qquad (10.61)$$

$$C = \begin{bmatrix} b_1 & b_2 & \cdots & b_{k-1} & b_k \end{bmatrix} \qquad (10.62)$$

The design of the gain matrix G of the full-state feedback control law

$$u_n = -G\hat{x}_n$$

by pole placement is achieved directly in terms of the estimated parameters a_i. Suppose the desired closed-loop characteristic polynomial for full-state feedback is

$$\Delta_c(z) = z^k + a_{c1}z^{k-1} + a_{c,k-1}z + a_{c,k} \qquad (10.63)$$

Then, because of the particular form of Γ, the gain matrix is

$$G(\hat{\theta}) = \begin{bmatrix} a_{c1} - \hat{a}_1 & a_{c2} - \hat{a}_2 & \cdots & ack - \hat{a}_k \end{bmatrix} \qquad (10.64)$$

Calculation of the gain matrix K of the observer,

$$\hat{x}_{n+1} = \Phi(\hat{\theta})\hat{x}_n + \Gamma(\hat{\theta})u_n + K[y_n - C(\hat{\theta})\hat{x}_n]$$

however, is not as simple. One approach would be to implement the Bass-Gura formula on-line:

$$K(\hat{\theta}) = [(OW)']^{-1} \begin{bmatrix} a_{o1} - \hat{a}_1 \\ a_{o2} - \hat{a}_2 \\ \cdots \\ a_{ok} - \hat{a}_k \end{bmatrix} \qquad (10.65)$$

where a_{oi} are the coefficients of the desired characteristic polynomial of the observer, and O and W are the matrices defined in Chapters 8 and 9:

$$O = \begin{bmatrix} C' & \Phi'C' & \cdots & \Phi'^{k-1}C' \end{bmatrix} \qquad (10.66)$$

$$W = \begin{bmatrix} 1 & \hat{a}_1 & \hat{a}_2 & \cdots & \hat{a}_{k-1} \\ 0 & 1 & \hat{a}_1 & \cdots & \hat{a}_{k-2} \\ \vdots & \vdots & \vdots & & \vdots \\ 0 & 0 & 0 & \cdots & 1 \end{bmatrix}$$

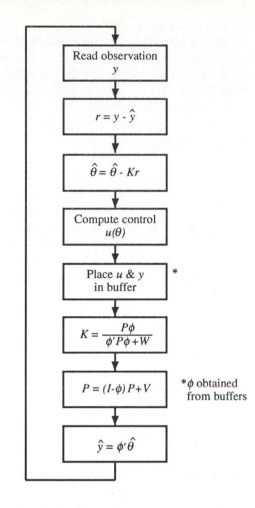

Figure 10.10 Flowchart of adaptive control algorithm based on recursive parameter estimation.

A flow chart illustrating the sequence of calculations of a control algorithm is shown in Figure 10.10. Each cycle of the algorithm begins with the reading of the observation y. (The subscript n is implicit.) Next, the residual $r = y - \hat{y}$ is computed using the predicted observation $\hat{y}$, which could have been determined at the previous step. The residual is then used to update the parameter estimate $\hat{\theta}$, which is used to calculate the present control u. As discussed above, any appropriate algorithm for calculating the control may be used, and the flow chart only indicates the need for performing the calculation. Since the update of the parameters requires k past values of the observation and the control, these quantities are placed in "ring buffers" each of length k. The vector ϕ is obtained by concatenating the contents of these buffers, starting at proper pointer locations, which are incremented at each step. Using the new vector ϕ, the new gain matrix $\tilde{K}$, the new covariance matrix $\tilde{P}$, and the predicted observation $\hat{y}$ are computed in sequence to conclude the cycle.

The algorithm requires vector and matrix operations: addition, subtraction, and multiplication. (Matrix inversion is not required for parameter estimation but the operation may be needed in the control algorithm.) Hence, for the algorithm to be generic, i.e., applicable to systems of arbitrary order k, "canned" packages to implement these operations would be beneficial. (See Problem 10.8.)

10.3   MODEL REFERENCE ADAPTIVE CONTROL

Suppose you have a process that you believe (from experience with similar processes or by intuition) can be controlled adequately by proportional plus integral (PI) control. You also have an idea of how the process ought to respond to a reference input y_r, with the PI control in use. But you don't know enough about the process to determine the proportional and integral gains. Simulation is not a viable tool, since you don't know enough about the process to formulate a design model. So you decide to adjust the gains adaptively. You might use the architecture shown in Figure 10.11, which is a typical configuration of a model-reference adaptive control system. The top part of the diagram has the form of a PI controller, except that multipliers are shown instead of fixed gains k_p and k_I. These quantities in the adaptive system are generated by the adaptive loop, which operates on the error

$$\epsilon = y_d - y \tag{10.67}$$

between the actual plant output y and the output y_d of a model. If the dynamics of the model are chosen correctly, it should be possible to drive this error toward zero. As the error approaches zero and the closed-loop plant behavior approaches that of the model, the signal that drives the output of the adaptive loop vanishes. Since the adaptively-generated gains should remain at their nonzero values, the integrators shown in the adaptive loop are required for proper operation. If the adaptive system is stable it should work as anticipated.

The major problem with this approach, as you might suspect, is determining whether the system is in fact stable. Because of the presence of the multipliers, the system is nonlinear. Other nonlinearities are usually included in processing the error ϵ. Moreover, the order of the complete system is the sum of the orders of the plant, the model, and any filter in the adaptive loop. This order can be quite large. So the stability analysis that you would have to perform in order to verify operation could be rather overwhelming. The worst of the problem, however, is that the plant dynamics are presumably unknown.

For a number of years these difficulties stood in the way of serious consideration of model-reference adaptive control. Theoretical developments during the 1970s improved the situation somewhat, but many issues still remain.

One of the issues is the choice of the model. If no restrictions are imposed on the model, why would you not make the model be a linear system with a transfer function of unity? The adaptive loop would generate large values for k_p and k_I and

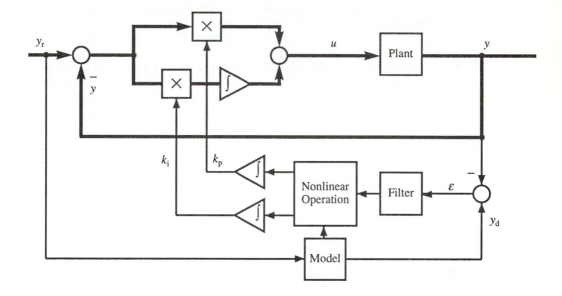

Figure 10.11 Model-reference adaptive control for adjusting proportional and integral gains.

thereby drive the transfer function of the upper closed-loop system to unity. Suppose you could come close to achieving this condition. What is the physical control u that would be needed? The control signal that achieves $y = y_r$ in the frequency domain is of course given by

$$U(s) = G^{-1}(s)Y_r(s)$$

where $G(s)$ is the transfer function of the plant. In words, the control is the inverse of the plant transfer function multiplied by the Laplace transform of the desired output. This relationship reveals two of the problems that might occur with this approach:

1. If the transfer function of the plant is non-minimum phase, i.e., it has zeros in the right half plane, $G^{-1}(s)$ has poles in the right half plane. This implies that, in order to achieve the desired output, $u(t)$ will have to contain increasing exponentials. These will ultimately saturate any real actuator and hence prevent the adaptive control from performing as intended. Moreover, unless the saturation is somehow accounted for, the system is vulnerable to instability.

2. If the degree of the denominator of $G(s)$ is greater than the degree of the numerator, as it almost always is, and the input contains discontinuities, the control signal will contain impulses, doublets, etc., reflecting that $G^{-1}(s)$ has a numerator of higher degree than the denominator, which implies differentiation of the

input. Again, since no physical actuator can produce this kind of input, the aim of the adaptive design will be frustrated.

The ramification of this analysis is that the transfer function $G_m(s)$ of the model must be compatible with the physical capabilities of the plant. To avoid increasing exponentials in the plant input, the transfer function of the model must have zeros in the right half plane at the same locations as they are in the actual plant. (As we will soon see, this is not difficult to achieve even if you don't know where the plant zeros are located.) To avoid differentiation of discontinuous function, the model must have a relative degree no smaller than that of the plant. You may want to impose other practical restrictions in the model, such as limiting the high-frequency response to avoid susceptibility to noise or unmodeled high-frequency resonances.

One way of obtaining an appropriate model for use in implementing the adaptive control is to assume that it is the transfer function of the closed-loop system obtained from a "nominal plant" model and a corresponding well-designed compensator. Specifically, if the plant has a nominal transfer function, say $G_0(s)$, and the compensator for this plant model has the transfer function $D_0(s)$, the reference model for purposes of design of the adaptive control would be

$$G_r(s) = H_0(s) = \frac{D_0(s)G_0(s)}{1 + D_0(s)G_0(s)} \tag{10.68}$$

Note that $G_r(s)$ has the zeros of $G_0(s)$ unless the compensator is designed to cancel them. Thus, barring a perverse compensator design that places poles at or near the right half plane zeros of $G_0(s)$, the requirement that the reference model contain the right-half plane zeros of the plant is automatically achieved, even if the location of the plant zeros is not known. In addition, if the denominator of the transfer function of the compensator has at least the same degree as its numerator, the reference model $G_r(s)$ will have a relative degree equal to the relative degree of the nominal plant. Hence use of a reference model of the form of (10.68) satisfies both of the practical requirements listed above.

Determination of the compensator and the calculation of the corresponding closed-loop transfer function $G_r(s)$ requires a nominal plant model, of course. Why, you might ask, should you use an adaptive control design if you know the transfer function of the plant? The question is reasonable and is one of the paradoxes of adaptive control: If you know enough about the plant to design the adaptive control algorithm, you don't need it. Notwithstanding the paradox, you might elect to use model- reference adaptive control when you know enough about the plant to establish its nominal model, but not enough to have confidence that the compensator designed using that model will perform adequately without adaptive features.

10.3.1   Classical Model-Reference Adaptive Control

A simple application of model reference adaptive control is to adjust the loop gain of a feedback control system. The configuration of such a system is shown in Figure

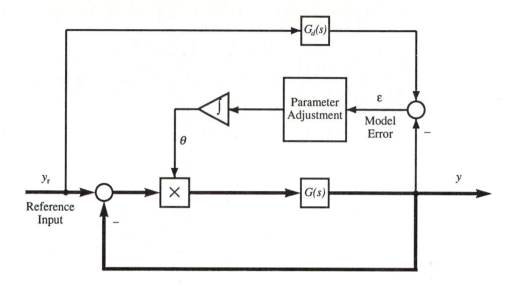

Figure 10.12   Model-reference control of loop gain

10.12. The loop gain K of a non-adaptive system is replaced by a multiplier, one input to which is the adaptation parameter θ, which the adaptive loop is required to determine.

The box in Figure 10.12 labled "Parameter Adjustment" denotes an appropriate algorithm. The output of this algorithm is input to an integrator the output of which is the desired parameter θ. Owing to the presence of the integrator, the output of the parameter adjustment algorithm should go to zero as the estimate of the parameter approaches the correct value.

Many parameter adjustment algorithms have been proposed over the years. One of the earliest of these is the "gradient" algorithm in which a performance measure $J(\epsilon)$ is defined, and the parameter is adjusted to reduce this performance measure. In particular, the input to the integrator can be defined by

$$\dot{\theta} = -\gamma \frac{\partial J}{\partial \theta}$$

where γ is an appropriate scale factor. If the performance criterion is quadratic:

$$J(\epsilon) = \epsilon^2(\theta)/2$$

then

$$\frac{\partial J}{\partial \theta} = \epsilon(\theta)\frac{\partial \epsilon}{\partial \theta} \tag{10.69}$$

The quantity ϵ needed to implement (10.69) is directly measured, but how can we get $\partial\epsilon/\partial\theta$ in terms of quantities that can be measured? Consider the relation for the error in the frequency domain:

$$J(\theta) = \left(H_m - \frac{\theta G}{1 + \theta G} \right) Y_m$$

where $G = G(s)$, the plant transfer function. (The complex frequency s is understood). Then

$$\frac{\partial J(\theta)}{\partial \theta} = -\frac{G}{(1 + \theta G)^2} Y_r \qquad (10.70)$$

But

$$\frac{G}{1 + \theta G} = \frac{Y}{\theta Y_r}$$

and

$$\frac{1}{1 + \theta G} = \frac{U}{\theta Y_r}$$

Hence (10.69) becomes

$$\frac{\partial J}{\partial \theta} = -\frac{YU}{\theta^2 Y_r} \qquad (10.71)$$

Absorbing the dependence of $\partial J/\partial \theta$ upon θ^2 into the scale factor γ, and reverting back to the time domain gives the adjustment algorithm:

$$\dot{\theta} = -\gamma \frac{yu}{y_r} \qquad (10.72)$$

This algorithm is based entirely upon measurable quantities; the only problem in implementation is the requirement for division by y_r. If y_r goes through zero only at discrete instants, the division by y_r is not a practical problem (although it creates a significant theoretical problem if you want to analyze the stability of the closed-loop system). Another practical problem is that the algorithm could generate a value of θ for which the main loop will be unstable. If there is enough *a priori* knowledge about the plant to specify a range of θ for which the main loop is stable, the algorithm can be altered to keep θ within this range. With a digital implementation, for example, the following algorithm will accomplish the task:

$$\theta_{n+1} = \begin{cases} \theta_{\max} & \text{for } z > \theta_{\max} \\ \theta_{\min} & \text{for } z < \theta_{\min} \\ z & \text{otherwise} \end{cases} \qquad (10.73)$$

where

$$z = \theta_n - \gamma \frac{y_n u_n}{(y_r)_n}$$

In view of the assumptions that were used in deriving (10.73), the most practical way of evaluating its performance is by simulation. For this purpose, of course, a hypothetical model (or a set of hypothetical models) of the plant is required, so you cannot really be sure how it will work until you try it on a real plant. The following example might reinforce your confidence that it can work.

Example 10.4   Model-reference adaptive control of instrument servo

In Example 10.3 we studied the design of an adaptive control law for an instrument servo based on parameter estimation. In this example we consider the same plant, but now use model-reference adaptive control. Here we assume that the motor time-constant is known and only the gain is uncertain. (This situation could occur in practice, for example, if the amplifier driving the motor is subject to drift.)

The reference model is taken as the closed-loop transfer function of the servo with the gain set at some nominal value, i.e.,

$$H_r(s) = \frac{K/s(s + \alpha)}{1 + K/s(s + \alpha)}$$

where K is the nominal forward loop gain and α is the known corner frequency of the servomotor.

Knowing that the second-order system will be stable for all positive gains, we can implement (10.73) with $\theta_{min} = 0$. But it might be desirable to avoid a loop gain of zero by setting θ_{min} to a small positive number. In this example the minimum value of θ was set at 1.0 and a sampling interval of 0.01 sec was used. In addition, the following parameters of the reference model were selected:

$$\alpha = 18.5., \quad K = 100.$$

These parameters give a step response with no overshoot. To evaluate the performance of the adaptive control algorithm, a square wave with a period of 4 sec was used as the reference input. A number of simulations were performed, the results of several of which are shown here.

In figure 10.13 the square wave response with and without adaptation are compared. Figure 10.13(a) shows the performance with the gain K kept constant at a value of 20. The step response is substantially slower than desired. In contrast, when the adaptive control operates with the adjustment parameter γ set at 0.5, Figure 10.13(b) shows the loop gain θ starting at the initial value of 20 and then, after dropping several times, recovering and converging to the correct value of 100 in about 6 sec. The difference between the reference output and the actual output is negligible thereafter.

What happens when the time-constant of the plant is different from that of the reference model? In this case, no adjustment of the parameter θ can make the closed-loop system respond exactly as the reference system does. The best that can be expected is that behavior of the closed-loop system comes reasonably close to that of the reference system. The simulation results appear to be favorable. When the adaptation is disabled, and the time constant of the plant reduced to 12 from 18.5 and θ intialized at 20, the closed loop system is still sluggish. as shown in Figure10.14(a). When the adaptation is enabled, however, using $\gamma = 0.5$, the transient response is significantly improved as the gain adapts to a more appropriate value (around 75). The peak error is substantially reduced as shown in Figure10.14(b).

10.4   STABILITY OF ADAPTIVE CONTROL ALGORITHMS

The foregoing example, Example 10.3 , shows that adaptive control can improve the performance of a control system. But we have not proved that the algorithms we have

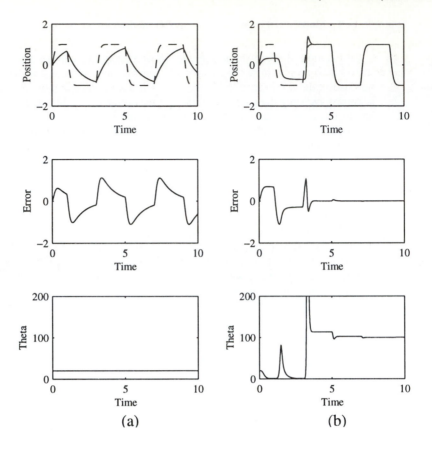

Figure 10.13 Model reference adaptive control of a servo. in observations. (a) Adaptation absent. (b) Adaptation present.

illustrated are stable. Since adaptive control systems entail complicated nonlinear dynamics, they are difficult to analyze. One of the properties that is difficult to establish analytically is stability. Reliance on simulations, moreover, is fraught with danger, because an adaptive control algorithm that appears to be stable may in fact be unstable, as the following celebrated example shows.

Example 10.5 Instability of an Adaptive Control System

The perverse behavior that can be exhibited by an adaptive control system was demonstrated by Rohrs, *et al.* [7]

The plant has the transfer function

$$G(s) = G_1(s)G_2(s) \tag{10.74}$$

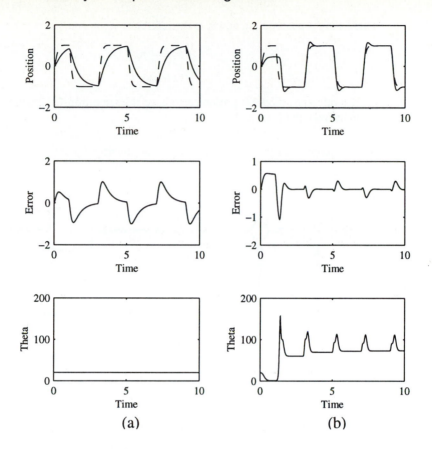

Figure 10.14 Model reference adaptive control of a servo with reference model not matched to plant. (a) Adaptation absent. (b) Adaptation present.

where

$$G_1(s) = \frac{2}{s+1}, \quad G_2(s) = \frac{229}{s^2 + 30s + 229} \qquad (10.75)$$

The poles of $G_2(s)$ are at $s = -15 \pm j2$; they are well damped and fairly remote from the dominant pole of the system at $s = -1$. Hence, if we didn't know of the existence of the poles of G_2 we might not be able to locate them. (And, if we did know of the location of these poles, why would we use adaptive control in the first place?)

Hence, it would seem reasonable to base the design of an adaptive control on the assumption that $G(s) = G_1(s)$. We adopt a model-reference adaptive control strategy

$$u = -\theta_1 y + \theta_2 y_r \qquad (10.76)$$

where y is the plant output and θ_r is the reference input, and

$$\dot{\theta}_1 = -\gamma y_c e \tag{10.77}$$

$$\dot{\theta}_2 = \gamma y e \tag{10.78}$$

The reference model is presumed to have the same transfer function as $G_1(s)$; the commanded input is y_c; and

$$e = y - y_m$$

is the difference between the model output and the actual plant output.

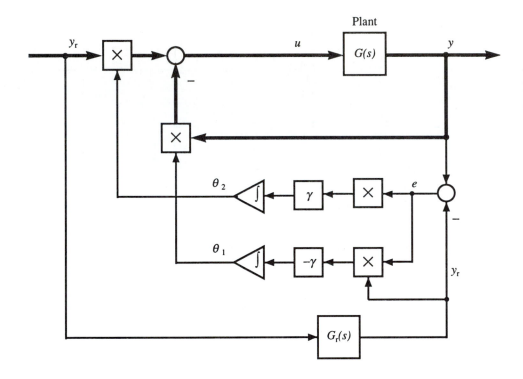

Figure 10.15   Block diagram of adaptive control system that can become unstable.

A block diagram of the system, given in Figure 10.15, shows that θ is a feedback gain and θ_2 is a feedforward gain. Considering θ_1 and θ_2 fixed, and omitting the contribution of $G_2(s)$, we see that the transfer function of the closed-loop system is

$$H(s) = \frac{2\theta_2}{s + 1 + 2\theta_1} \tag{10.79}$$

Since the closed-loop plant is first order, it is stable for any fixed θ_1. Moreover, with θ_1 and θ_2 properly adjusted, it is possible to achieve any first-order transfer function. With Rohrs, *et al.* we take

$$H_r(s) = \frac{3}{s+2} \tag{10.80}$$

which is achieved when

$$\theta_1 = 1, \quad , \theta_2 = 1.5 \tag{10.81}$$

For future reference, note the following:

1. The d-c gain of $H(s)$ is $2\theta_2/(1 + 2\theta_1)$, and the d-c gain of $H_r(s)$ is 3/2. Thus it is possible to match the d-c gains of H and H_r by making

$$\theta_2 = \theta_1 + .5 \tag{10.82}$$

Any pair of θs on this line will match d-c gains.

2. When the complete transfer function of the plant $G(s)$ is used instead of $G_1(s)$, the closed-loop transfer function of the system (still with θ_i = const.) is

$$H(s) = \frac{458\theta_2}{s^3 + 31s^2 + 259s + 229 + 458\theta_1} \tag{10.83}$$

This transfer function, being third-order, is stable for only a finite range of θ_1, namely

$$-0.5 < \theta_1 < 17.03 \tag{10.84}$$

If the parameter θ_1 should ever leave this range, the closed-loop system would become unstable.

If we knew $G_2(s)$, we could limit θ_1 to the range (10.84). But of course, we are assuming that we don't know it. So we have no way of knowing how to bound θ_1. Thus, if the adaptation strategy (10.77), (10.78) were to lead to a value of θ_1 outside the permissible range, we could expect the system to go unstable. (Because θ_1 is not constant, even if it were inside this range the system might be unstable.)

This is precisely the behavior that was discovered by Rohrs, *et al.* To evoke this unstable behavior, they used a reference input

$$y_r(t) = C_1 + C_2 \sin \omega t$$

with the frequency ω chosen to be the frequency at which (10.83) becomes zero. This frequency is given by

$$\omega = \sqrt{259} = 16.09$$

The results of two simulations of the overall system, with $\gamma = 1$ and with the initial conditions:

$$\theta_1(0) = 0.5, \quad \theta_2(0) = 2.5$$

and with

$$C_1 = 0.3, \quad C_2 = 1.85$$

are shown in Figure 10.16 for a step input, and in Figure 10.17 for an input of a step plus a sinusoid at the frequency ω given above.

The behavior for the step input is acceptable, although the response is oscillatory owing to the presence of the unmodeled dynamics. The parameters θ_1 and θ_2 converge to the values given in (10.81).

For the step-plus-sinusoid reference input, however, the system is unstable. The output and the parameter estimates diverge rapidly after about 15 sec. (The actual time histories shown in these figures is somewhat different from those presented in [7] and from those given in [5]; the results depend on the values of γ and the initial conditions.)

Similar unsatisfactory behavior is observed when there is sinusoidal noise in the measurement of the plant output.

A detailed analysis of this example occupies about a dozen pages of [5] where it is argued that the culprit is the absence of persistent excitation. This lack renders it impossible to determine both θ_1 and θ_2. (Note that the computed parameters tend to track in accordance with (10.82).) But the unmodeled dynamics is surely an important factor, since the instability occurs only in the presence of the unmodeled dynamics.

One lesson of this example is that apparently simple applications can have unanticipated complications.

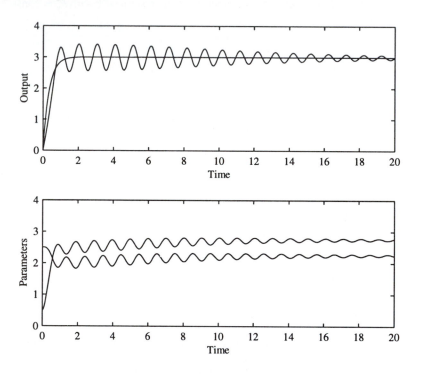

Figure 10.16   Behavior of model reference adaptive control for plant with unmodeled dynamics and step reference input.

The foregoing example illustrates some issues of concern in the stability of adaptive control systems. They include:

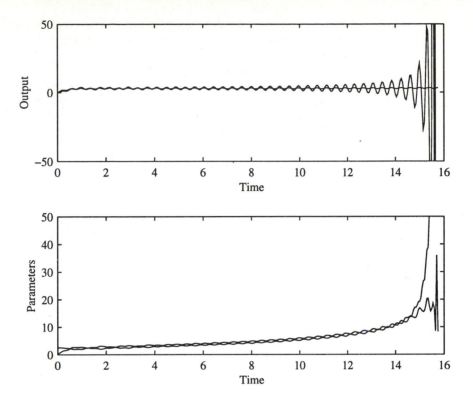

Figure 10.17 Behavior of model reference adaptive control for plant with unmodeled dynamics and sinusoidal reference input.

1. Stability of the main control loop

2. Stability of the adaptive loop

3. Convergence of parameter estimates to true values

One can readily appreciate that 1 and 2 are not equivalent issues. Suppose you know that the main control loop for your plant is stable for all positive values of loop gain, including time-varying gain. (This is a great deal of knowledge about the plant. How you would come to have so much knowledge is another matter we we won't go into now.) If so, any method of gain adjustment that assures that the gain always remains positive will also assure the stability of the main control loop. But it tells you nothing about the stability of the adaptive loop. The latter may produce a gain that oscillates between known (positive) limits, that wanders apparently aimlessly (chaotically?) through positive values, or that diverges to $+\infty$. These types of behavior would classify the adaptive loop as unstable, but not the main loop.

In addressing these issues, the concepts of *positive-real functions* and *persistent excitation* have been found useful.

10.4.1 Strictly Positive-Real (SPR) Functions

Analyzing the stability of adaptive control systems has been greatly facilitated by techniques based on the concept of *(strictly) positive-real functions*. Recall from Chapter 3 (Section 3.2.1) that a positive-real function is one which has no poles and zeros in the right half of the s plane. A *strictly* positive-real function is one that has no poles or zeros on the imaginary axis or in the right half plane. Many functions that would qualify as positive real would not qualify as strictly positive real: the transfer functions of common *ideal* electrical network elements (capacitors, inductors) are positive-real but not strictly positive-real.

The key idea in using SPR functions for analyzing stability of adaptive control systems is to manipulate the block diagram of the system so that it has the configuration of a feedback system comprising a linear SPR system in a closed-loop with a zero-memory passive system (i.e., a system in which energy is always dissipated) as shown in Figure 10.18. Recall from Chapter 3 that a passive, zero-memory nonlinearity has the property that

$$y\phi(y) \geq 0$$

for all inputs y to the nonlinearity.

If the adaptive control algorithm with a given plant can be shown to be equivalent to the feedback system of Figure 10.18, the stability of the system is assured. The argument for stability follows along the lines of the argument leading up to the circle criterion, as given in Chapter 3: Using the Kalman-Yacubovich lemma, a state-space representation of the strictly positive-real plant is developed and using this representation a quadratic candidate Liapunov function is defined such that its derivative is guaranteed to be negative definite whenever $y\phi(y)$ is positive. (One reason why stability of the main loop is never at issue when unmodeled dynamics are absent in Example 10.5 is that the plant is then strictly positive-real.)

The restriction that adaptive algorithm and corresponding plant can be represented in the form of the feedback system of Figure 10.18 is formidable. The contortions used to effect this representation are exemplified by Chapter 5 of Åström and Wittenmark [5].

"Simple" Adaptive Control   A technique for getting around the SPR requirement on the plant is to deal with a modified plant, consisting of the original plant and a feedforward compensator, as shown in Figure 10.19(a). The transfer function C(s) of the compensator is chosen such that the modified plant, with transfer function

$$\tilde{G}(s) = G(s) + C(s) \tag{10.85}$$

is SPR. Since feedforward compensation is rarely feasible in practice, the implementation is accomplished using the configuration illustrated in Figure 10.19(b). It is easy

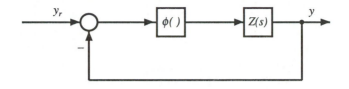

Figure 10.18 Feedback configuration for proving stability of adaptive control algorithms.

to see that Figure 10.19(a) and Figure 10.19(b) are equivalent with regard to stability (with a reference input of zero.) This design technique has been studied extensively by Kaufman, Bar-Kana, and Sobel who report [8] excellent results in a variety of practical applications that have been investigated by simulation. These authors call plants that can be compensated in this manner "almost strictly positive real" (ASPR). They have given their design technique the appellation "simple adaptive control," although in practice the control law may turn out to be anything but simple.

One of the key issues is knowing enough about the plant to be able to define an appropriate compensator transfer function $C(s)$. One way of accomplishing this, of course, is to make $C(s)$ a strictly positive-real function with a very high gain so that the parallel combination is dominated by $C(s)$. This will work in principle but may not be satisfactory in practice, because the adaptive feature will be all but washed out by the feedback path around it as the implementation of Figure 10.19(b) clearly shows. To successfully apply the method you have to make $C(s)$ only large enough to ensure that $\bar{G}(s)$ is SPR. To do this, you must know a good deal about the plant transfer function $G(s)$.

10.4.2 Persistent Excitation

An issue of adaptive control is the convergence of the parameter estimate $\hat{\theta}$ to its "correct" value θ. (In adaptive control techniques based on estimation of parameters of the plant, the correct value is the true value of the parameter; in model-reference adaptive control, the correct value is that which minimizes the error.)

One might argue that the failure of a parameter estimate to converge to its correct value is a subsidiary issue as long as the adaptive control law works (even with the incorrect parameter estimate). This reasoning might be valid if you can ascertain that the algorithm works for every input to which the system might be subjected. But in most applications the performance of the algorithm is tested only with a limited set of inputs. Satisfactory performance for this set of inputs is not a reliable indicator that its performance will be satisfactory for inputs outside the test set. Hence, you might

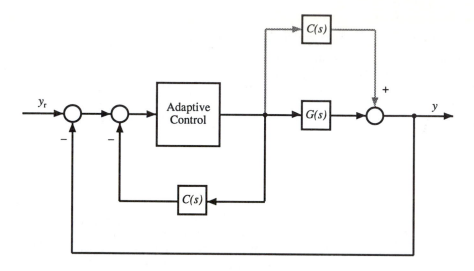

Figure 10.19   Compensation to make plant strictly positive real.

verify that an adaptive control algorithm appears to be satisfactory for some inputs although it fails to estimate a parameter, and find its performance unacceptable for other inputs. It may perform acceptably for some inputs without accurately estimating the parameter in question because its performance in *for this set of inputs* could well be insensitive to the parameter in question. It can be justifiably argued that if the performance is acceptable for *every* input the parameter in question is truly irrelevant to the operation of the system and one need not even bother trying to estimate the parameter. Except for this extreme case, it would be hazardous to use an adaptive control algorithm that fails to estimate a parameter that it is designed to estimate.

Åström [5] calls the input to a parameter estimation algorithm *persistently-exciting of order n* if the algorithm assures the estimation of n parameters. In general, the nature of both the algorithm and the input would determine the order of the persistent excitation. (In Example 10.5 , a sinusoidal reference signal is persistently exciting only of order 1 and hence does not allow both parameters to be identified.)

In the parameter estimation technique discussed in Section 10.2.2 the order of persistence of excitation is the rank of a matrix M.

10.5 NOTES

Note 10.1 History

In contrast to other areas of control theory that have waxed and waned over the years, interest in adaptive control has been strong for well over a quarter century. In a recent survey, Jacobs [9] identified an average of over 100 papers dealing with the subject from the mid 1960s to 1980. If anything, the number of papers on adaptive control has been on the increase in recent years. An tutorial article by Åström [10] includes 369 references, with probably an equal number having been omitted.

The field of adaptive control has been fraught with controversy. One of the fundamental issues has been the very definition of the term *adaptive control*. It is generally agreed that if you look at a control algorithm that no one told you is supposed to be adaptive, you would find it very difficult, if not impossible, to determine that it was adaptive. It would just look like another nonlinear control algorithm. This difficulty was recognized early on by J.B. Truxal, who proposed the following definition: "An adaptive system is any physical system which has been designed with an adaptive viewpoint."[11]. By this definition, any control algorithm is adaptive that its author so declares. In conformity with this permissive definition, most of the contributions to the early compendium in which Truxal's definition appears deal with almost anything but adaptive control.

Truxal's definition proved convenient over the years, permitting investigators who were not working on adaptive control to say they were when it was fashionable to do so, and conversely, permitting those who were really working on adaptive control to call it something else when adaptive control was out of fashion.

The fashion was often dictated by the agencies that provided funding for the research. There was a period of time during which research planners in the U.S. Air Force pinned great hopes on adaptive control. Abetted by Truxal's elastic definition, it was prudent to pursue research in adaptive control to improve one's opportunities to gain funding from the Air Force. After a fatal accident, when adaptive control became an Air Force *bete noir*, it was more prudent to work on nonlinear control, or parameter estimation, or some other topic.

Adaptive control did not share in the reputation for mathematical rigor that other areas of control theory acquired during the 1960s and 70s. In contradistinction to the latter, it was widely held that adaptive control techniques were too ad hoc and that their stability and performance were suspect. The status of adaptive control theory improved substantially with the discovery of rigorous proofs of the stability of certain adaptive control algorithms during the mid 1970s. Adaptive control theory now has a rigorous foundation. [12], [13].

Whether the complexity of adaptive control is justified by the payoff in terms of system performance remains an issue. There have been few practical instances in which adaptive control has convincingly outperformed systems that could have been designed by other methods. Nevertheless, the advantages of adaptive control—not having to know much about the process you want to control, and tolerance to large parameter changes—have made it attractive for many industrial applications. Commercial adaptive controllers have been on the market for a number of years [5] and have been successfully used in many applications.

10.6   PROBLEMS

Problem 10.1   Gain-scheduled control of cargo-moving gantry

Compare the performance of the gain-scheduled control of Example 10.1 with a fixed gain control law, the gains of which are calculated for the average cable length.

Problem 10.2   Robot arm with integral control

Consider the robot arm that was the subject of an adaptive control design in Example 10.2 . Investigate the use of a non-adaptive control law that provides integral action and compare performance with that shown in Example 10.2

Problem 10.3   Adaptive control of belt-driven robot arm

A belt-driven robot arm is governed by the following equations:

$$\dot{x}_1 = x_3 \tag{10.86}$$

$$\dot{x}_2 = x_4 \tag{10.87}$$

$$\dot{x}_3 = cx_1 - \theta(x_1 - x_2)/J_1 \tag{10.88}$$

$$\dot{x}_4 = \theta(x_1 - x_2)/J_2 - bx_4 + u \tag{10.89}$$

where x_1 and x_2 denote the angles of the drive shaft and of the arm, respectively, b is the damping, $c = g/l$ (where l is the effective length of the arm), and θ is the effective spring rate of the belt, the parameter to be estimated. Investigate the use of the parameter estimation technique of Section 10.2.2 to estimate the spring coefficient θ, and use the estimated value of θ for design of an adaptive feedback control law.

Simulate the performance with for several appropriate values of inertia, damping and spring rate.

Problem 10.4   Adaptive control of belt-driven robot arm (Continued)

Consider the belt-driven robot arm of Problem 10.3. Modify the parameter estimation algorithm to estimate the parameter c as well as θ and simulate the performance of the algorithm with and without noise.

Problem 10.5   Adaptive control of instrument servo

Continue the investigation of the parameter estimation design of Example 10.3 , exploring the issues raised in the example.

Problem 10.6   Recursive least squares

Show that the recursive least-squares solution of (10.47) leads to (10.48).

Problem 10.7   Covariance matrix in recursive least squares

Show that if the noise covariance matrices V_n and W_n are both zero, the gain matrix K_n in (10.56) would go to zero in at most $2k$ steps, thereby preventing the parameter estimates from further changing.

Problem 10.8   Implementation of recursive parameter estimation algorithm

Prepare the computer code to implement the ARMA model recursive parameter estimation algorithm outlined in Section 10.2.3.

Problem 10.9   Instability and Persistence of Excitation

(a) Does the lack of persistent excitation in Example 10.5 depend on the fact that the reference input is sinusoidal? Investigate the effect of using a square-wave input, which is richer in harmonics.

(b) Investigate the effect of random noise in measuring the output of the plant.

10.7   REFERENCES

[1] B. Friedland, *Control System Design: An Introduction to State-Space Methods*, McGraw-Hill, New York, 1986.

[2] B. Friedland and J. Richman, and D. Williams, "On the 'Adiabatic Approximation' for Design of Control Laws for Linear, Time-Varying Systems," IEEE Trans. on Automatic Control, Vol. AC-32, No. 1, pp. 62–63, January 1987.

[3] B. Friedland, "A Simple Nonlinear Observer for Estimating Parameters in Dynamic Systems," Proc. 1993 IFAC Congress, Sydney, Australia, July 18-23, 1993.

[4] B. Friedland, S. Mentzelopoulou, and Y.-J. Park, "Friction Estimation in Multi-Mass System," Proc. 1992 IEEE Conf. on Decision and Control,San Francisco, CA, June 1993.

[5] K.J. Åström and B. Wittenmark, *Adaptive Control*, Addison-Wesley Publishing Co., Reading, MA, 1989.

[6] G.J. Bierman, *Factorization Methods for Discrete Sequential Estimation*, Academic Press, New York, 1977.

[7] C. Rohrs, L. Valavani, M. Athans, and G. Stein,"Robustness of Continuous-Time Adaptive Control Algorithms in the Presence of Unmodeled Dynamics," IEEE Trans. on Automatic Control, Vol. AC- 30, No. 9, September 1985, pp. 881-889.

[8] H. Kaufman, I. Bar-Kana, and K. Sobel, *Direct Adaptive Control Algorithms*, Springer-Verlag, New York, 1994.

[9] O.L. Jacobs, "Introduction to Adaptive Control," in *Self-Tuning and Adaptive Control: Theory and Applications*, C.J. Harris and S.A. Billings (eds.) Peter Peregrinus, Ltd., London, 1980.

[10] K.J. Åström, "Adaptive Feedback Control, " Proceedings of the IEEE, Vol. 75, No. 2, pp. 185–217, February 1987.

[11] J.G. Truxal, "The Concept of Adaptive Control," Chapter 1 in E. Mishkin and L. Braun (eds.) *Adaptive Control Systems*, McGraw- Hill Book Co., New York, 1961.

[12] G.C. Goodwin and K.S. Sin, *Adaptive Filtering Prediction and Control*, Prentice-Hall, Inc., Englewood Cliffs, NJ, 1984.

[13] K.S. Narendra and A.M. Annaswamy, *Stable Adaptive Control*, Prentice-Hall, Inc., Englewood Cliffs, NJ, 1991.

BIBLIOGRAPHY

Books

Apostol, T.M., *Mathematical Analysis*, Addison-Wesley Publishing Co., Reading, MA, 1957.

Armstrong-Hélouvry, B., *Control of Machines with Friction*, Kluwer Academic Publishers, Boston, MA, 1991.

Åström, K.J. and Wittenmark, B., *Adaptive Control*, Addison-Wesley Publishing Co., Reading, MA, 1989.

Athans, M., and Falb, P.L., *Optimal Control*, McGraw-Hill Book Co., New York, 1966.

Atherton, D., *Nonlinear Control Engineering*, 2nd ed., Van Nostrand Reinhold, New York, 1982.

Bellman, R., *Dynamic Programming*, Princeton University Press, Princeton, NJ, 1957.

Bierman, G.J., *Factorization Methods for Discrete Sequential Estimation*, Academic Press, New York, 1977.

Bliss, G.A., *Lectures on the Calculus of Variations*, The University of Chigago Press, Chicago, IL, 1946.

Booker, H.G., *An Approach to Electrical Science*, McGraw-Hill, New York, 1959.

Bryson, A.E. Jr. and Ho, Y.-C.,*Applied Optimal Control*, Blaisdell Publishing Co., Waltham, MA, 1969.

Cauer, W., *Synthesis of Linear Communication Networks,* McGraw- Hill, New York, 1958.

Friedland, B., *Control System Design: An Introduction to State Space Methods*, McGraw-Hill, New York, 1986.

Gantmacher, F.R., *Matrix Theory*, Chelsea Publishing Co., New York, 1959.

Gleick, J., *Chaos: Making a New Science*, Viking Press, New York, 1987.

Goodwin, G.C. and Sin, K.S., *Adaptive Filtering Prediction and Control*, Prentice-Hall, Inc., Englewood Cliffs, NJ, 1984.

Hsu, J.C. and Meyer, A.U., *Modern Control Principles and Applications*, McGraw-Hill, New York, 1968.

Isidori, A., *Nonlinear Control Systems: An Introduction,* 2nd Ed., Springer-Verlag, New York, 1989.

Itkis, U., *Control Systems of Variable Structure,* John Wiley and Sons, New York, 1976.

James, H.M., Nichols, N.B., and Phillips, R.S., *Theory of Servomechnaisms* (MIT Radiation Laboratory Series, Volume 25), McGraw-Hill, New York, 1947.

Jury, E.I., *Sampled-Data Control Systems*, John Wiley and Sons, Inc., New York, 1958.

Kaufman, H., Bar-Kana, I., and Sobel, K., *Direct Adaptive Control Algorithms*, Springer-Verlag, New York, 1994.

Khalil, H.K., *Nonlinear Systems*, New York: Macmillan Publishing Co., 1992.

Lefschetz, S., *Stability of Nonlinear Control Systems,* Academic Press, New York, 1965.

Liapunov, A. M., *Problème Général de la Stabilité du Mouvement*, Princeton University Press, Princeton, NJ, 1947. (Reprint of Russian monograph of 1892.)

Minorsky, N., *Introduction to Nonlinear Mechanics*, J.W. Edwards, Ann Arbor, MI, 1947.

Mohler, R., *Nonlinear Systems* 2 Volumes, Englewood Cliffs, NJ: Prentice-Hall, 1991

Narendra, K.S., and Annaswamy, A.M., *Stable Adaptive Control*, Prentice-Hall, Inc., Englewood Cliffs, NJ, 1991.

Poncaré, H. *Oeuvres*, Gauthier-Villars, Paris, 1928.

Pontryagin, L.S.,Boltyanskii, V., Gamkrelidze, R., and Mischchenko, E., *The Mathematical Theory of Optimal Processes*, Interscience Publishers, Inc. New York, 1962.

Ragazzini, J.R. and Franklin, G.F., *Sampled-Data Control Systems*, McGraw-Hill, New York, 1958.

Routh, E.J., *A Treatise on the Stability of a Given State of Motion*, MacMillan, London, 1877.

Schwarz, R.J. and Friedland, B., *Linear Systems*, McGraw-Hill, New York, 1965.

Slotine, J.-J.E. and Li, W., *Applied Nonlinear Control*, Prentice- Hall, Inc., Englewood Cliffs, NJ, 1991.

Smith, O.J.M., *Feedback Control Systems*, McGraw-Hill, New York, 1958.

Sorenson, H.W., (ed.) *Kalman Filtering: Theory and Applications* IEEE Press, New York, 1985.

Utkin, V.I., *Sliding Modes and their Application in Variable Structure Systems* (in Russian), Nauka, Moscow, 1974.

Vidyasagar, M., *Nonlinear Systems Analysis,* Prentice-Hall, Inc., Englewood Cliffs, NJ, 1978.

Papers and Reports

Aizerman, M.A., "On a Problem Concerning Stability in the Large of Dynamical Systems," Uspekhi Mat. Nauk (USSR), Vol. 4, pp. 187-188, 1949.

Aronowitz, F., "The Laser Gyro," in M. Ross (Ed.) *Laser Applications* Vol. 1, pp. 133-200, Academic Press, New York, 1971.

Aström, K.J., "Adaptive Feedback Control, " Proceedings of the IEEE, Vol. 75, No. 2, pp. 185–217, February 1987.

Athans, M., "The Matrix Minimum Principle," Information and Control, Vol. 11, pp.592–606, November 1967.

Axelsson, J.P., *Modelling and Control of Fermentation Processes,* Doctoral Dissertation, Department of Automatic Control, Lund Institute of Technology, Lund, Sweden, March 1989.

Bass, R.W. and Gura, I., "High-Order System Design Via State-Space Considerations," Proc. Joint Automatic Control Conference, Troy, NY, , pp. 311–318, June 1965.

Bendixson, I. "Sur les courbes définies par des équations différentielles," Acta Mathematica, Vol. 24, pp 1–88, 1901.

Bode, H.W. and Shannon, C.E., "A Simplified Derivation of Linear Least Square Smoothing and Prediction Theory," Proc. IRE, Vol. 38, pp. 417–425, April 1959.

Brandenburg, G., Hertle, H., and Zeiselmair, K., "Dynamic Influence and Partial Compensation of Coulomb Friction in a Position and Speed Controlled Elastic Two-Mass System," Preprints, 10th IFAC World Congress, Vol. 3., pp. 91-99, Munich, Germany, July 27–31, 1987.

Canudas de Wit, C., Aström, K.J., and Braun, K.,"Adaptive Friction Compensation," IEEE Jour. of Robotics and Automation, Vol. RA-3, No. 6., pp. 681–685, December 1987.

Canudas de Wit, C.,Olsson, H., and Aström, K.J., and Lischinsky, P., "A New Model for Control of Systems with Friction," IEEE Trans. on Automatic Control, Vol. 40, No. 3, pp. 419–425, March 1995.

Dahl, P.R., "A Solid Friction Model," AFO 4695-67-C-0158, The Aerospace Corporation, El Segundo, CA, 1968.

Fitts, R.E., "Two Counterexamples to Aizerman's Conjecture," IEEE Trans. on Automatic Control, vol AC-11, no. 3, pp.553-556, July 1966.

Friedland, B., "Steady-State Behavior of Kalman Filter with Discrete- and Continuous-Time Observations," IEEE Trans. on Automatic Control, Vol. AC- 25, No. 5, pp. 988–992, October 1980.

Friedland, B., "On the Properties of Reduced-Order Kalman Filters," IEEE Trans. on Automatic Control, Vol. 34, No. 3, pp. 321–324, March 1989.

Friedland, B., "A Simple Nonlinear Observer for Estimating Paramters in Dynamic Systems," Proc. 1993 IFAC Congress, Sydney, Australia, July 18-23, 1993.

Friedland, B., Hutton, M.F., Williams, C., and Ljung, B., "Design of Servo for Gyro Test Table Using Linear Optimum Control Thoery," IEEE Trans. on Automatic Control, Vol. AC-21, No. 2, pp. 293–296, April 1976.

Friedland, B., Mentzelopoulou, S., and Park, Y.-J., "Friction Estimation in Multi-Mass System," Proc. 1992 IEEE Conf. on Decision and Control, San Francisco, CA, June 1993.

Friedland, B. and Mentzelopoulou, S., "On Adaptive Friction Compensation Without Velocity Estimation," Proc. 1992 Conference on Control Applications, Vol. 2, pp. 1076–1081, Dayton, OH, September 1992.

Friedland, B. and Park, Y.-J., "On Adaptive Friction Compensation," IEEE Trans. on Automatic Control, Vol. 37, No. 10., pp. 1609–1612, 1992.

Friedland, B., Richman, J., and Williams, D., "On the 'Adiabatic Approximation' for Design of Control Laws for Linear, Time-Varying Systems," IEEE Trans. on Automatic Control, Vol. AC-32, No. 1, pp. 62–63, January 1987.

Gunckel, T.L. III and Franklin, G.F., "A General Solution for Linear Sampled Data Control," J. Basic Engineering (Trans. ASME) Vol. 85D. No.1., pp. 197–201, January 1963.

Haessig, D.A. Jr. and Friedland, B., "On the Modelling and Simulation of Friction," J. Dynamic Systems, Measurements, and Control, (Trans. ASME) Vol. 113, pp. 354–362, September 1991.

Harnoy, A., Friedland, B., and Rachoor, H., "Modeling and Simulation of Elastic Forces in Lubricated Bearings for Precise Motion Control," Wear, Vol. 172, pp. 155–165, 1994.

Hong, S.C. and Park, M.H., "Microprocessor-Based High-Efficiency Drive of a DC Motor," IEEE Trans. on Industrial Electronics, Vol.IE-34, No. 4, pp.433-440, November 1984.

Hurwitz, A. "Über die Bedingungen, unter welchen eine Gleichung nur Wurzeln mit negativen reelen Teilen besitzt," Math. Ann. Vol. 146, pp.273– 284, 1895.

Hutton M. and Friedland, B., "Routh Approximations for Reducing Order of Linear, Time Invariant Systems," IEEE Trans. on Automatic Control, Vol. AC-20, No. 3., pp. 329–337, June 1975.

Jacobs, O.L., "Introduction to Adaptive Control," in *Self-Tuning and Adaptive Control: Theory and Applications*, C.J. Harris and S.A. Billings (eds.) Peter Peregrinus, Ltd., London, 1980.

Johnson, C.D. and Wonham, W.M., "On a Problem of Letov in Optimal Control," Proceedings, 1964 Joint Automatic Control Conference, Stanford, CA, pp. 317-325, June 24-26, 1964.

Jury, E.I. and Blanshard, D., "A Stability Test for Linear Discrete Systems in Table Form," Proc. IRE, Vol. 49, pp. 1947–1948, 1961.

Kalman, R.E. "Physical and Mathematical Mechanisms of Instability in Nonlinear Control Systems," Trans. ASME, Vol. 79, pp. 553–566, 1957.

Kalman, R.E. "A New Approach to Linear Filtering and Prediction Problems," J. Basic Engineering (Trans. ASME), Vol. 82D, No. 1, pp. 35–45, March 1960.

Kalman, R.E. "Lyapunov Functions for the Problem of Lur'e in Automatic Control," Proc. National Academy of Sciences, USA, Vol. 48, pp. 201-205, 1963.

Kalman, R.E. "When is a Linear Control System Optimal?" J. Basic Engineering (Trans. ASME) Vol. 86D, No.1, pp. 51-60, March 1964.

Kalman R.E. and Bertram, J.E., "Control System Analysis and Design via the Second Method of Lyapunov," J. Basic Engineering (Trans. ASME), Vol. 82, pp. 371–399, 1960.

Karnopp, D. "Computer Simulation of Slip-Stick Friction in Mechanical Dynamic Systems," J. Dynamic Systems, Measurements, and Control, (Trans. ASME) Vol. 107, pp. 100-103, 1985.

LeMay, J.R. "Recoverable and Reachable Zones for Control Systems with Linear Plants and Bounded Controller Outputs," Proceedings, 1964 Joint Automatic Control Conference, Stanford, CA,, pp. 305–312, June 24-26, 1964.

Liu,Y., Anderson, B.D.O., and Ly, U.-Y., "Coprime Factorization Controller Reduction with Bezout Identity Induced Frequency Weighting," Automatica, Vol. 26, No. 2, March 1990.

Lorenz, E.N. "Deterministic Nonperiodic Flow," J. Atmospheric Science, Vol. 20, pp. 130–141, 1963.

McShane, E.J. "On Multipliers for Lagrange Problems," Am. J. Math, Vol. 61, pp. 809–819, 1939.

Mentzelopoulou, S. and Friedland, B., "Experimental Evaluation of Friction Estimation and Compensation Techniques," Proc. 1994 American Control Conference, Baltimore, MD, pp. 3132–3136, June 1994.

Meyer, G. and Cicolani, L.,"A Formal Structure for Advanced Automatic Flight Control Systems," NASA Technical Note D-7940, 1975.

Meyer, G. and Cicolani, L., "Application of Nonlinear Systems Inverses to Automatic Flight Control Design—System Concepts and Flight Evaluations," In *Theory and Applications of Optimal Control in Aerospace Systems*, NATO AGARDograph No.251, July 1981.

Moore, B.C., "Principal Component Analysis in Linear Systems: Controllability, Observability, and Model Reduction," IEEE Trans. on Automatic Control, Vol. AC-26, No. 1., pp. 17–27, January 1981.

Parks, P.C., "A New Proof of the Routh-Hurwitz Stability Criterion Using the 'Second Method' of Lyapunov," Proc. Cambridge Philosophical Society, Vol. 58, Pt 4., pp. 694-720, 1962.

Rohrs, C., Valavani, L. , Athans, M. , and Stein, G., "Robustness of Continuous-Time Adaptive Control Algorithms in the Presence of Unmodeled Dynamics," IEEE Trans. on Automatic Control, Vol. AC-30, No. 9, pp. 881-889, September 1985.

Rouelle D. and Takens, F., "On the Nature of Turbulence," Communications on Mathematical Physics, Vol. 20, pp.167-192, 1971.

Schäfer, U., and G. Brandenburg, "Model Reference Position Control of an Elastic Two-Mass System with Compensation of Coulomb Friction," Proc. 1993 American Control Conference, Vol. 2., pp. 1937–1941, San Francisco, CA, June 1993.

Simon, H.A. "Dynamic Programming Under Uncertainty with a Quadratic Criterion Function," Econometrica, Vol. 24, 74–81, 1956.

Truxal, J.G. "The Concept of Adaptive Control," Chapter 1 in E. Mishkin and L. Braun (eds.) *Adaptive Control Systems*, McGraw-Hill, New York, 1961.

Youla, D.C., Jabr, H.A., and Bongiorno, J.J. Jr., "Modern Wiener-Hopf Design of Optimal Controllers, Part II, The Multivariable Case, " IEEE Trans. on Automatic Control, Vol. AC-21, pp.319-338, 1976.

Ziegler, J.G. and N.B. Nichols, "Optimum Settings for Automatic Controllers," Trans. ASME, Vol. 64, No. 8, pp. 759–768, 1942.

INDEX